Emerging Power Converters for Renewable Energy and Electric Vehicles

T0187729

Emerging Power Converters for Renewable Energy and Electric Vehicles

Modeling, Design, and Control

Edited by

Md. Rabiul Islam
Md. Rakibuzzaman Shah
Mohd. Hasan Ali

CRC Press
Taylor & Francis Group
Boca Raton London New York

CRC Press is an imprint of the
Taylor & Francis Group, an **informa** business

MATLAB® is a trademark of The MathWorks, Inc. and is used with permission. The MathWorks does not warrant the accuracy of the text or exercises in this book. This book's use or discussion of MATLAB® software or related products does not constitute endorsement or sponsorship by The MathWorks of a particular pedagogical approach or particular use of the MATLAB® software.

CRC Press
Boca Raton and London

First edition published 2021
by CRC Press
6000 Broken Sound Parkway NW, Suite 300, Boca Raton, FL 33487-2742
and by CRC Press
2 Park Square, Milton Park, Abingdon, Oxon, OX14 4RN
© 2021 Taylor & Francis Group, LLC

CRC Press is an imprint of Taylor & Francis Group, LLC

ISBN: 978-0-367-52803-4 (hbk)
ISBN: 978-0-367-52814-0 (pbk)
ISBN: 978-1-003-05847-2 (ebk)

Typeset in Times
by KnowledgeWorks Global Ltd.

Contents

Acronyms

CHAPTER 1

CHB	Cascaded H-bridge
DAB	Dual active bridge
DPS	Dual phase shift
EMS	Energy management system
EPRI	Electrical Power Research Institute
ESS	Energy storage system
FER	Front-end rectifier
FREEDM	Future renewable electric energy delivery
HFML	High-frequency magnetic link
IGBT	Insulated gate bipolar transistor
IoT	Internet of Things
LV	Low voltage
LVAC	Low-voltage AC
LVDC	Low-voltage DC
MAB	Multiple active bridge
MG	Microgrid
MLC	Multilevel converter
MOSFET	Metal–oxide–semiconductor field effect transistor
MV	Medium voltage
MVAC	Medium-voltage AC
MVDC	Medium-voltage DC
NPC	Neutral point clamped
PAB	Penta-active bridge
PWM	Pulse width modulation
QAB	Quad-active bridge
SPS	Single-phase shift
SST	Solid-state transformer
TAB	Triple active bridge
UNIFLEX	Universal and flexible power management system

CHAPTER 2

APOD	Alternate phase disposition
BCPWM	Bus-clamping pulse width modulation
CHB	Cascaded H-bridge
CSVPWM	Conventional space-vector pulse width modulation
DAB	Dual active bridge
IGBT	Insulated gate bipolar transistor
IPD	In-phase disposition

MMC	Modular multilevel converter
PET	Power electronic transformer
POD	Phase opposition disposition
PWM	Pulse width modulation
SDBCPWM	Sixty-degree bus-clamping pulse width modulation
SDPWM	Sixty-degree pulse width modulation
SPWM	Sinusoidal pulse width modulation
STATCOM	Static synchronous compensator
TDBCPWM	Thirty-degree bus-clamping pulse width modulation
THD	Total harmonic distortion
THPWM	Third harmonic injected PWM
THSDBCPWM	Third harmonic injected sixty-degree bus-clamping pulse width modulation
THTDBCPWM	Third harmonic injected thirty-degree bus-clamping pulse width modulation
TRPWM	Trapezoidal pulse width modulation

CHAPTER 3

AC	Alternating current
BP	Bipolar pad
CC	Capacitor-capacitor
CLC	Capacitor-inductor-capacitor
CSI	Current source inverter
DC	Direct current
DCM	Discontinous conduction mode
DWC	Dynamic wireless charging
EMF	Electric and magnetic fields
EMI	Electromagnetic interference
EV	Electric vehicle
FBMC	Full-bridge matrix converter
FBVSI	Full-bridge voltage source inverter
GaN	Gallium nitride
GIT	Gate injection transistor
ICNIRP	International Commission on Non-Ionizing Radiation Protection
IEC	International Electrotechnical Commission
IEEE	Institute of Electrical and Electronics Engineers
IGBT	Insulated gate bipolar transistor
IPT	Inductive power transfer
ISO	International Standards Organization
LCC	Inductor-capacitor-capacitor
LCCL	Inductor-capacitor-capacitor-inductor
LCL	Inductor-capacitor-inductor
MC	Matrix converters
MOSFET	Metal–oxide–semiconductor field-effect transistor
MTTF	Mean time to failure

PCB	Printed circuit board
PF	Power factor
PFC	Power factor correction
PP	Parallel-parallel
PS	Parallel-series
PWM	Pulse width modulation
RMS	Root mean squares
S2V	Solar to vehicle
SAE	Society of Automotive Engineers
SiC	Silicon carbide
SP	Series-parallel
SRC	Sinusoidal ripple charging
SS	Series-series
THD	Total harmonic distortion
UL	Underwriters Laboratories
V2G	Vehicle to grid
VA	Voltage-ampere
VSI	Voltage source inverter
WPT	Wireless power transfer
ZCS	Zero-current switching

CHAPTER 4

AC	Alternating current
BEV	Battery-powered electric vehicle
DC	Direct current
EDLC	Electric double-layer capacitors
EIS	Electrochemical impedance spectroscopy
ESR	Equivalent series resistance
ESS	Energy storage system
EV	Electric vehicle
FC	Fuel cells
HESS	Hybrid energy storage system
HEV	Hybrid electric vehicle
HWFET	Highway fuel economy test
Li-ion	Lithium-ion
LPF	Low-pass filter
MOSFET	Metal–oxide–semiconductor field-effect transistor
NEDC	New European driving cycle
PE	Power electronics
PHEV	Plug-in hybrid electric vehicle
PI	Proportional-integral
SC	Supercapacitors
SoC	State of charge
SoH	State of health
UC	Ultracapacitors

CHAPTER 5

BDC	Bidirectional DC–DC converter
BLIL	Bridgeless interleaved
CCCM	Critical continuous conduction mode
CCM	Continuous conduction mode
CS	Charging station
DCM	Discontinues conduction mode
DCVM	Discontinuous capacitor voltage mode
DEG	Diesel engine generator
DPT	Direct power transfer
DT	Distribution transformer
EMI	Electromagnetic interference
EV	Electric vehicle
EWC	Enabling window control
FPGA	Field Programmable Gate Array
G2V	Grid to vehicle
HSC	High-speed controller
IEC	International Electrotechnical Commission
IGBT	Insulated gate bipolar transistor
LED	Light-emitting diode
MOSFET	Metal–oxide–semiconductor field effect transistor
PFC	Power factor correction
PHEV	Plug-in hybrid electric vehicle
PQ	Power quality
PWM	Pulse width modulation
RCD	Resistor-capacitor-diode
RFI	Radio frequency interference
SEPIC	Single-ended primary inductor converter
SOC	State of charge
SPWM	Sinusoidal pulse width modulation
THD	Total harmonic distortion
UPS	Uninterruptible power supply
V2B	Vehicle to building
V2G	Vehicle to grid
V2H	Vehicle to home
V2O	Vehicle to office electric vehicle
VSI	Voltage source inverter
ZCS	Zero current switching
ZVS	Zero voltage switching
ZVT	Zero voltage transition

CHAPTER 6

$\mathbf{\Delta V_{dc}}$	Smoothing index of the coordinated control loop (V)
B:	External stiffness (N.m/rad.s)

$C_p(\lambda, \beta)$	Power coefficient which is a function of the tip speed ratio λ and blade pitch angle β (o)
DFIG:	Doubly fed induction generator
dq0	Park transformations
E	Kinetic energy (J)
ε_p	Tracking Error in Power in the Sliding Mode Control Algorithm (W)
FFC	Full-scale frequency converter
f_g:	Grid frequency (Hz)
f_g^*	Reference grid frequency (Hz)
GSC	Grid side converter
H	Inertia constant of wind turbine (J/VA)
i_g	Grid current (A)
J	Moment of inertia (kgm^2)
K	External damping factors (N.m/rad.s)
λ_{opt}	Optimal tip-speed ratio
λ_r	Rotor flux linkage (wb)
λ_s	Stator flux linkage (wb)
L_g	Grid filter inductance (H)
L_{lr}	Rotor inductance (H)
L_{ls}	Stator inductance (H)
L_m	Mutual inductance (H)
L_r	Rotor self-inductance (H)
L_s	Stator self-inductance (H)
LVRT	Low-voltage ride-through
MMC	Modular multilevel converter
MPC	Model predictive control
MPPT	Maximum power point tracking
ω	Synchronous speed of the dq reference frame
ω_m	Rotational speed of the wind turbine blade (rad/s)
ω_{rl}	Angular slip frequency of the generator (rad/s)
ω_s	Synchronous speed of the generator (rad/s)
P	Active or real power (W)
P	Air density (kg/m^3)
P_a	Aerodynamic power (W)
P_c^*	Power smoothing command (W)
P_g	Power delivered to the grid from the generator (W)
P_g^*	Reference power (W)
PLL	Phase locked loop
PMSG	Permanent magnet synchronous generator
P_w	Mechanical power extracted from wind turbine (W)
Q	Reactive power (var)
R	Radius of the wind turbine blade (m)
R_g	Grid filter resistance (Ω)
ROCOF	Rate of change of frequency (Hz)

R_r	Rotor resistance (Ω)
R_s	Stator resistance (Ω)
RSC	Rotor side converter
S	Nominal apparent power (VA)
SCIG	Squirrel cage induction generator
SMC	Sliding mode control
SSI	Sub-synchronous interaction
T_a	Aerodynamic torque (N.m)
T_{em}	Generator electromagnetic torque (N.m)
Θ	Transformation angle (o)
T_{hs}	High-speed torque (N.m)
T_{ls}	Low-speed torque (N.m)
T_Θ	Park transformation matrix
v_1	GSC output voltage (V)
V_{dc}	DC link voltage (V)
v_g	Grid voltage (V)
VSC:	Voltage source converter
VS-WECS	Variable-speed wind energy conversion system
V_w	Wind speed (m/s)
WECS	Wind energy conversion system
WRIG	Wound rotor induction generator

CHAPTER 7

AHP	Analytic hierarchy process
AM	Amplitude modulation
ANFIS	Adaptive neuro-fuzzy interference system
ANNs	Artificial neural networks
BTA	Boosting tree algorithm
CCFL	Cluster center fuzzy logic
CFF	Current form factors
CHAID	Chi-square automatic interaction detector
CMU	Converter monitoring unit
CWT	Continuous wavelet transform
DFIG	Double fed induction generator
DWT	Discrete wavelet transform
EDM	Electrical discharge machining
EESG	Electrically excited synchronous generator
EKF	Extended Kalman filter
ELM	Extreme learning machine
EMF	Electromotive force
EMP	Electromagnetic power
EPVA	Extended Park's Vector Approach
FEM	Finite element model
FESR	Front-end speed regulation
FFT	Fast Fourier transform

FIS	Fuzzy interference system
FSCs	Fault signature components
gmean	Geometric mean
GP	Gaussian process
HHT	Hilbert–Huang transform
IDFT	Localized discrete Fourier-transform
IF	Instantaneous frequency
IoT	Internet of Things
K-NN	k-nearest neighbor
LSSVM	Least square support vector machine
LVRT	Low-voltage ride-through
MAE	Mean absolute error
MAPE	Mean absolute percentage error
MF	Membership functions
MSB	Modulation signal bispectrum
NAS	Network-attached storage
NN	Neural network
PCA	Principal component analysis
PMSG	Permanent magnet synchronous generator
PWM	Pulse width modulation
QPC	Quadratic phase coupling
RFA	Random forest algorithm
RMS	Root mean squared
RWISC	Rotor winding inter-turn short-circuit
SCADA	Supervisory control and data acquisition
SCIG	Squirrel cage induction generator
SD	Standard deviation
SNR	Signal-to-noise ratio
SPRT	Sequential probabilistic ratio test
SSCI	Sub-synchronous control interactions
STFT	Short-term Fourier transform
STFT	Short-time Fourier transform
SVDD	Support vector data description
SVM	Support vector machine
VP	Virtual power
WECS	Wind energy conversion system
WRIMs	Wound rotor induction motors
WT	Wind turbine
WTG	Wind turbine generator
WTGU	Wind turbine generator unit

CHAPTER 9

1LG	Single line-to-ground
2LG	Double line-to-ground
2LL	Line-to-line

3LG	Three phase-to-ground
ANFIS	Adaptive neuro-fuzzy inference system
ANN	Artificial neural network
FACTS	Flexible AC transmission systems
FLC	Fuzzy logic controller
PI	Proportional-integral
PID	Proportional-integral-derivative
PSS	Power system stabilizer
SMES	Superconducting magnetic energy storage
SSSC	Static synchronous series compensator
STATCOM	Static synchronous compensator
SVC	Static VAR compensator
TCBR	Thyristor controlled braking resistor
TCSC	Thyristor controlled series capacitor
THD	Total harmonic deviation
TSC	Thyristor switched capacitor
UPFC	Unified power flow controller

CHAPTER 10

BDC	Bidirectional DC–DC converter (BDC)
BESS	Battery energy storage system
DERs	Distributed energy resources
DFIG	Doubly fed induction generator
DG	Diesel generator
ES	Energy storage
FRT	Fault ride-through
GSC	Grid side converter
HESS	Hybrid energy storage system
HMG	Hybrid AC/DC microgrid
ILC	Interlinking converter
LPF	Low-pass filter
MAS	Multi-agent systems
PV	Photovoltaic
PWM	Pulse width modulation
RSC	Rotor side converter
SPWM	Sinusoidal PWM
VSCs	Voltage source converters

SYMBOLS

CHAPTER 5

C, C1 & C2	Capacitor values
Ci	Inverter-side filter capacitor
CL	DC link capacitor

Cr	Resonant capacitor
Cs	Source-side capacitor
D	Duty cycle
Δi	Ripple current
ΔVDC	Voltage ripple at DC bus
η	Efficiency
Fnw	Conversion gain
fs	Switching frequency
iau	Auxiliary inductor current
Iau,p	Current at the primary side of the auxiliary inductor
iin	Input current
iL	Inductor current
IL.pk	Inductor peak current
ILm,peak	Peak value of leakage current due to mutual inductance Lm1
Iout	Inverter output current
Iout.crit	Critical value of output current
is	Current at supply side
l	Inductance ratio
L, LA & LB	Inductor values
Laux	Auxiliary inductor
Lb	Battery side capacitor
Leq	Equivalent inductance value
Li	Inverter side filter inductor
Lm	Mutual inductance
Ls	Supply side inductor
Mcrit	Critical value of gain
Mmin	Minimum value of gain
P11	Power at auxiliary stage
Pcore_loss	Core loss
PEV	Power consumption by EV charger
Phalfline	Power transferred in the half line cycle
Pin	Input power
Pnon-EV	Power consumption by residential loads
Po	Output power
Po1	Power output at the first stage
Po2	Power output at the second stage
Pout,max	Maximum output power
Ps	Power consumption of overall system
RL	Load resistance
t, to, t1 & t2	Switching intervals
thold	Hold-up time
Ts	Switching period
Vau	Auxiliary inductor voltage
VB.min	Minimum battery voltage
Vcm	Capacitor's maximum voltage
Vcut-off	Battery cut-off voltage

Vdc	DC link voltage
VDC.min	Minimum DC link voltage
Vfull-charge	Battery voltage at full charge
Vin	Input voltage
Vo	Output voltage
Vout	Inverter output voltage
Vs	Supply voltage
Z0,max	Maximum characteristics impedance

CHAPTER 7

acc$_i$	Accuracy of class *i*
Br	Rotor flux density waves
Bs	Stator flux density waves
Cbearing	Parasitic capacitances for the bearing
Ccap	Parasitic capacitances for the insulation cap
C_{fj}	Power distribution coefficient of the *j*th faulty turbine
Crf	Parasitic capacitances for the rotor core to stator frame
Crwf	Parasitic capacitances for the rotor winding to stator frame
Crwr	Parasitic capacitances for the rotor winding to rotor core
Crwsr	Parasitic capacitances for the rotor winding to stator winding
Cswf	Parasitic capacitances for the stator winding to stator frame
Cswr	Parasitic capacitances for the rotor core to stator winding
CWT_{local}	Matrix of wavelet coefficients
D_c	Pitch or cage diameter
E(X)	Statistical expectation operator of the random variable X
F ()	Fourier transform of the signal
fc	Carrier wave frequencies
ffault	Fault characteristic frequency of the bearing
fi, fo, fb, fc	Characteristic frequencies of the inner raceway, outer raceway, ball, and cage in a mechanical bearing
fm	Modulation wave frequencies
fr	Rotating frequency of the bearing
fs	Fundamental frequency
fs	Fundamental frequency of the stator current
g0	Mean air-gap length without eccentricity
I0, Ik	Amplitude of the fundamental current and harmonic current
ia, ib, ic	Three-phase currents in the stator windings
Ia,l, Ib,l, Ic,l	Leakage current of phase A, phase B, and phase C
ic	IGBT collector-emitter on-state current
Ii±	Amplitude of the modulation harmonics
ijv	Average absolute value of the phase current
Lbrush	Equivalent inductance of the grounding brush
lm	Effective lamination length
N_1	Number of the healthy wind turbines
N_2	Number of the faulty wind turbines

NB	Number of balls in the bearing
ω	Angular frequency
ωs	Stator current frequency of the generator
p	Number of pole-pairs
P	Power
P_{fj}	Power output of the jth faulty turbine working in a power-reduced mode
P_{hi}	Power output of the ith healthy turbine operating in the baseline mode
φ0	Phase of the fundamental current
φ0 ± φi	Phase of the modulation harmonics
φν	Phase of torque variation due to bearing faults
Rbrush	Equivalent resistance of the grounding brush
s	Slip ratio
σ_n^2	Noise variance
σ_f^2	Signal variance
T0	Torque produced by wind power
θr	Rotor position
Tν	Amplitude of torque variation due to bearing faults
Uag, Uab, Uac	Amplitudes of voltage vectors between phase A and ground, phase B, and phase C
Va, Vb, Vc	Phase output voltages of the converter
vceon	IGBT collector-emitter on-state voltage
Vcom	Common mode voltage
x_*	Input of system ·
X(f)	Discrete Fourier transform (DFT) of a discrete time current signal x(k)
y_*	Unknown predicted output
Zpg	Winding impedances of phase to ground
Zpp	Winding impedances of phase to phase
Zr	Number of rotor slot

Preface

With the rapid penetration of renewable energy, modern power networking is becoming decentralized, and bidirectional power flow is possible. Power electronic loads, including electric vehicles (EVs), are highly sophisticated, nonlinear, and unpredictable. Power electronics load such as EVs have tremendous potential to cause severe power quality problems in future renewable energy integrated power systems. The rapid growth in grid-connected renewable power generation, EVs, and power quality problems has provided the impetus for new power converter technologies such as multilevel, multiphase, bidirectional solid-state transformer (SST) converters, and other high-voltage DC and AC devices with advanced switching and control techniques.

Recent advances in power semiconductor devices (ultra-fast recovery and silicon carbide), soft magnetic materials (amorphous or nanocrystalline), and controllers have led to the development of grid-connected high-power-density power converters for renewable energy generation and EV applications. However, the design process of the grid-connected converter involves multiphysics problems that entail complex trade-offs between reliability, efficiency, and cost. Therefore, extensive multiphysics research in the fields of grid-connected power converters, switching, and control for compensation of grid power quality problems; protection, as well as condition monitoring, of grid-connected converters, and magnetics are needed to develop next-generation power converters for grid applications.

This book collects recent advancements in energy conversion with high-power-density power converters with an emphasis on control methods for bidirectional power transfer and compensation of grid power interaction in mixed AC/DC systems; AC and DC system stability; magnetic design for high-frequency, high-power-density systems with advanced soft magnetic materials; modeling and simulation of mixed AC/DC systems, switching strategies for enhanced efficiency; and reliability for sustainable grid integration.

The chapter on **Modeling, Design, and Control of Solid-State Transformer for Grid Integration of Renewable Sources** focuses on the modeling and control issues related to the design of the emerging solid-state transformer (SST) for the integration of renewable energy sources to the distribution grid. The state-of-the-art topologies of SST and their applications in integrating renewable energy sources are covered. The emerging solid-state switching devices and power converter topologies suitable for SST operation are discussed in terms of their power ratings and associated thermal loss scenarios.

Recently, medium-voltage power converters that do not require power frequency transformers and line filters are becoming increasingly popular. This chapter on **Magnetic Linked Power Converter for Transformer-Less Direct Grid Integration of Renewable Generation Systems** aims to review different power converter topologies with different switching and control strategies for the grid integration of renewable energy sources. This chapter also presents current research activities and possible future directions for the power converter topologies in a grid-connected renewable energy power conversion system.

The chapter on **Power Electronics for Wireless Charging of Future Electric Vehicles** deals with the operation, fundamentals, industrial standards, and power conversion architectures for inductive power transfer (IPT)-based battery charging for EV applications. The state-of-the-art power electronic converter topologies utilized in IPT-based charging systems are reviewed in this chapter.

The chapter on **Power Electronics-Based Solutions for Supercapacitors in Electric Vehicles** deals with the fundamental details of SCs such as construction, classification, electrical characteristics, charge/discharge behavior for various types of loads, and modeling approaches. However, there are some concerns associated with individual SCs, such as voltage imbalance and unutilized energy when utilized for various applications.

EVs, especially e-cars, are becoming the major attraction to smart grid technologies and building energy management techniques due to their vehicle-to-grid (V2G), vehicle-to-home (V2H), vehicle-to-building (V2B), and vehicle-to-office (V2O) functionalities. However, the primary application of an EV charger is battery charging with power factor correction (PFC) and galvanic isolation. As these chargers are an integral part of the EV with bidirectional power flow, their power density, efficiency, reliability, and lifetime are vital in improving the performance of the EV. Hence, the challenges in EV chargers and their recent developments are needed to be discussed comprehensively. The chapter on **Front-end Power Converter Topologies for Plug-In Electric Vehicles** provides a summary of front-end AC–DC converter topologies and DC–DC converters on their operation and design considerations to provide a broad scope of these converters.

Modern and sophisticated wind farms are required to be capable of meeting intriguing energy demands; technical implications as well as stable operational characteristics imposed by the regulatory commission; related authoritative bodies; and system infrastructure. Thereby, manifestations of technically prudent control methodologies suitable for integration in wind energy systems are important for penetrating this renewable energy source. The chapter entitled **Advanced and Comprehensive Control Methods in Wind Energy Systems** deals with the comprehensive discussions on various control methodologies, such as pitch control methods, inertial control, direct current vector control, sliding mode control, model predictive control systems, and coordinated control.

With the increasing popularity of wind power generation, diagnostic and monitoring technologies for offshore wind turbines have become more and more important. Among them, related technologies based on converters have attracted more attention. The chapter on **Converter-Based Advanced Diagnostic and Monitoring Technologies for Offshore Wind Turbines** reviews the important progress in this field from device (generator and converter) to system level in recent years and makes comprehensive comparison and analysis.

Renewable energy integrated industrial-scale DC microgrids could play a significant role in minimizing global emission and providing clean energy to the community. However, before implementing this new technology on a commercial scale, some of the key technical challenges, such as choosing a suitable topology and impact of various disturbances in an industrial-scale microgrid need to be understood better and examined thoroughly. The chapter on a **Comprehensive Stability Analysis** of **Multi-Converter-Based DC Microgrids** primarily provides an overview of the design aspects of an industrial-scale DC microgrid.

Microgrids and smart grids are gradually evolving from conventional power systems. Remote microgrid systems deliver reliable, renewable power cost-effectively to places without access to an electrical grid. To ensure constant voltage and frequency to the consumers, stability of remote microgrids should be improved and maintained at a desired level. The chapter on **Stability of Remote Microgrids: Control of Power Converters** deals with the nonlinearity that arises due to the balanced and unbalanced faults occurring at different points in the microgrid system. To improve the stability due to nonlinearity issues, the use of intelligent controller-based FACTs devices has been explained. Also, suitable techniques to deal with time delay issues have been discussed.

Mixed AC/DC power systems, especially a hybrid AC/DC microgrid (HMG), is getting increasing attention in the power community worldwide. The purpose of operating these grids is to supply energy from distributed energy resources (DERs), such as wind turbines, photovoltaic (PV) arrays, diesel generators, etc., to the loads independent of the main utility grid in a relatively small area. The chapter on **Mixed AC/DC System Stability under Uncertainty** deals with an introduction to the hybrid microgrid system. Then it discusses the structure of HMGs, including their topologies; grid components; power converters, such as DC–DC converters and voltage source converters (VSCs); and various control strategies, such as centralized/distributed control, primary control (droop-control, voltage/current control loops, power sharing loops), secondary/tertiary control (voltage/frequency restoration), DC–DC converter control schemes, and interlinking converter (ILC) control.

Editors

Md. Rabiul Islam
School of Electrical, Computer and Telecommunications Engineering
Faculty of Engineering and Information Sciences
University of Wollongong
New South Wales
Australia

Md. Rakibuzzaman Shah
School of Engineering, Information Technology and Physical Sciences
Federation University Australia
Victoria
Australia

Mohd. Hasan Ali
Department of Electrical and Computer Engineering
University of Memphis
Memphis, Tennessee

About the Editors

Md. Rabiul Islam received his PhD degree in electrical engineering from University of Technology Sydney (UTS), Sydney, Australia, in 2014. He was appointed a lecturer at RUET in 2005 and promoted to full professor in 2017. In early 2018, he joined at the School of Electrical, Computer, and Telecommunications Engineering (SECTE), University of Wollongong (UOW), Wollongong, Australia. He is a Senior Member of IEEE. His research interests are in the fields of power electronic converters, renewable energy technologies, power quality, electrical machines, electric vehicles, and smart grid. He has authored or co-authored more than 200 papers. including 50 *IEEE Transactions/IEEE* Journal papers. He has written or edited 5 technical books published by Springer and Taylor & Francis Group. Dr. Islam has received more than 20 awards for his outstanding publications, including 2 Best Paper Awards from *IEEE Transactions on Energy Conversion* in 2020. He was also awarded several prestigious fellowships/scholarships, including the Australian Government Endeavour Research Fellowship, University of Queensland Postdoctoral Research Fellowship (Chancellor Fellow), Japanese Government Monbukagakusho Scholarship, Australian Government International Postgraduate Research Scholarship, University of Technology Sydney President Scholarship, and the Asian Development Bank Scholarship. He has received funding from government and industries, including the Australian Research Council Discovery Project 2020 (AUD 487,629.00) titled "A Next Generation Smart Solid-State Transformer for Power Grid Applications". He was also involved in many innovative projects, such as the Australian Research Council discovery project entitled "An Optimal Electrical Drive System for Plug-in Hybrid Electric Vehicles" and the Australian Renewable Energy Agency project (AUD10.5 million) entitled "Smart Sodium Storage Solution". He is serving as an Editor for *IEEE Transactions on Energy Conversion* and *IEEE Power Engineering Letters*, and Associate Editor for *IEEE Access*. He has also served as a Guest Editor for *IEEE Transactions on Energy Conversion*, *IEEE Transactions on Applied Superconductivity*, and *IET Electric Power Applications*.

Md. Rakibuzzaman Shah received his PhD degree from the University of Queensland, Brisbane, QLD, Australia. He is a senior lecturer in smart power systems engineering at the School of Science Engineering and Information Technology, Federation University Australia (FedUni Australia). Prior to joining FedUni Australia, he had worked with the University of Manchester, the University of Queensland, and Central Queensland University. He has experience working at, and consulting with, DNOs and TSOs on individual projects and has done collaborative work on a large number of projects (EPSRC project on Multi-terminal HVDC, Scottish and Southern Energy Multi-infeed HVDC) - primarily on the dynamic impact of integrating new technologies and power electronics into large systems. He is an active member of the IEEE and CIGRE. He has more than 50 international publications and has spoken at leading power system conferences around the world. His research interest include future power grids (i.e., renewable energy integration, wide-area control), asynchronous grid connection through VSC-HVDC, power system stability and dynamics, application of data mining in power systems, application of control theory in power systems, distribution system energy management and low carbon energy systems.

Mohd. Hasan Ali received his PhD degree in electrical and electronic engineering from Kitami Institute of Technology, Kitami, Japan, in 2004. He is currently an Associate Professor with the Electrical and Computer Engineering Department at the University of Memphis, Tennessee, USA, where he leads the Electric Power and Energy Systems (EPES) Laboratory. His research interests include advanced power systems, smart-grid and microgrid systems, renewable energy systems, energy storage systems, and cybersecurity issues in modern power grids. Dr. Ali has more than 185 publications including 2 books, 4 book chapters, 2 patents, 59 top ranked journal papers, 96 peer-reviewed international conference papers and 20 national conference papers. He serves as an Editor of the *IEEE Transactions on Sustainable Energy* and *IET Generation, Transmission and Distribution* (GTD) journals. Dr. Ali is a Senior Member of the IEEE Power and Energy Society (PES). He is also the Chair of the PES of the IEEE Memphis Section.

Contributors

Vassilios G. Agelidis
Technical University of Denmark
Lyngby, Denmark

Xi Chen
Huazhong University of Science and
Technology
Wuhan, China

Dhiman Chowdhury
Department of Electrical Engineering
University of South Carolina
Columbia, South Carolina

Yashwanth Dasari
Institute of Technology
University of Ontario
Oshawa, Canada

Sagnika Ghosh
Tennessee State University
Nashville, Tennessee

Mohammad Habibullah
School of Information Technology and
Electrical Engineering
The University of Queensland
Brisbane, Australia

Phuoc Sang Huynh
Institute of Technology
University of Ontario
Oshawa, Canada

Md. Moinul Islam
Electric Reliability Council of Texas
Austin, Texas

Morteza Daviran Keshavarzi
University of Memphis
Memphis, Tennessee

Yi Liu
Huazhong University of Science and
Technology
Wuhan, China

A. M. Mahfuz-Ur-Rahman
University of Wollongong
Wollongong, NSW, Australia

N. Mithulananthan
School of Information Technology and
Electrical Engineering
The University of Queensland
Brisbane, Australia

Kashem M. Muttaqi
University of Wollongong
Wollongong, NSW, Australia

Md. Ashib Rahman
University of Wollongong
Wollongong, NSW, Australia

Deepak Ronanki
Indian Institute of Technology
Roorkee

Chandra Sekar S.
Motilal Nehru National Institute of
Technology, Allahabad
Prayagraj, India

Rahul Sharma
School of Information Technology
and Electrical Engineering
The University of Queensland
Brisbane, Australia

Asheesh K. Singh
Motilal Nehru National Institute of
Technology, Allahabad
Prayagraj, India

Sri Niwas Singh
Indian Institute of Technology
Kanpur, India

Danny Sutanto
University of Wollongong
Wollongong, NSW, Australia

Sheldon S. Williamson
Institute of Technology
University of Ontario
Oshawa, Canada

1 Modeling, Design, and Control of Solid-State Transformer for Grid Integration of Renewable Sources

Md. Ashib Rahman, Md. Rabiul Islam,
Kashem M. Muttaqi, and Danny Sutanto
University of Wollongong

CONTENTS

1.1 INTRODUCTION

The use of a smart solid-state transformer (SST), instead of the traditional distribution transformer, has been envisioned as a potential solution for the integration of renewable energy sources, such as distributed generators, battery energy storage systems (ESS), and electric vehicle charging stations to the distribution grid. The internet of things (IoT)-based SST framework can perform the same function as an embedded energy router in a power grid. Such a framework removes the complexities associated with the interconnection of the traditional distribution grid and distributed microgrid (MG) (AC, DC, or hybrid) units from the viewpoints of reliability, controller performance, and communication functionalities. In comparison with the traditional distribution transformer, the smart SST exhibits several unique functionalities, such as improvement of the grid power quality, regulation of the voltage and power factor, support of the grid reactive power, provision of real-time communication, and intelligent management of the energy flow.

The modern SST design requires multidisciplinary expertise from the field of communication, power electronics, magnetics, and control to harness the aforementioned benefits. Due to the integration of different types of linear and nonlinear local loads and the zonal DC or AC MG systems with the SST via power and communication network, as shown in Figure 1.1, a complicated energy scheduling and optimization algorithm should be adopted to achieve a balance between the energy supply and the load demand, and to prioritize the load. Moreover, the power grid may suffer from several power quality problems, like sag, swell, transients, and harmonics. Controller functionalities of an SST vary according to the types of the loads and their energy demands. Therefore, a proper design of the controller architecture and their coordination are crucial for the reliable operation of the SST-based distribution grids.

An SST integrated with renewable energy and ESS or with AC/DC MG system has wide operating capabilities. An intelligent energy management system (EMS) should be designed for the power supply reliability and economic load dispatch to put less burden on the utility grid. The designed EMS should handle renewable energy intermittency, smooth charging and discharging of the ESS, and ensure seamless operation of the SST in grid-forming and grid-feeding modes. The designed EMS should demonstrate superior performance under various power management scenarios. Moreover, the next-generation smart SST should efficiently handle several internal and external faults and should facilitate fault ride-through capability. Each active switch in SST is a potential source of the fault and quick fault detection and isolation helps in reducing the possibility of the converter failure. Certain protective measures must be undertaken against several unwanted situations, such as extreme overvoltage stresses on the medium-voltage (MV) side, switching transients, medium- and low-voltage (LV) short circuit, non-ideal load, overcurrent requirements, and grid voltage sag/swell. A novel on-line communication network-based fault isolation system is required so that the fault can be detected and isolated within the shortest possible time.

1.2 TOPOLOGIES, CONVERTERS, AND SWITCHES OF THE SST

The history of SSTs goes back to 1970 when a technique of power conversion using a high-frequency magnetic link (HFML) was patented [1] and a basic theoretical concept of thyristor-based power electronic transformer [2] was proposed by W. McMurray. The proposed model was a single-stage direct AC–AC converter through an HFML mimicking the working principle (voltage step-up and step-down) of the traditional distribution transformer. In 1980 and 1995, AC–AC converter-based SST topologies were implemented by the Naval Construction Battalion Centre [3] and the Electrical Power Research Institute (EPRI) [4], respectively. However, SST topologies had low power handling capabilities due to the unavailability of the high power solid-state switching devices and therefore, not suitable for traditional

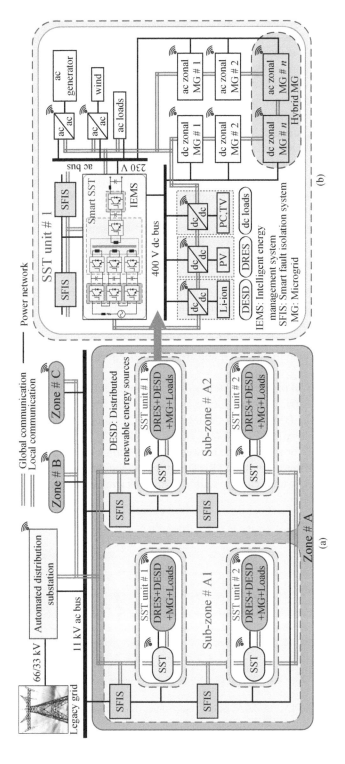

FIGURE 1.1 SST-based future power distribution system: (a) SST-based zonal power distribution system and (b) zoomed view of SST unit # 2.

grid integration. Since 1970, several SST topologies have been proposed [5, 6] based on the application areas, working principles, types of switches utilized, and number of power conversion stages.

Extensive research works related to the implementation of novel SST topologies, control functionalities, and the energy-efficient operation are going on all over the world by different research groups. In 2005, a multilevel intelligent universal transformer was proposed by the researchers of EPRI for DC fast charging application [7], as shown in Figure 1.2. The diode clamped converters were used as a front-end rectifier (FER) and DC–DC (single side of HFML) converter. Silicon-insulated gate bipolar transistor (Si-IGBT) modules were utilized at the load side DC–AC converter. Traditional ferrite core was used to design the HFML and operated at 20 kHz. In 2011, an advanced power electronic transformer (UNIFLEX) was proposed for the universal and flexible power management system [8], shown in Figure 1.3. It was a three-stage topology, where cascaded multilevel converter (MLC) for the FER and several full-bridge isolated DC–DC converters were adopted. The advanced power electronic transformer stepped down the 3.3 kV voltage level to 415 V. The amorphous magnetic-material-based HFML was adopted for the isolated DC–DC stage. A power electronic transformer topology was designed by ABB for traction application [9]. The topology adopted the 6.5-kV Si-IGBT-based H-bridge cells and a nanocrystalline magnetic-material-based HFML for the power transformation. For the DC–DC conversion, the LLC resonant converter was adopted without any dedicated control loop to operate it at its resonance frequency to achieve maximum efficiency.

Several successful projects on designing and implementing different SST topologies have been carried out at Future Renewable Electric Energy Delivery and Management (FREEDM) System Center in North Carolina State University, USA. The first SST topology (FREEDM Gen I) was a 3.6 kV-120 V, 10-kVA SST-enabled Green Energy Hub for an interface between the traditional utility grid and the residential power distribution system [10, 11] with advanced bidirectional power flow capability. The renewable energy source, ESS, electric vehicle charging station,

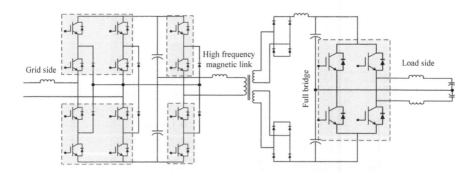

FIGURE 1.2 A 2.4-kV, 20-kVA power electronic transformer from Electrical Power Research Institute [7].

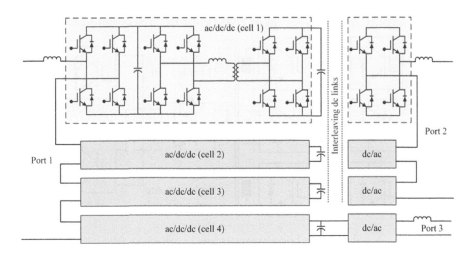

FIGURE 1.3 A 3.3-kV, 300-kVA power electronic transformer for a unified and flexible power management system [8].

and LV AC/DC load can be connected at the 400-V DC or 120-V AC ports. The LV grid will not be affected by the disturbances at the distribution grid due to the isolated DC–DC stages. Also, unpredictable and unscheduled power oscillations from renewable energy sources connected to the LV side will not pollute the distribution grid. Therefore, the renewable-based distribution grid becomes stable and resilient. However, the efficiency of the system was only 88%. The efficiencies of the FREEDM Gen-II SST prototype [12] based on the 10-kV 10-A silicon carbide metal–oxide–semiconductor field-effect transistor (SiC MOSFET) and the FREEDM Gen-III SST prototype [13, 14] based on the 15-kV 10-A SiC MOSFET increased to 95.5% and 97.5%, respectively. To reduce the number of the HFMLs, researches have proposed novel multi-winding HFML-based SST topology [15, 16], as shown in Figure 1.4. The multi-winding HFML can be operated as a multiple active bridge (MAB) isolated DC–DC converter within the SST. In this way, the number of HFMLs and the overall complexity of designing the SST are reduced.

Among several existing SST topologies, the three-stage SST topology is the most popular due to the availability of both low-voltage AC (LVAC) and low-voltage DC (LVDC) ports along with an isolated HFML DC–DC conversion stage. The converters for each stage vary according to the power requirements, switching functions, control dynamics, fault-tolerant capabilities, modularity, and switching vector redundancy. The three-phase three-level neutral-point clamped (NPC) converters are industrially matured and found application in the FER stage of SSTs. The DC-bus voltage balancing is easier in case of three-level NPC converters in comparison to the other cascaded MLC topologies. A modular MLC is another popular choice for the FER stage of the SST [17]. It facilitates an MV DC link, redundancy, and advanced control. However, several half-bridge modules and bulky smoothing capacitors are

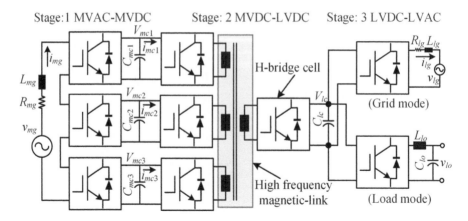

FIGURE 1.4 A multiple active bridge converter-based single-phase solid-state transformer topology with full H-bridge cells [16].

required, thus increasing the system cost and complexity. Another modular topology for FER stage of the SST is the cascaded H-bridge (CHB) MLC [11], which facilitates simple control system architecture, fault-tolerant capability, and redundancy in the switching space vectors. For the HFML-based DC–DC stage, the utilization of half-bridge or full-bridge modules is the common practice. In three-phase HFML-based topologies, three-level NPC converters or matrix converters have also been applied and investigated [18]. The third stage is the LV DC–AC stage. Several LV inverter topologies, such as H-bridge, NPC, mixed-mode H-bridge with bidirectional power flow capability exist in the literature. The inverter should handle the intermittency of the renewable energy generation system, random power fluctuations, mode identification system integration, and disturbance rejection capability for the LV grid implementation.

For the switching devices of the earlier topologies of the SST, commercially available Si IGBTs with antiparallel Si diodes with the voltage-handling capability of 1.2–6.5 kV have been extensively utilized. The recommended operating frequency of the IGBTs is less than 10 kHz to avoid a substantial switching energy loss. Therefore, they are not suitable for high-frequency DC–DC stage of the SST to reduce the size of the HFML. To deal with the high switching losses at higher frequencies, the SiC MOSFET devices have come into the picture. SiC MOSFET devices of 1.2 kV have been applied in the SST for more than 40 kHz switching operation and have a more reduced loss profile than the Si IGBTs at higher frequencies. Commercially available SiC MOSFETs have a maximum voltage rating of 1.7 kV and current rating of 115 A, as shown in Figure 1.5(a). There are several manufacturing companies, like CREE, Wolfspeed, Semikron, etc., which continue their effort to commercialize high-power SiC MOSFET devices. For example, SEMITRANS, SKiM 93, SEMiX 3 Press-Fit modules can handle 10–350 kW power with a maximum 600 A chip current. However, LV rated switches require

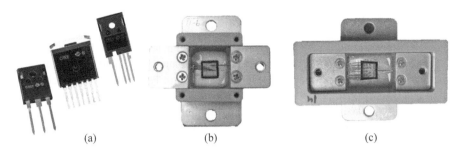

(a) (b) (c)

FIGURE 1.5 (a) Commercially available 1.2 kV and 1.7 kV Si MOSFETs from CREE/ Wolfspeed, (b) prototype 10-kV SiC MOSFET, and (c) prototype 15-kV SiC MOSFET [19].

a cascaded connection for the direct grid integration of the SST, which makes the system costly, heavy, and less efficient. To tackle this issue, the 10 kV and 15 kV, 10-A SiC MOSFET devices have been developed and characterized, as shown in Figures 1.5(b) and (c), respectively.

However, they are not commercially available until now. They have been applied for the SST and tested in the laboratory for 12–40 kHz switching frequency operation. With these high-power-rated switching devices, the number of components of the SST drastically reduces enabling high-frequency operation at a low loss and high power density operation. However, one disadvantage of these MOSFET devices is that they show high conduction losses at high-power operation, which tends to decrease the efficiency of the SST. The IGBTs have lower conduction losses at high-power applications due to the drift region conductivity modulation methods. Therefore, the 15-kV SiC IGBT has been developed [20] for high-power applications, although not fully characterized and not commercially available in the market. It is evident from the above discussion that the high-power SiC MOSFETs and SiC IGBTs will be available in the market soon, which will make the SST an energy-efficient renewable energy hub with grid power quality improvement capabilities. This might become a potential alternative of the traditional transformer.

1.3 HIGH-FREQUENCY MAGNETIC LINK FOR THE SST

Among the three stages, the magnetically isolated dc–dc converter stage faces some strict design challenges due to the inclusion of the HFML. The design process of the HFML involves multiphysics problems that entail complex tradeoffs between electrical and magnetic properties including efficiency. Moreover, the performance of the HFML depends on the switching characteristics of power switching devices and various excitation signals. Therefore, extensive multiphysics research in the field of design and optimization of HFMLs is needed to develop next-generation technologies. The core loss in the HFML tends to increase as it is commonly operated at high frequency with non-sinusoidal excitation signals. The higher the frequency, the smaller the size of the HFML will be. Moreover, for a high-density power transfer with a smaller sized core, the maximum flux density of the core should be high enough to prevent core saturation as well as thermal heating. To deal with those issues, several advanced magnetic materials, such as amorphous and nanocrystalline materials, are often used due to their low core loss profile at higher frequencies and high flux density characteristics.

For the manufacturing of the traditional distribution transformer core, the silicon steel ($Fe_{97}Si_3$, $Fe_{93.5}Si_{6.5}$) and ferrite (MnZn, NiZn) have been extensively adopted [5]. The silicon steel material has the highest magnetic flux density (2.0 T); however, it has higher core loss property. The soft ferrite material has low specific flux density (0.26–0.55 T) which results in an increased size of the HFML. The amorphous ($Fe_{76}(SiB)_{24}$, $Co_{73}(SiB)_{27}$) magnetic material has comparable flux density (1.56 T) and shows high power loss at higher frequencies [21]. The nanocrystalline (FeCuNbSiB) magnetic material has lowest core loss and comparable flux density (1.23–1.45 T) [5] and therefore, it is the most suitable candidate in terms of flux density and core loss for the development of energy-efficient, compact SST. It is to be noted that the cost of the nanocrystalline material is comparatively high in comparison to the other existing magnetic materials.

In addition to the electromagnetic performance, the availability of the magnetic materials (especially the size of the ribbon) and the complexity to design the core in the laboratory need to be considered in selecting a suitable core shape. The selection of the core shape is as important as the selection of the HFML core size. The shape and size optimization can be carried out based on the electromagnetic field analysis for the efficient design of the HFML core. To select a suitable core shape, six basic shapes are considered for investigation. The top view of these shapes with their dimensions is shown in Figure 1.6. The cross-sectional view of these shapes with their side views and dimensions are depicted in Figure 1.7. The volume of all these shapes are considered to be 140 cm^3 and their cross-sectional area is 5 cm^2.

The magnetic flux distribution performances of different core shapes are illustrated in Figure 1.8. The simulation results from the ANSYS software environment show that the magnetic flux distribution is not uniform for all other shapes except for the shape shown in Figures 1.8(a) and (f). Figures 1.8(b)–(d) show the core shapes in which the magnetic flux density is more near the edges. Due to the ununiformed distribution of the flux, the core loss increases. Moreover, the distortion of the output voltage of the HFML is observed for the core in which the flux is not properly distributed. For each of the shapes, the secondary load is varied to find the maximum power rating, and the magnetic flux distribution and distortion in the induced voltage are observed.

The torus shape is found to be the best to develop the core of the HFML, as the shape shows a good conducting magnetic flux. Although vendors may design this complex shape while condensing the molten magnetic alloy, the special requirement increases the cost and development time of this type of core. Moreover, the predesigned core may not match the requirement, which may affect system efficiency. Vendors usually provide amorphous and nanocrystalline magnetic alloy as a ribbon with a thickness of about 20 μm. It is really difficult to develop a torus shape core

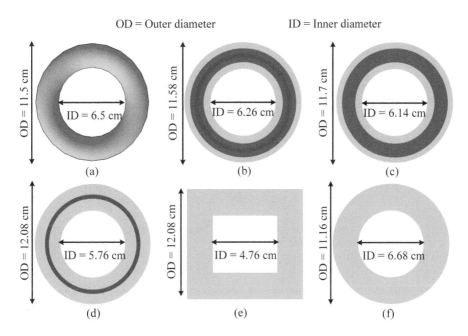

FIGURE 1.6 Top views of the cores with different shapes: (a) torus, (b) octagonal cross-section toroidal core, (c) hexagonal cross-section toroidal core, (d) twisted rectangular cross-section toroidal core, (e) rectangular cross-section square-shaped core, and (f) rectangular cross-section hollow cylindrical core.

in a laboratory environment with such thin sheets or ribbons. In the laboratory, the magnetic alloy ribbon can be wrapped on a cylindrical frame with a suitable adhesive for the construction of the HFML core. Therefore, the second option is selected, i.e., the hollow cylindrical core with a rectangular cross-sectional area, as shown in Figure 1.8(f), due to its simple and effective construction. A systematic analysis can

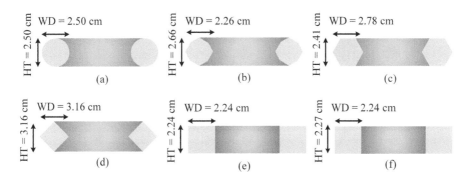

FIGURE 1.7 Cross-sections and side views of the cores with the corresponding dimensions: (a) torus, (b) octagonal cross-section toroidal core, (c) hexagonal cross-section toroidal core, (d) twisted rectangular cross-section toroidal core, (e) rectangular cross-section square-shaped core, and (f) rectangular cross-section hollow cylindrical core.

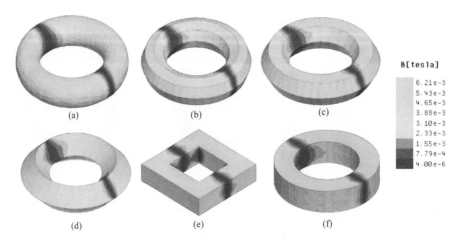

FIGURE 1.8 Magnetic flux distributions in the cores: (a) torus, (b) octagonal cross-section toroidal core, (c) hexagonal cross-section toroidal core, (d) twisted rectangular cross-section toroidal core, (e) rectangular cross-section square-shaped core, and (f) rectangular cross-section hollow cylindrical core [22].

be carried out in the ANSYS software environment to optimize the size of the core by following the human intervened genetic algorithm [22]. The dimensions of the optimized prototype core are shown in Figure 1.9.

To design and characterize the optimized 3-kVA HFML under non-sinusoidal excitations, 20 μm thick and 2.5 cm wide amorphous and nanocrystalline magnetic alloy ribbons are collected. A platform is designed in the laboratory to develop a core

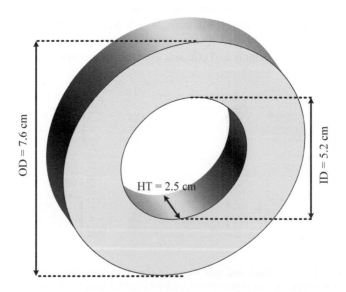

FIGURE 1.9 Dimensions of the optimized rectangular cross-section hollow cylindrical core.

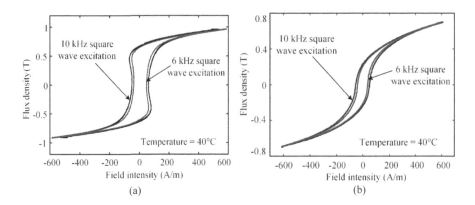

FIGURE 1.10 Hysteresis (B-H) loops under different frequency square wave excitations for: (a) amorphous core and (b) nanocrystalline core [22, 23].

with micrometer thickness ribbon. The developed cores are wound with Litz wires in a single-layer arrangement, which in turn reduce the skin and proximity effects in high-frequency applications. It is important to characterize magnetic materials under non-sinusoidal excitations for the design optimization of the HFML, as vendors usually provide core loss data for sinusoidal excitations. The HFMLs are tested in the laboratory under a square wave excitation generated by a high-frequency inverter. The magnetic field intensity is calculated from the measured excitation current and the magnetic field density is calculated from the measured induced voltage in the secondary winding.

Figures 1.10(a) and (b) show the measured B–H loops of the amorphous and the nanocrystalline magnetic materials, respectively, under 6 and 10 kHz square wave excitations at 40°C. The area of the B–H loops for the amorphous alloy is much bigger than those obtained from the nanocrystalline alloy. That means that the core loss is higher for the amorphous material than that of the nanocrystalline material with similar excitation. The power analyzer Tektronix PA4000 is used to measure the losses under square wave excitations of different amplitude and frequency. Figures 1.11(a) and (b) compare the loss characteristics of the amorphous alloy-based HFML with the nanocrystalline alloy-based HFML. It can be seen that the amorphous alloy-based core provides higher core loss than the nanocrystalline alloy-based core.

1.4 MODELING OF MULTIPLE ACTIVE BRIDGE DC–DC CONVERTER FOR THE SST APPLICATION

The utilization of the traditional HFML-based dual active bridge (DAB) DC–DC converter for the SST application is a common practice. However, due to the cascaded connection of the switching modules for direct grid integration, several DAB converters are required thus increasing the number of HFMLs. Although this technique increases the modularity characteristics of the system, the utilization of several

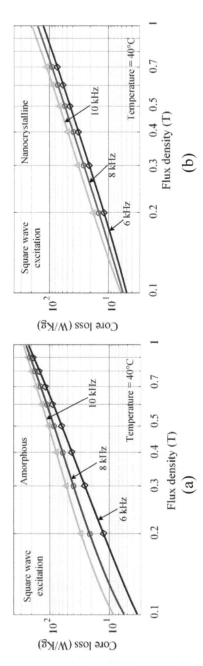

FIGURE 1.11   Core loss characteristics under different frequency square wave excitations for: (a) amorphous core and (b) nanocrystalline core [22, 23].

HFMLs increases the SST design complexities and the possibility of parameter mismatches. Therefore, multiple active bridge (MAB) DC–DC converter concept has been proposed for next-generation SST, where multiple H-bridges share the same HFML [24–26]. This configuration reduces the number of HFMLs and increases cross-coupling power transfer capability. The detailed modeling and the control technique of the MAB converter are investigated for SST application. The small and large signal average models of the converter are developed to better understand the control dynamics.

The detailed model of the MAB DC–DC converter is shown in Figure 1.12, which shows that the MV side has four active bridges with four constant and equal DC voltages V_2, V_3, V_4, and V_5. The windings of the HFML are numbered as "1," "2," "3," "4," and "5." The subscript numbers of the parameters relate to the corresponding windings. The windings turn ratios are denoted by n_2, n_3, n_4, and n_5, and the MV side capacitors are expressed as C_2, C_3, C_4, and C_5. R and L denote the corresponding winding resistance and leakage inductance, respectively. The bridge switches and winding currents are symbolized by "s" and "i," respectively. At the LV side of HFML, the constant DC voltage V_1 can be replaced by V_o, which will be considered as

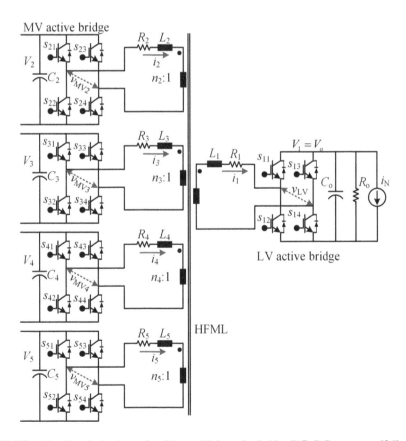

FIGURE 1.12 Detailed schematic of the multiple active bridge DC–DC converter [26].

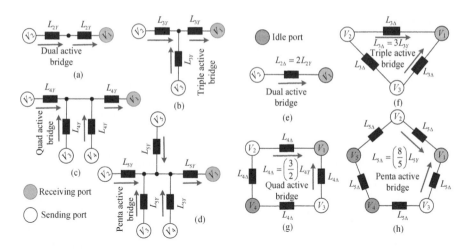

FIGURE 1.13 (a)–(d) Star equivalent models and (e)–(h) delta equivalent models of the multiple active bridge converters [26].

the converter output side. Further, the smoothing capacitor C_o, load R_o, and current source i_N are added at the LV side.

To investigate the power flow characteristics, the MAB converter is simplified by an equivalent star configuration, as shown in Figures 1.13(a)–(d), where the symbol L_Y with the numbered subscripts defines the leakage inductances of the windings of the corresponding converters. To simplify the analysis further, the star equivalent models of each type of converter are converted to the equivalent delta models, as shown in Figure 1.13(e)–(h). The delta equivalent inductances L_Δ with the associated subscript numbers for the type of the converter are expressed in terms of the star equivalent inductances. From the mathematical point of view, one new type of port is introduced, i.e., the idle port. This port does not transfer power and remains idle. For the normal DAB converter, the transferred power P_{D12}, from winding "2" to "1" for the single-phase shift (SPS) control (with unity turns ratio) can be written as:

$$P_{D12} = \left(\frac{1}{4}\right)\frac{V_1 V_2}{f_{sw} L_{2Y}} D(1-D) \tag{1.1}$$

where, D defines the duty ratio, as shown in Figure 1.14(a) for the SPS control, and f_{sw} is the switching frequency. If the LV side voltage source is replaced by the resistive load R_o and a current source, as shown in Figure 1.12, then the voltage V_1 can be obtained by using (1.2):

$$V_1 = \left(\frac{1}{4}\right)\frac{V_2 R_o}{f_{sw} L_{2Y}} D(1-D) \tag{1.2}$$

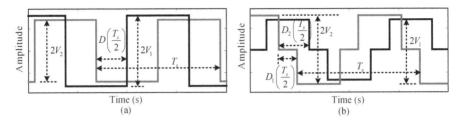

FIGURE 1.14 High-frequency magnetic link voltage waveforms of the ports "1" and "2" under (a) single-phase shift modulation and (b) dual-phase shift modulation.

If the dual-phase shift control (DPS) is applied, then the transferred power from winding "2" to "1" can be written as:

$$P_{D12} = \begin{cases} \left(\dfrac{1}{4}\right)\dfrac{V_1 V_2}{f_{sw} L_{2Y}}[D_2(1-D_2)-\dfrac{1}{2}D_1^2], \ 0 \le D_1 \le D_2 < 1 \\[2mm] \left(\dfrac{1}{4}\right)\dfrac{V_1 V_2}{f_{sw} L_{2Y}}[D_2(1-D_1-\dfrac{1}{2}D_2)], \ 0 \le D_2 < D_1 < 1 \end{cases} \tag{1.3}$$

where D_1 and D_2 denote the inner phase shift and the outer phase shift ratio, respectively, as shown in Figure 1.14(b). The processed power of port "1" and the transferred power to port "1" are the same for the DAB converter. For the triple active bridge (TAB) converter, the transferred power from port "2" and "3" to port "1" for the SPS control will be:

$$P_{T12} = \left(\frac{1}{6}\right)\frac{V_1 V_2}{f_{sw} L_{2Y}}D(1-D) \tag{1.4}$$

$$P_{T13} = \left(\frac{1}{6}\right)\frac{V_1 V_3}{f_{sw} L_{2Y}}D(1-D) \tag{1.5}$$

For the DPS control scheme, the transferred power P_{T12} can be modified as:

$$P_{T12} = \begin{cases} \left(\dfrac{1}{6}\right)\dfrac{V_1 V_2}{f_{sw} L_{2Y}}[D_2(1-D_2)-\dfrac{1}{2}D_1^2], \ 0 \le D_1 \le D_2 < 1 \\[2mm] \left(\dfrac{1}{6}\right)\dfrac{V_1 V_2}{f_{sw} L_{2Y}}[D_2(1-D_1-\dfrac{1}{2}D_2)], \ 0 \le D_2 < D_1 < 1 \end{cases} \tag{1.6}$$

The transferred power P_{T13} can be obtained by replacing the voltage V_2 in (1.6) by V_3. It should be noted that the only difference between the transferred power with SPS and DPS scheme is the duty ratio-based parameters. Therefore, for the rest of the analysis, the discussion is limited only for the SPS scheme. The processed power of port "1" for the TAB converter can be expressed as:

$$P_{T1} = \left(\frac{1}{6}\right)\frac{V_1(V_2 + V_3)}{f_{sw}L_{3Y}}D(1-D)$$

(1.7)

Similarly, the processed power of port "1" for the quad active bridge (QAB) and the penta-active bridge (PAB) converter can be written as:

$$P_{Q1} = \left(\frac{3}{16}\right)\frac{V_1(V_2 + V_3)}{f_{sw}L_{4Y}}D(1-D)$$

(1.8)

$$P_{P1} = \left(\frac{1}{5}\right)\frac{V_1(V_2 + V_3)}{f_{sw}L_{5Y}}D(1-D)$$

(1.9)

To develop the lossless average model of the PAB converter operating in SPS control technique, all the voltages and all the leakage inductances are assumed equal, i.e., the currents in all the windings will be the same. The dynamic equations can be written as:

$$\frac{di_1(t)}{dt} = \frac{1}{5L_{5Y}}\left[s_2(t)v_2(t) - s_1(t)v_o(t)\right]$$

(1.10)

$$\frac{dv_o(t)}{dt} = -\frac{v_o(t)}{R_oC_o} + \left(\frac{4}{C_o}\right)s_1(t)i_1(t) - \frac{i_N}{C_o}$$

(1.11)

where $s_1(t)$ and $s_2(t)$ are the switching functions for the bridge "1" and bridge "2," respectively. According to the generalized averaging framework developed in reference [27], Equations (1.10) and (1.11) can be separated into their associated zeroth order and first-order current (real and imaginary) components as:

$$\frac{d\langle i_1\rangle_{1R}}{dt} = \frac{1}{5L_{5Y}}\left[\begin{array}{c}\langle s_2\rangle_0\langle v_2\rangle_{1R} + \langle s_2\rangle_{1R}\langle v_2\rangle_0 - \\ \langle s_1\rangle_0\langle v_o\rangle_{1R} - \langle s_1\rangle_{1R}\langle v_o\rangle_0\end{array}\right] + \omega_s\langle i_1\rangle_{1I}$$

(1.12)

$$\frac{d\langle i_1\rangle_{1I}}{dt} = \frac{1}{5L_{5Y}}\left[\begin{array}{c}\langle s_2\rangle_0\langle v_2\rangle_{1I} + \langle s_2\rangle_{1I}\langle v_2\rangle_0 - \\ \langle s_1\rangle_0\langle v_o\rangle_{1I} - \langle s_1\rangle_{1I}\langle v_o\rangle_0\end{array}\right] - \omega_s\langle i_1\rangle_{1R}$$

(1.13)

$$\frac{d\langle v_o\rangle_0}{dt} = -\frac{\langle v_o\rangle_0}{R_oC_o} + \frac{4}{C_o}\left[\begin{array}{c}\langle s_1\rangle_0\langle i_1\rangle_0 + 2\langle s_1\rangle_{1R}\langle i_1\rangle_{1R} + \\ 2\langle s_1\rangle_{1I}\langle i_1\rangle_{1I}\end{array}\right] - \frac{\langle i_N\rangle_0}{C_o}$$

(1.14)

To avoid any saturation of the HFML, the zeroth values of the switching functions are designed to be zero, i.e., $\langle s_1\rangle_0 = \langle s_2\rangle_0 = 0$. Now, Equations (1.12)–(1.14) can be modified as:

$$\frac{d\langle i_1\rangle_{1R}}{dt} = \frac{1}{5L_{5Y}}\left[\langle s_2\rangle_{1R}\langle v_2\rangle_0 - \langle s_1\rangle_{1R}\langle v_o\rangle_0\right] + \omega_s\langle i_1\rangle_{1I}$$

(1.15)

$$\frac{d\langle i_1\rangle_{1I}}{dt} = \frac{1}{5L_{5Y}}\left[\langle s_2\rangle_{1I}\langle v_2\rangle_0 - \langle s_1\rangle_{1I}\langle v_0\rangle_0\right] - \omega_s\langle i_1\rangle_{1R} \tag{1.16}$$

$$\frac{d\langle v_o\rangle_0}{dt} = -\frac{\langle v_o\rangle_0}{R_oC_o} + \frac{8}{C_o}\left[\langle s_1\rangle_{1R}\langle i_1\rangle_{1R} + \langle s_1\rangle_{1I}\langle i_1\rangle_{1I}\right] - \frac{\langle i_N\rangle_0}{C_o} \tag{1.17}$$

Using Equations (1.15)–(1.17), Equation (1.18) can be derived from Equations (1.10) and (1.11), assuming that $\langle i_N\rangle_0 = I_N$ and $\langle v_o\rangle_0 = V_o$.

$$\frac{d\vec{x_L}}{dt} = A_L\vec{x_L} + B_L\vec{u_L} \tag{1.18}$$

where

$$\vec{x_L} = \begin{bmatrix} \langle v_o\rangle_0 & \langle i_1\rangle_{1R} & \langle i_1\rangle_{1I} \end{bmatrix}^T$$

$$\vec{u_L} = \begin{bmatrix} V_2 & I_N \end{bmatrix}^T$$

$$A_L = \begin{bmatrix} -\dfrac{1}{R_0C_0} & \left(\dfrac{8}{C_0}\right)\langle s_1\rangle_{1R} & \left(\dfrac{8}{C_0}\right)\langle s_1\rangle_{1I} \\[3mm] -\left(\dfrac{1}{5L_{5Y}}\right)\langle s_1\rangle_{1R} & 0 & \omega_s \\[3mm] -\left(\dfrac{1}{5L_{5Y}}\right)\langle s_1\rangle_{1I} & -\omega_s & 0 \end{bmatrix}$$

$$B_L = \begin{bmatrix} 0 & -\dfrac{1}{C_0} \\[3mm] \left(\dfrac{1}{5L_{5Y}}\right)\langle s_2\rangle_{1R} & 0 \\[3mm] \left(\dfrac{1}{5L_{5Y}}\right)\langle s_2\rangle_{1I} & 0 \end{bmatrix}$$

Here, the subscripts "R" and "I" define the real and imaginary components, respectively. The average of the switching functions, after including the necessary corrections as mentioned in [28], can be expressed by Equations (1.19)–(1.22):

$$\langle s_2\rangle_{1R} = 0 \tag{1.19}$$

$$\langle s_2\rangle_{1I} = -\frac{2}{\pi} \tag{1.20}$$

$$\langle s_1 \rangle_{1R} = -D(1-D)\left(\frac{\pi^2}{4}\right) \tag{1.21}$$

$$\langle s_1 \rangle_{1I} = -\left(\frac{2}{\pi}\right)\cos\left\{\sin^{-1}\left[D(1-D)\left(\frac{\pi^3}{8}\right)\right]\right\} \tag{1.22}$$

After substituting the values of the switching functions found from Equations (1.19)–(1.22) in Equation (1.18), the large-signal average model of the MAB will be formed. The system shifts from the steady state to the transient condition if there are any small perturbations and therefore, the small-signal perturbation-based transfer function between the control input and output needs to be developed. The small-signal model can be represented as:

$$\frac{d\vec{x_s}}{dt} = A_L \vec{x_s} + \vec{B_s}\Delta d \tag{1.23}$$

where

$$\vec{x_s} = \begin{bmatrix} \Delta v_{o0} & \Delta i_{1R} & \Delta i_{1I} \end{bmatrix}^T$$

$$\vec{B_s} = \begin{bmatrix} \left(\frac{16}{C_0}\right)\left[D(1-D)\left(\frac{\pi^3}{8}\right)I_{1I} - \cos\left\{\sin^{-1}\left[D(1-D)\left(\frac{\pi^3}{8}\right)\right]\right\}\right]I_{1R} \\[2em] \frac{2}{5}\left(\frac{V_{o0}}{L_{5Y}}\right)\cos\left\{\sin^{-1}\left[D(1-D)\left(\frac{\pi^3}{8}\right)\right]\right\} \\[2em] -\frac{2}{5}\left(\frac{V_{o0}}{L_{5Y}}\right)D(1-D)\left(\frac{\pi^3}{8}\right) \end{bmatrix}$$

where symbol "Δ" represents the small perturbations. Figure 1.15 demonstrates the step-by-step categorical task flow for the derivation and the implementation of the large-signal average model of the MAB converter. With the help of Equations (1.10)–(1.22), the time-domain state-space model can be developed. The time-domain state-space model is transformed into the frequency-domain transfer function model by using the numerical values from Table 1.1 and the corresponding MATLAB® function. After that, the step responses of the model are derived.

For the validation of the developed large-signal average model, three types of models are considered for comparison. The first one is the output voltage equation for the PAB converter that can be derived from Equation (1.9). This model is kept as a reference and the two other models are compared with it, as shown in Figure 1.16. The figure shows that the improved model perfectly tracks the reference voltage when compared with that from the generalized average model. Based on the developed small-signal model, the duty ratio-based voltage balance controller for the PAB

Step 1. Develop the time domain dynamic model of the MAB converter based on (1.10) and (1.11);

$\downarrow$

Step 2. Derive the average model of the switching functions based on (1.19)–(1.22);

$\downarrow$

Step 3. Represent the zero and first order components in time domain state-space form based on (1.18);

$\downarrow$

Step 4. Insert the numerical values of the variables from Table 1.1 into (1.18) for the time domain large signal model of the MAB converter; and,

$\downarrow$

Step 5. Transform the state-space model to frequency-domain transfer function model using MATLAB to obtain the step response, as shown in Figure 1.16

FIGURE 1.15   Step-by-step procedure to implement the large-signal model of the multiple active bridge converter.

converter is designed, as shown in Figure 1.17(a). The main objective of the MV side controller is to correct the output voltage if there is any voltage unbalance at the input. The controller architecture produces four duty signals for driving the four active bridges of PAB. The duty signal for the LV side bridge is kept constant and considered as the reference. The output results are compared with different proportional-integral (PI) controller parameters, as shown in Figure 1.17(b). The reference voltage is set to 180 V and the time step is set as 0.1 second. Figure 1.17(b) shows that with the increase of the integral parameter, the response becomes oscillatory and any further increase will make the system unstable.

TABLE 1.1

Simulation Parameters

Parameter	Values
Rated power	10 kVA
Rated input AC voltage	4.8 kV
Rated LV side voltage	531.54 V
Rated output resistance, R_o	47 Ω
Output capacitor, C_o	500 μF
Rated MV side AC voltage	1.9 kV
Leakage inductance	420 μH
Winding resistance	0.64 Ω
Nominal duty ratio, D	0.5
Switching frequency, f_{sw}	40 kHz

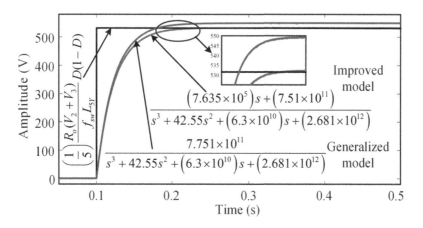

FIGURE 1.16 Improved performance of the derived large-signal average model for the penta-active bridge converter.

1.5 MULTISTAGE CONTROL ARCHITECTURE FOR THE SST

In this study, for the converters of a three-stage MAB converter-based SST, shown in Figure 1.4, a data-driven coordinated controller architecture is designed to improve the grid power quality in terms of the grid voltage distortions, sag, and swell. An improved dual-loop current controller and an inter-module decoupled voltage balance controller are designed for the MV stage of the SST. For a 30-kVA, 1- kV/400 V SST, three H-bridge modules with the 10-kV SiC devices can be connected in series to form the 6.35-kV (single-phase) MVAC with the nominal 3.74-kV MVDC link voltage. The LVDC link (400 V) can be used to integrate the distributed renewable energy sources, ESSs, DC loads, and DC MGs. Finally, the 400-V LVDC voltage is converted to 230-V (single-phase) LVAC voltage for integrating AC loads and AC MGs.

There are several control data and several distinct control objectives that need to be regulated based on the specific SST topology and application. The SST topology considered in this paper has the following control data: (1) LVAC reactive power, Q_{lg}; (2) LVAC active power, P_{lg}; (3) MVAC grid current, i_{mg}; (4) LVAC voltage, v_{lo}; (5) LVAC current, i_{lg}; (6) LVDC link voltage, V_{lc}; (7) MVAC reactive power, Q_{mg}; (8) MVDC link voltage, V_{mc}; and (9) MVDC link current, i_{mc}. In this study, the discussion is limited to the controller coordination of the SST converters when the SST is connected to the MVAC grid. The control coordination in the islanded mode is out of the scope of this study, as the system will then behave as a normal distributed AC or DC MG system.

Figure 1.18 shows the coordinated controller architecture. The architecture has four distinct control layers which are managed in a coordinated manner. The third control layer is the core control layer where decisions are made on which controller combination will be activated. For the controller design, the synchronously rotating *dq* frame-based control technique is adopted. The average differential model of the

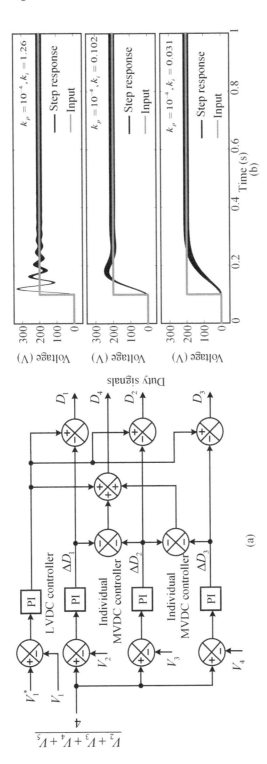

FIGURE 1.17 (a) Low-voltage DC controller with individual medium-voltage DC regulators for the penta-active bridge DC–DC converter and (b) step-response of the designed controller for different control parameters.

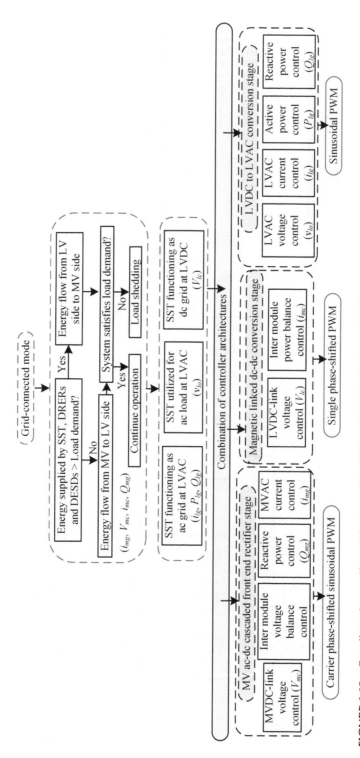

FIGURE 1.18 Coordinated controller architecture of the SST-enabled distribution grid [16].

FER in terms of the real and imaginary components can be written as Equations (1.24)–(1.26):

$$L_{mg}\frac{di_{mg}}{dt} = -R_{mg}i_{mg} + v_{mg} - \sum_{N=1}^{3}V_{mcN}d_{mN} \tag{1.24}$$

$$L_{mg}\frac{di_{ig}}{dt} = -R_{mg}i_{ig} + v_{ig} - \sum_{N=1}^{3}V_{mcN}d_{iN} \tag{1.25}$$

$$i_{mg}d_{mN} + i_{ig}d_{iN} = C_{mcN}\frac{dV_{mcN}}{dt} + \frac{V_{mcN}}{R_{mcN}}(N = 1,2,3) \tag{1.26}$$

where, in the real axis, i_{mg} is the grid current, V_{mcN} is the MVDC-link voltage, v_{mg} is the grid voltage, L_{mg} is the inductance, R_{mg} is the grid equivalent resistance, C_{mcN} is the DC link capacitance, and d_{mN} is the switching function. In the imaginary axis, the subscripts, "m" of the current, the voltage, and the switching function are replaced by the subscript "i". After transforming into a dq reference frame, Equations (1.24)–(1.26) can be written as:

$$\begin{bmatrix} \dfrac{di_{mgd}}{dt} \\ \dfrac{di_{mgq}}{dt} \end{bmatrix} = \begin{bmatrix} -\dfrac{R_{mg}}{L_{mg}} & \omega \\ -\omega & -\dfrac{R_{mg}}{L_{mg}} \end{bmatrix}\begin{bmatrix} i_{mgd} \\ i_{mgq} \end{bmatrix} + \begin{bmatrix} \dfrac{v_{mgd}}{L_{mg}} \\ \dfrac{v_{mgq}}{L_{mg}} \end{bmatrix} - \begin{bmatrix} \displaystyle\sum_{N=1}^{3}\dfrac{V_{mcN}d_{mdN}}{L_{mg}} \\ \displaystyle\sum_{N=1}^{3}\dfrac{V_{mcN}d_{mqN}}{L_{mg}} \end{bmatrix} \tag{1.27}$$

$$i_{mgd}d_{mdN} + i_{mgq}d_{mqN} = C_{mcN}\frac{dV_{mcN}}{dt} + \frac{V_{mcN}}{R_{mcN}}(N = 1,2,3) \tag{1.28}$$

where, the subscripts, "d" and "q", represent the d- and the q-axis components of the corresponding parameters, respectively. For the voltage imbalance mitigation among each H-bridge cells, a separate voltage balance controller is designed. According to reference [29], for the complete decoupling between the main system controller and the inter-module voltage balance controller, Equations (1.29) and (1.30) must be satisfied, where Δv_{mcN} is the difference between the reference and the measured voltage, and V_{mc_nom} is the nominal MVDC-link voltage.

$$\sum_{N=1}^{3}\Delta v_{mcN} = 0 \tag{1.29}$$

$$\sum_{N=1}^{3}V_{mcN} = \sum_{N=1}^{3}V_{mc\_nomN} = 3V_{mc\_nom} \tag{1.30}$$

To satisfy the conditions of Equations (1.29) and (1.30), after small-signal perturbations, the condition of Equation (1.31) should hold true, where d_{md} is the duty cycle for the nominal MVDC link voltage output and Δd_{mdN} is the small variation due to Δv_{mcN}. From Equation (1.31), (1.32) can be derived, which avoids the interaction between the system voltage controller and the voltage balance controller. Based on Equation (1.32), the voltage balance controller is designed, as shown in Figure 1.19, where the proportional and integral gains of the PI controllers are set as 0.1 and 0.05, respectively.

$$\sum_{N=1}^{3}\left(V_{mcN}+\Delta v_{mcN}\right)\left(d_{mdN}+\Delta d_{mdN}\right)=3V_{mc\_nom}d_{md} \tag{1.31}$$

$$\Delta d_{md3}=\frac{\displaystyle\sum_{N=1}^{3}\Delta v_{mcN}d_{md}-\sum_{N=1}^{2}V_{mcN}\Delta d_{mdN}}{V_{mc3}} \tag{1.32}$$

It should be noted that instead of the traditional PI-based current controller, a dual loop current control architecture is designed due to its superior harmonic suppressing capability [30]. The closed-loop d-axis current transfer function model can be derived as:

$$\frac{i_{mgd}}{i_{mgd\_ref}}=\frac{C_{PI}(s)P_{m}(s)\left[1+C_{P}(s)P_{s}(s)\right]}{\left[1+C_{P}(s)P_{m}(s)\right]+C_{PI}(s)P_{m}(s)\left[1+C_{P}(s)P_{s}(s)\right]} \tag{1.33}$$

where $C_{PI}(s)$ and $C_{P}(s)$ are the PI and P controller model, respectively. $P_{m}(s)$ and $P_{s}(s)$ define the main and the pseudo plant models, respectively. The $P_{s}(s)$ can be expressed by Equation (1.35). If it is assumed that the $P_{s}(s)$ mimics the $P_{m}(s)$ exactly, then Equation (1.34) can be simplified and expressed by (1.35).

$$P_{s}(s)=\frac{1}{sL_{mg}+R_{mg}} \tag{1.34}$$

$$\frac{i_{mgd}}{i_{mgd\_ref}}=\frac{C_{PI}(s)P_{m}(s)}{1+C_{PI}(s)P_{m}(s)} \tag{1.35}$$

The proportional and the integral gains of the dual-loop current controller are selected as 100 and 700, respectively, to get a bandwidth of 160 Hz. The proportional gain of the P controller model is selected such that the disturbance rejection loop bandwidth becomes 390 Hz, which does not exceed the Nyquist rate of the switching frequency. The voltage controller is designed in a traditional way, i.e., the voltage controller dynamics are kept much slower (10 Hz) than that of the current controller. In a balanced steady-state condition, each MVDC port of the DC–DC stage transfers an equal amount of power to the LVDC port. The power flow from each MVDC port to the LVDC port can be derived as:

$$P_{ml1}=\frac{L_{mc1}'V_{lc}L_{m3}'L_{m4}'D_{ml1}(1-D_{ml1})}{2f_{sw}\left[L_{m2}'\left(L_{l}L_{m3}'+L_{l}L_{m4}'+L_{m3}'L_{m4}'\right)+L_{l}L_{m3}'L_{m4}'\right]} \tag{1.36}$$

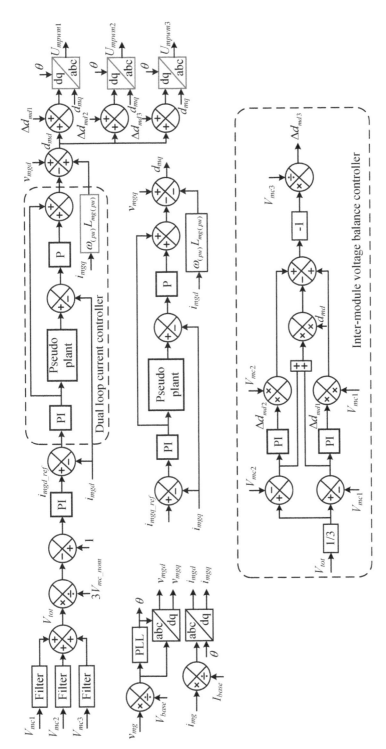

FIGURE 1.19 Controller architecture of the cascaded front-end rectifier [16].

$$L'_{m2} = \left(\frac{N_l}{N_{m2}}\right)^2 L_{m2}, L'_{m3} = \left(\frac{N_l}{N_{m3}}\right)^2 L_{m3}, L'_{m4} = \left(\frac{N_l}{N_{m4}}\right)^2 L_{m4} \tag{1.37}$$

$$V'_{mc1} = \left(\frac{N_l}{N_{m1}}\right) V_{mc1} \tag{1.38}$$

Here, all the parameters are referred to as the LVDC port. P_{ml1} is the supplied power from the MVDC port "1" to the LVDC port, D_{ml} is the duty cycle, and V_{lc} is the LVDC link voltage. L_l defines the winding leakage inductance at the LVDC side and L_m with the numbered subscripts, define the three corresponding winding leakage inductances at the MVDC side. N_l is the number of turns of the LV winding and N_m with the numbered subscripts define the number of turns of the three corresponding MV windings. Now, if I_{mcN} defines the DC input current of the Nth port at the MVDC side, the power equation can be expressed as:

$$\sum_{N=1}^{3} P_{mlN} + P_{loss} = \sum_{N=1}^{3} V_{mcN} I_{mcN} \tag{1.39}$$

where, P_{loss} is the total power loss of the DC–DC stage. The controller of the FER stage keeps the MVDC link voltages constant, i.e., $V_{mc1} = V_{mc2} = V_{mc3} = V_{mc}$. Therefore, according to Equation (1.40), the summation of the input current will be constant according to the power rating of the DC–DC stage.

$$\sum_{N=1}^{3} I_{mcN} = \frac{\sum_{N=1}^{3} P_{mlN} + P_{loss}}{V_{mc}} \tag{1.40}$$

This property helps to design a power balance controller by controlling the current, as shown in Figure 1.20. The controller ensures equal amount of power flow in the magnetic-linked H-bridges at the FER.

As previously stated, the LVAC port may be operated as a grid mode or as a load mode. In the grid mode, the LV inverter should supply the regulated active and reactive power according to the grid demand along with a pollution-free grid current. The dynamic dq model of the LVAC stage operating as a grid mode can be expressed as:

$$\begin{bmatrix} \dfrac{di_{lgd}}{dt} \\[2mm] \dfrac{di_{lgq}}{dt} \end{bmatrix} = \begin{bmatrix} -\dfrac{R_{lg}}{L_{lg}} & \omega \\[2mm] -\omega & -\dfrac{R_{mg}}{L_{mg}} \end{bmatrix} \begin{bmatrix} i_{lgd} \\[2mm] i_{lgq} \end{bmatrix} + \begin{bmatrix} \dfrac{v_{lgd}}{L_{lg}} \\[2mm] \dfrac{v_{lgq}}{L_{lg}} \end{bmatrix} - \begin{bmatrix} \dfrac{V_{lc}d_{ld}}{L_{lg}} \\[2mm] \dfrac{V_{lc}d_{lq}}{L_{lg}} \end{bmatrix} \tag{1.41}$$

where, i_{lg} is the LVAC grid current, V_{lc} is the LVDClink voltage, v_{lg} is the LVAC grid voltage, L_{lg} and R_{lg} are the LVAC grid equivalent inductance and resistance, respectively, and d_l is the LV inverter switching function. Based on Equation (1.41),

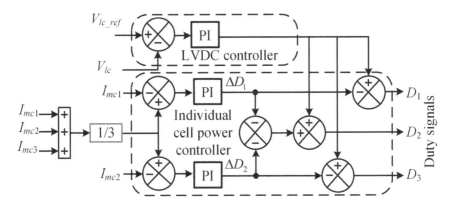

FIGURE 1.20 Power and voltage balance controller high-frequency magnetic link DC–DC stage [16].

the dual loop current controller with the outer loop active and reactive power controller is designed in a cascaded manner. The controller architecture is similar to the one shown in Figure 1.19. The only difference is that there is no inter-module voltage balance controller and the current reference components are generated from the active and reactive power controller instead of the voltage controller. The current controller PI values are chosen to have a bandwidth of 198 Hz and the P-value is chosen to eliminate any frequency distortions from the grid current. The proportional and integral values of the active and reactive power controllers are selected to have a controller bandwidth of 7.2 Hz. In the case of accessing the LVAC stage like a load mode, the load will be connected via a second-order low-pass LC filter. The main purpose of the controller will be to supply a well-regulated AC voltage. The output voltage with the filter dynamics in the *dq* rotating frame can be derived as:

$$
\begin{bmatrix} \dfrac{dv_{lod}}{dt} \\ \dfrac{dv_{loq}}{dt} \end{bmatrix} = \begin{bmatrix} 0 & \omega \\ -\omega & 0 \end{bmatrix} \begin{bmatrix} v_{lod} \\ v_{loq} \end{bmatrix} + \frac{1}{C_{lo}} \begin{bmatrix} i_{lld} - i_{lod} \\ i_{llq} - i_{loq} \end{bmatrix} \tag{1.42}
$$

where, v_{lo} is the load voltage, i_{ll} is the inverter current, i_{lo} is the load current, L_{lo} is the filter inductance, and C_{lo} is the filter capacitance. The controller design methodology is similar to the one designed for the grid mode controller, except this time, the load voltage is controlled in the outer control loop, as shown in Figure 1.21.

A sophisticated processor and hardware circuitry are required for the control coordination and the controller design for an SST to manage several control data with the shortest possible time scale. Moreover, the embedded controller should incorporate suitable communication technology for the seamless operation of the SST with the grid and transmit the necessary state estimated feedback data to the command center. In this respect, the entire control architecture is implemented in the dSPACE MicroLabBox-based embedded platform. The Xilinx Kintex-7 XC7K325T

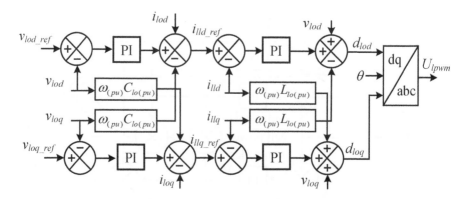

FIGURE 1.21   Controller for the low-voltage DC–AC stage operating as a load mode [16].

FPGA can be operated along with the real-time processor to satisfy special input/ output requirements. The FPGA control logics are implemented by the dedicated Simulink® RTI FPGA programming block sets. The PC with the Control Desk 6.2 is operated as a real-time interface for monitoring, observing the system status, and taking actions accordingly.

A simulation is performed to test the performance of the designed controller architecture under grid voltage distortions, swell, and sag with the parameters mentioned in Table 1.2. To demonstrate the performance of the coordinated controller under grid voltage abnormalities, it is assumed that the DC MG is self-sustaining and does not fetch or feed power from or to the SST. In this case, the LVAC side load demand is entirely met up by the SST, i.e., the power flow direction will be from the MVAC to the LVAC side.

TABLE 1.2
Simulation Parameters for Controller Implementation

Parameter	Values
Nominal power ratings	30 kVA, 11 kV/400 V
Nominal MVDC and LVDC-link voltage	3.74 kV, 400 V
Switching frequency	2.5 kHz
MVAC grid inductance, L_{mg} and resistance, R_{mg}	100 mH, 0.7 Ω
MVDC- and LVDC-link capacitances, C_{mc} and C_{lc}	2200 μF, 2200 μF
LVAC grid inductance, L_{lg} and resistance, R_{lg}	50 mH, 0.7 Ω
LVAC output capacitance, C_{lo} and inductance, L_{lo}	2200 μF, 0.23 mH
Magnetic link winding inductances (MV side)	420 μH
Magnetic link winding inductance (LV side)	180 μH
Magnetic link turns ratio	9.5:1
Magnetic link operating frequency	20 kHz

Abbreviations: LVDC, low-voltage DC; MVDC, mediun-voltage DC.

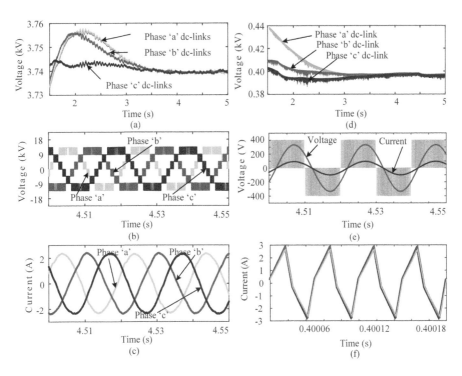

FIGURE 1.22 The performance of the SST under balanced grid condition: (a) MVDC link voltage; each phase contains three MVDC link voltages, (b) MV PWM voltages before filtering circuit, (c) MVAC grid currents, (d) LVDC link voltages; each phase contains single LVDC link, (e) LV PWM voltage before filtering with corresponding LVAC voltage and current, when LVAC port is functioning as load mode, and (f) HFML currents. MVDC, medium-voltage DC; LVDC, low-voltage DC; PWM, pulse width modulation; HFML, high-frequency magnetic link.

Figure 1.22 shows the performance of the SST under balanced conditions when the power is flowing from the MVAC side to the LVAC side. The system reaches steady state in 3.5 second. The controller of the FER ensures the constant and balanced MVDC link voltages at 3.74 kV, shown in Figure 1.22(a), along with a pollution-free grid current of 1.57 A, shown in Figure 1.22(c). Figure 1.22(b) shows the three-phase seven-level MV pulse width modulated (PWM) waveforms before filtering with a switching frequency of 2.5 kHz. Figure 1.22(d) illustrates the balanced LVDC link voltages at 400 V, ensured by the voltage balance controller of the HFML DC–DC stage. Figure 1.22(e) shows the two-level PWM voltage waveform along with the filtered LVAC voltage and current waveforms, when the grid is accessed as a load via the second-order LC filter. The controller shown in Figure 1.21 ensures that the LVAC port can supply a constant AC voltage of 230 V to the load. Figure 1.22(f) demonstrates the 20-kHz HFML current waveforms under a SPS modulation. The power balance controller of Figure 1.20 ensures that all the three bridges of the FER supply an equal amount of power to the LV side.

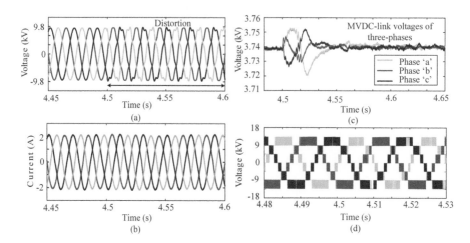

FIGURE 1.23 The performance of the solid-state transformer under grid voltage distortions starting at 4.5 second: (a) distorted MVAC grid voltages, (b) MVAC currents, (c) MVDC link voltages, and (d) MV PWM voltages. MVAC, medium-voltage AC; MVDC, medium-voltage DC; PWM, pulse width modulation.

Figures 1.23–1.26 demonstrate the controller performance under grid voltage abnormalities. Figure 1.23(a) shows that the grid voltages are intentionally distorted by adding 0.06 pu. fifth and 0.05 pu. seventh harmonics. The distortion starts at 4.5 second. Figure 1.23(b) shows that the designed dual-loop current controller can suppress the unwanted harmonics from the MVAC currents. Figure 1.23(c) reveals that the MVDC link voltages oscillate from the stable condition for 0.05 second and eventually

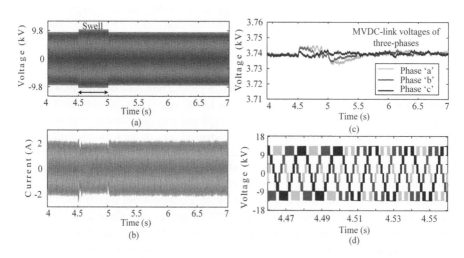

FIGURE 1.24 The performance of the solid-state transformer under grid voltage swell starting at 4.5 second: (a) MVAC grid voltages, (b) MVAC currents, (c) MVDC link voltages, and (d) MV PWM voltages. MVAC, medium-voltage AC; MVDC, medium-voltage DC; PWM, pulse width modulation.

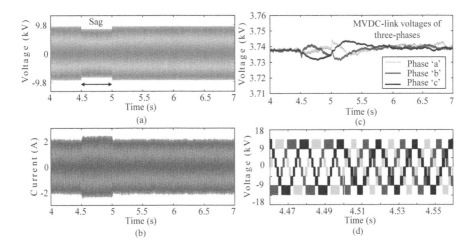

FIGURE 1.25 The performance of the solid-state transformer under grid voltage sag starting at 4.5 second: (a) MVAC grid voltages, (b) MVAC currents, (c) MVDC link voltages, and (d) MV PWM voltages. MVAC, medium-voltage AC; MVDC, medium-voltage DC; PWM, pulse width modulation.

they track the nominal value of 3.74 kV even if the distortion persists. The SST performance is further investigated under the voltage swell conditions. A 10% grid voltage swell is introduced at 4.5 second till 5 second, as shown in Figure 1.24(a). The MVAC current decreases due to the voltage sag and it shifts the MVDC link voltages from the nominal condition, as shown in Figures 1.24(b) and (c), respectively. However, the system eventually reaches to the balanced condition within 0.05 second after removing the voltage swell. A 10% voltage sag is also introduced in the grid voltage for 0.5 second starting at 4.5 second, shown in Figure 1.25(a). This time, the controller increases the MVAC current to compensate for the effect of the voltage sag and tries to keep the MVDC link voltages at their nominal conditions, shown in Figures 1.25(b) and (c), respectively. Figure 1.26 shows the LVDC link voltages under momentary grid voltage abnormalities. Figure 1.26(a) shows that the distortion has no considerable effect on the LVDC link voltage. The LVDC link voltages face momentary 2.5% voltage oscillations from the stable condition due to the voltage swell, as shown in Figure 1.26(b). Figure 1.26(c) shows that the designed control system prevents any considerable effect on the LVDC link voltages due to the grid voltage sag.

For the experimental validation of the proposed controller architecture, a 2.5-kVA scale-down laboratory prototype of the single-phase SST is implemented. The MVAC grid voltage is chosen as 400 V and the MVDC link voltages of the FER are chosen as 250 V with the modulation index of 0.8. The LVDC link voltage is kept constant at 200 V. Figure 1.27 demonstrates the balanced steady-state performance of the SST. Figure 1.27(a) shows the seven-level PWM voltage at the MV side, the MVAC current, and the constant MVDC link voltage. Figure 1.27(b) shows the constant LVDC link voltage, the HFML primary and secondary currents with the high-frequency inverter voltages at the MV side.

The performance of the designed controller is also tested with grid voltage distortion, grid voltage swell, and grid voltage sag. Figure 1.28(a) shows the case where the

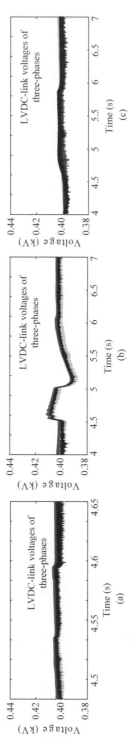

FIGURE 1.26 Low-voltage DC link voltages under grid voltage abnormalities starting from 4.5 second: (a) grid voltage distortions, (b) grid voltage swell, and (c) grid voltage sag.

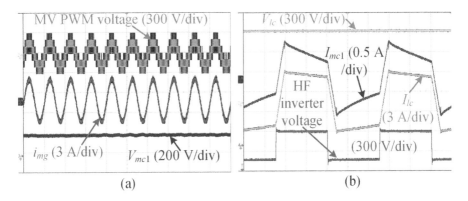

FIGURE 1.27 Experimental test results of the solid-state transformer under balanced steady-state condition: (a) MV PWM voltage, MVAC current, and MVDC link voltage and (b) LVDC link voltage, HFML currents, and high-frequency inverter voltage [16]. MVDC, medium-voltage DC; LVDC, low-voltage DC; PWM, pulse width modulation; HFML, high-frequency magnetic link.

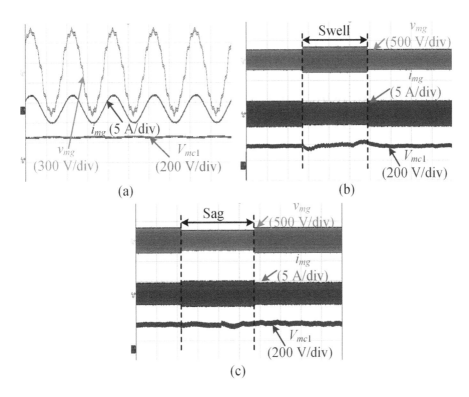

FIGURE 1.28 MVAC voltage, MVAC current, and MVDC link voltage of the SST under abnormal grid conditions: (a) grid voltage distortion, (b) grid voltage swell, and (c) grid voltage sag [16]. MVAC, medium-voltage AC; MVDC, medium-voltage DC; SST, solid-state transformer.

grid voltage is polluted with the fifth and seventh harmonics. It can be seen that the dual-loop controller of the FER eliminates the harmonics from the MVAC current and injects pure sinusoidal grid current. Moreover, the MVDC link voltage is also kept constant by the MVDC link voltage balance controller. Figure 1.28(b) shows the SST performance when a voltage swell occurs at the grid voltage. The controller decreases the grid current accordingly to maintain the power balance. A slight deviation at the MVDC link voltage is observed at the start of the voltage swell. Figure 1.28(c) shows how the system behaves under grid voltage sag. The controller increases the grid current to eliminate the impact of the voltage sag. The MVDC link voltage waveforms also show a satisfactory performance under the voltage sag condition.

1.6 CONCLUSIONS

This chapter discusses the potential of the SSTs to replace the bulky traditional distribution transformers and the diverse applications of the SSTs in the renewable energy-based grid integration and grid power quality improvement functionalities. The types of power electronic converters adopted for the efficient and reliable operation of the SSTs are discussed. The applicability of the commercially available switches in SST to operate at a high-frequency with low loss characteristics is described. Moreover, this chapter sheds lights on the upcoming HV switching devices to simplify the design and implementation of the next-generation SST. The detailed design process and fabrication procedure for developing an optimized HFML in the laboratory are covered for a highly efficient isolated DC–DC stage of the SST. The modeling and analysis of multistage controller design for the SST are mentioned for the fast and stable regulation of the DC link voltages and power routing yet solving several grid power quality problems, like grid voltage sag, swell, and distortions.

REFERENCES

1. W. McMurray, "Power converter circuits having a high-frequency link," *U.S. Patent 3517300*, June 23, 1970.
2. W. McMurray, "The thyristor electronic transformer: A power converter using a high frequency link," *IEEE Trans. Ind. Gen. Appl.*, vol. IGA-7, no. 4, pp. 451–457, July 1971.
3. J. L. Brooks, "Solid state transformer concept development," In *Naval Material Command. Port Hueneme, Civil Eng. Lab*, Naval Construction Battalion Center, 1980.
4. P. Reischl, "Proof of the principle of the solid-state transformer and the AC/AC switch mode regulator," San Jose State Univ., San Jose, CA, *EPRI TR-105 067*, 1995.
5. M. A. Hannan *et al.*, "State of the art of solid-state transformers: advanced topologies, implementation issues, recent progress and improvements," *IEEE Access*, vol. 8, pp. 19113–19132, 2020.
6. F. Ruiz, M. A. Perez, J. R. Espinosa, T. Gajowik, S. Stynski, and M. Malinowski, "Surveying solid-state transformer structures and controls: providing highly efficient and controllable power flow in distribution grids," *IEEE Ind. Electron. Mag.*, vol. 14, no. 1, pp. 56–70, Mar. 2020.
7. J.-S. Lai, A. Maitra, A. Mansoor, and F. Goodman, "Multilevel intelligent universal transformer for medium voltage applications," *Conf. Rec. IEEE/IAS Annu. Meet.* vol. 3, pp. 1893–1899, Oct. 2005.

8. S. Bifaretti, P. Zanchetta, A. Watson, L. Tarisciotti, and J. C. Clare, "Advanced power electronic conversion and control system for universal and flexible power management," *IEEE Trans. Smart Grid*, vol. 2, no. 2, pp. 231–243, 2011.

9. D. Dujic, C. Zhao, A. Mester, J. K. Steinke, M. Weiss, S. Lewdeni-Schmid, T. Chaudhuri, and P. Stefanutti, "Power electronic traction transformer-low voltage prototype," *IEEE Trans. Power Electron.*, vol. 28, no. 12, pp. 5522–5534, 2013.

10. T. Zhao, G. Wang, S. Bhattacharya, and A. Q. Huang, "Voltage and power balance control for a cascaded H-bridge converter-based solid-state transformer," *IEEE Trans. Power Electron.*, vol. 28, no. 4, pp. 1523–1532, Apr. 2013.

11. X. She, X. Yu, F. Wang, and A. Q. Huang, "Design and demonstration of a 3.6-kV–120-V/10-kVA solid-state transformer for smart grid application," *IEEE Trans. Power Electron.*, vol. 29, no. 8, pp. 3982–3996, Aug. 2014.

12. F. Wang, Gangyao, A. Huang, W. Yu, and X. Ni, "A 3.6kV high performance solid state transformer based on 13kV SiC MOSFET," In *2014 IEEE 5th International Symposium on Power Electronics for Distributed Generation Systems (PEDG)*, Galway, pp. 1–8, 2014.

13. F. Wang, G. Wang, A. Huang, W. Yu, and X. Ni, "Design and operation of A 3.6kV high performance solid state transformer based on 13kV SiC MOSFET and JBS diode," In *2014 IEEE Energy Conversion Congress and Exposition (ECCE), Pittsburgh*, PA, pp. 4553–4560, 2014.

14. A. Q. Huang, L. Wang, Q. Tian, Q. Zhu, D. Chen, and Y. Wensong, "Medium voltage solid state transformers based on 15 kV SiC MOSFET and JBS diode," In *IECON 2016 - 42nd Annual Conference of the IEEE Industrial Electronics Society*, Florence, pp. 6996–7002, 2016.

15. L. F. Costa, G. Buticchi, and M. Liserre, "Quad-active-bridge DC–DC converter as cross-link for medium-voltage modular inverters," *IEEE Trans. Ind. Appl.*, vol. 53, no. 2, pp. 1243–1253, Mar.–Apr. 2017.

16. M. A. Rahman, M. R. Islam, K. M. Muttaqi, and D. Sutanto, "Data driven coordinated control of converters in a smart solid state transformer for reliable and automated distribution grids," *IEEE Trans. Ind. Appl.*, (Early Access), Feb. 2020.

17. M. Liserre, G. Buticchi, M. Andresen, G. De Carne, L. F. Costa, and Z.-X. Zou, "The smart transformer: Impact on the electric grid and technology challenges," *EEE Ind. Electron. Mag.*, vol. 10, no. 2, pp. 46–58, Jun. 2016.

18. Y. Liu, Y. Liu, H. Abu-Rub, and B. Ge, "Model predictive control of matrix converter based solid state transformer," In *Proceedings of IEEE International Conference of Industrial Technology (ICIT)*, pp. 1248–1253, Mar. 2016.

19. V. Pala, *et al.*, "10 kV and 15 kV silicon carbide power MOSFETs for next-generation energy conversion and transmission systems," In *Proceedings of IEEE Energy Conversion Congress and Exposition (ECCE), Pittsburgh*, PA, pp. 449–454, 2014, doi: 10.1109/ECCE.2014.6953428.

20. A. Kadavelugu, et al., "Characterization of 15 kV SiC n-IGBT and its application considerations for high power converters," In *Proc. IEEE Energy Conversion Congress and Exposition*, Denver, CO, pp. 2528–2535, 2013, doi: 10.1109/ECCE.2013.6647027.

21. M. R. Islam, G. Lei, Y. Guo, and J. Zhu, "Optimal design of high-frequency magnetic links for power converters used in grid-connected renewable energy systems," *IEEE Trans. Magnetics*, vol. 50, no. 11, pp. 1–4, Nov. 2014, Art no. 2006204, doi: 10.1109/TMAG.2014.2329939.

22. M. R. Islam, O. Farrok, M. A. Rahman, M. R. Kiran, K. M. Muttaqi, and D. Sutanto, "Design and characterisation of advanced magnetic material-based core for isolated power converters used in wave energy generation systems," *IET Electric Power Appl.*, vol. 14, no. 5, pp. 733–741, 5 2020.

23. M. A. Rahman, M. R. Islam, K. M. Muttaqi, and D. Sutanto, "Characterization of amorphous magnetic materials under high-frequency nonsinusoidal excitations," *AIP Adv.*, vol. 9, pp. 035004, 2019.

24. M. A. Rahman, M. R. Islam, K. M. Muttaqi, Y. G. Guo, J. G. Zhu, D. Sutanto, and G. Lei, "A modified carrier-based advanced modulation technique for improved switching performance of magnetic-linked medium-voltage converters," *IEEE Trans. Ind. Appl.*, vol. 55, no. 2, pp. 2088–2098, Mar.–Apr. 2019.

25. M. R. Islam, M. A. Rahman, P. C. Sarker, K. M. Muttaqi, and D. Sutanto, "Investigation of the magnetic response of a nanocrystalline high-frequency magnetic link with multi-input excitations," *IEEE Trans. Appl. Supercond.*, vol. 29, no. 2, Mar. 2019, Art no. 0602205.

26. M. A. Rahman, M. R. Islam, K. M. Muttaqi, and D. Sutanto, "Modeling and control of SiC-based high-frequency magnetic linked converter for next generation solid state transformers," *IEEE Trans. Energy Convers.*, vol. 35, no. 1, pp. 549–559, Mar. 2020.

27. H. Qin and J. W. Kimball, "Generalized average modeling of dual active bridge DC-DC converter," *IEEE Trans. Power Electron.*, vol. 27, no. 4, pp. 2078–2084, Apr. 2012.

28. J. A. Mueller and J. W. Kimball, "An improved generalized average model of DC–DC dual active bridge converters," *IEEE Trans. Power Electron.*, vol. 33, no. 11, pp. 9975–9988, Nov. 2018.

29. X. She, A. Q. Huang, T. Zhao, and G. Wang, "Coupling effect reduction of a voltage-balancing controller in single-phase cascaded multilevel converters," *IEEE Trans. Power Electron.*, vol. 27, no. 8, pp. 3530–3543, Aug. 2012.

30. S. Gulur, V. M. Iyer, and S. Bhattacharya, "A dual-loop current control structure with improved disturbance rejection for grid-connected converters," *IEEE Trans. Power Electron.*, vol. 34, no. 10, pp. 10233–10244, Oct. 2019.

2 Magnetic Linked Power Converter for Transformer-Less Direct Grid Integration of Renewable Generation Systems

*A. M. Mahfuz-Ur-Rahman, Md. Rabiul Islam,
Kashem M. Muttaqi, and Danny Sutanto*
University of Wollongong

CONTENTS

2.1   INTRODUCTION

Most countries have already set their renewable energy targets to fulfill their increasing national energy demand and keep their environment sustainable. Renewable energy is non-pollutant, environment friendly, low maintenance, and will never run out.

Although the upfront cost of the renewable energy plant is relatively high and energy storage may be required due to its intermittent nature, it has more benefits than drawbacks.

Power electronic converters are the key elements that interconnect renewable energy sources with the power grid and provide the conditioning operation of the energy conversion system [1–5]. Traditionally, two-level inverters along with power frequency transformer-based power conversion systems are commonly used for the grid integration of renewable energy sources. However, the use of the power frequency step-up transformers and the filter circuits increases the total loss by 50% and the system volume by 40% [6]. Therefore, medium-voltage power converters, that can integrate the renewable energy sources directly with the power grid without requiring the heavy and bulky power frequency step-up transformers and the filter circuits, are becoming increasingly popular.

Recently, multilevel converters have become fashionable for use as medium-voltage power converters. These multilevel power converters do not require step-up transformers and line filters for grid integration of renewable energy sources with the electrical power grid, but they require multiple isolated balanced DC power sources, galvanic isolation, and more electronic components as the voltage level is increased.

High-frequency magnetic linked power converters are increasingly being used due to their compactness and better efficiency compared to the traditional power frequency transformer-based power conversion systems. Moreover, these high-frequency magnetic linked power converters when used in conjunction with the medium-voltage power converters, such as the multilevel power converters, can offer galvanic isolation and can generate the isolated balanced DC power sources required by the medium-voltage multilevel power converters.

In this chapter, an extensive review of the current research outcomes and the possible future directions of research to develop magnetically linked power converter technologies are presented.

This chapter is structured as follows: Section 2.2 presents two-level inverter topologies for grid integration of renewable energy sources; Section 2.3 describes multilevel inverters for transformer-less grid integration of renewable energy sources; Section 2.4 presents magnetic linked power converters for grid integration of renewable energy sources; Section 2.5 describes switching techniques for magnetic linked power converters; Section 2.6 presents the modeling of inverter losses; and finally Section 2.7 concludes this chapter.

2.2 TWO-LEVEL INVERTERS FOR GRID INTEGRATION OF RENEWABLE ENERGY SOURCES

Two-level power converters are very popular and commonly used for the grid integration of renewable energy sources to the power grid. Figure 2.1 shows half-bridge single-phase, full-bridge single-phase, and three-phase two-level inverter topologies.

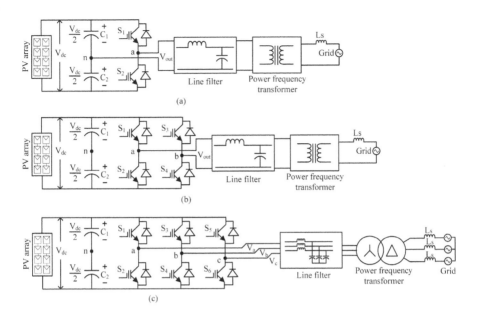

(a)

(b)

(c)

FIGURE 2.1 Traditional power converter topologies for grid integration of renewable energy sources: (a) Half-bridge single-phase, (b) full-bridge single-phase, and (c) three-phase two-level inverter topologies.

The half-bridge single-phase two-level inverter topology consists of two electronic switches (S_1 and S_2), two DC-link capacitors (C_1 and C_2), and a DC voltage source (V_{dc}). Here, the two electronic switches are complimentary in fashion. Therefore, when S_1 is on, then S_2 must be off and vice versa. When S_1 is on, then the voltage across the C_1 capacitor appears at the output terminal which is $+V_{dc}/2$. Similarly, when S_2 is on, then the voltage across the C_2 capacitor appears at the output terminal which is $-V_{dc}/2$. By using the pulse width modulation (PWM) technique, the average value at the output terminal can be varied between $+V_{dc}/2$ and $-V_{dc}/2$. Traditionally, a sinusoidal reference signal is compared with a high-frequency triangular carrier signal and the corresponding gate pulses are generated for the electronic switches to achieve sinusoidal voltage across the output terminals. The output voltage of the half-bridge single-phase inverter can be expressed as follows:

$$V_{out} = \sum_{n=1,3,5,\ldots}^{\infty} \frac{2V_{dc}}{n\pi} \sin n\omega t$$

$$= 0 \quad \text{for } n = 2,4,6,\ldots$$

(2.1)

$$V_{out\_m} = \frac{mV_{dc}}{2}$$

(2.2)

where V_{out} is the instantaneous output voltage of the half-bridge inverter, V_{out_m} is the peak value of the inverter output voltage, ω is the frequency of the inverter output voltage, and m is the modulation index.

The full-bridge single-phase two-level inverter topology generates the same output waveform using four electronic switches. The main difference is the DC bus voltage utilization. The full-bridge inverter can utilize the DC bus voltage up to 100% and the half-bridge inverter can utilize the DC bus voltage up to 50% of the input DC voltage. For the full-bridge single-phase two-level inverter topology, the two terminals in the output are connected to the middle points of the left-hand leg and the right-hand leg of the bridge circuit, respectively. Here, S_1 and S_4 need to be turned on to obtain $+V_{dc}$ at the output terminal and S_2 and S_3 need to be turned on to obtain $-V_{dc}$ at the output terminal.

Figure 2.2 shows the sinusoidal PWM technique with the corresponding output voltage waveforms for the full-bridge single-phase two-level inverter. A sinusoidal reference signal is compared with the high-frequency triangular carrier and the corresponding insulated gate bipolar transistors (IGBTs) are switched accordingly. Figure 2.2 also shows the leg voltage waveforms, V_{an} and V_{bn}, for the full-bridge single-phase two-level inverter. The output terminal voltage waveform of the half-bridge single-phase two-level inverter is the same as the output terminal voltage waveform of the full-bridge single-phase inverter except for the amplitude of the

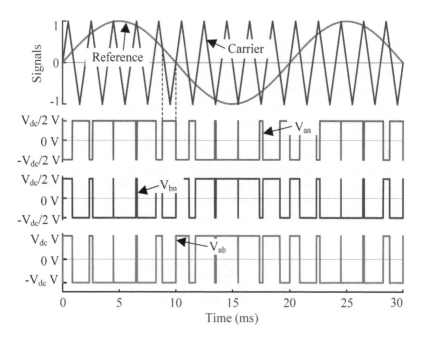

FIGURE 2.2 Sinusoidal pulse width modulation technique showing the output voltage waveforms for the full-bridge single-phase two-level inverter.

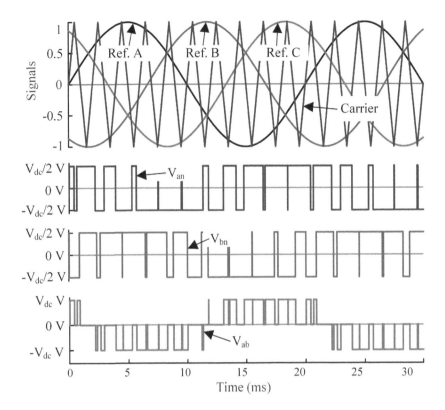

FIGURE 2.3   Sinusoidal pulse width modulation technique and the output voltage of the three-phase two-level inverter.

voltage. The peak value of the output voltage of the half-bridge single-phase two-level inverter is half of the peak value of the output voltage of the full-bridge single-phase two-level inverter.

A two-level three-phase inverter, as shown in Figure 2.1(c), consists of three legs where each leg consists of two electronic switches. Figure 2.3 shows the sinusoidal PWM technique and the output voltage waveforms of the three-phase two-level inverter.

In the three-phase two-level inverter, the peak voltage of the line-to-line voltage is the same as the DC bus voltage. Here, three sinusoidal signals are used as a reference and one single high-frequency triangular signal is used as a carrier signal. The three sinusoidal reference signals are used to drive the three legs independently.

Figure 2.4 shows the frequency spectrum of the output voltages of the full-bridge single-phase two-level inverter and the three-phase two-level inverter for a carrier frequency of 1 kHz and the modulation index of 1. The total harmonic distortion (THD) of the output voltage of a full-bridge single-phase two-level inverter is around 99.62% and the THD of the line voltage output of a three-phase two-level inverter is around 68.52%.

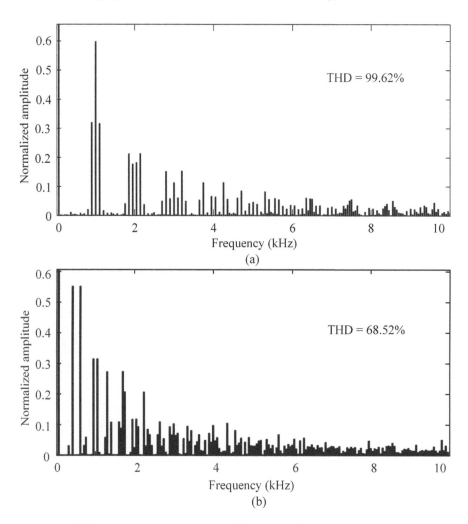

FIGURE 2.4 The frequency spectrum of (a) the output voltage of the full-bridge single-phase two-level inverter and (b) the output line voltage of the three-phase two-level inverter.

2.3 MULTILEVEL INVERTERS FOR TRANSFORMER-LESS GRID INTEGRATION OF RENEWABLE ENERGY SOURCES

The issue with the two-level inverter topologies is the high THD of the output voltage waveform due to which larger size of the filter is required. Multilevel inverters offer better output quality compared to that of the two-level inverters. The main conventional multilevel inverter topologies are the cascaded H-bridge (CHB), the neutral point clamp (NPC), and the flying-capacitor multilevel inverter topologies. As the number of levels increases in the multilevel inverter topologies, the number of devices increases significantly.

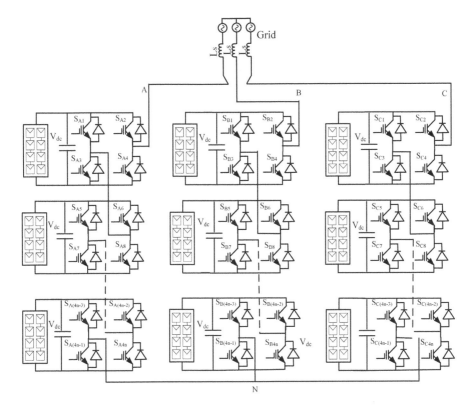

FIGURE 2.5 H-bridge multilevel inverter topology.

Figure 2.5 shows the H-bridge multilevel inverter topology. A seven-level three-phase CHB multilevel inverter topology requires 9 DC power sources and 36 switching devices [7]. Here, a single H-bridge unit can develop three voltage levels $+V_{dc}$, $-V_{dc}$, and 0. Thus by cascading the H-bridge converters, a higher output voltage can be obtained. In the CHB multilevel inverter topology, the advantage is that the number of possible output voltage level is more than twice the number of DC sources. If there are s DC sources, then the number of levels will be 2s+1. Another advantage is that the series combination of the H-bridge circuits allows the topology layout to be modularized for packaging. This enables the manufacturing procedure to be completed more quickly and cheaply. The main disadvantage of the H-bridge multilevel inverter topology is that a separate and balanced DC voltage source is needed for each of the H-bridge circuits.

Figure 2.6 shows the seven-level NPC multilevel inverter topology. A seven-level three-phase NPC multilevel inverter topology requires 36 switching devices, 30 diodes, 18 DC-link capacitors, and 1 DC power source [7]. In general, this topology requires 1 DC power source, $(2m-2)\times3$ switching devices, $(m-1)\times3$ DC-link capacitors, and $(2m-4)\times3$ clamping diodes to develop a three-phase m-level inverter.

The main advantage of the NPC multilevel inverter is that the capacitance requirement of this power converter topology is low because all the phases share a common DC bus voltage. Therefore, a back-to-back power converter topology is not only

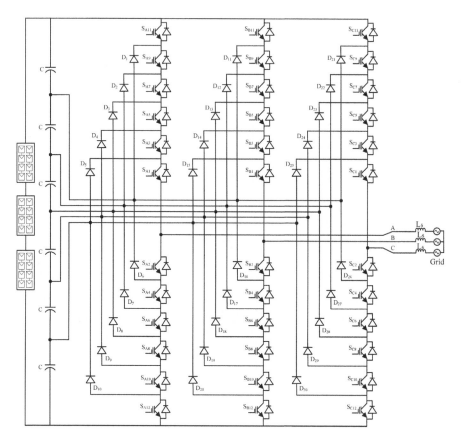

FIGURE 2.6   Seven-level neutral point clamp multilevel inverter topology.

possible but also realistic in a high-voltage back-to-back integration or an adjustable speed drive. Here, the capacitors can be pre-charged as a group. The efficiency of the NPC multilevel inverter is relatively high. The main disadvantage of this kind of inverter is that the active power flow is critical for a single power inverter due to the overcharging or discharging tendency of the intermediate DC levels without precise monitoring and control. Moreover, in this topology, the number of clamping diodes requirement is quadratically related to the number of levels which can make the system bulky for a higher number of voltage levels.

Figure 2.7 shows the seven-level flying capacitor multilevel inverter topology. The structure of the flying capacitor multilevel inverter is similar to the NPC multilevel inverter except that the inverter in this case requires capacitors instead of clamping diodes. In this topology, the DC side capacitors form a ladder structure, where the voltage across each capacitor differs from that of the next capacitor. The voltage change between the two nearby capacitor legs determines the number of the voltage steps in the output waveform. The m-level three-phase flying capacitor multilevel inverter topology requires 1 DC source, $(2m-2)\times3$ switching devices, $3\times(m-1)$ DC-link capacitors, and $3\times(m-1) \times (m-2)/2$ auxiliary capacitors.

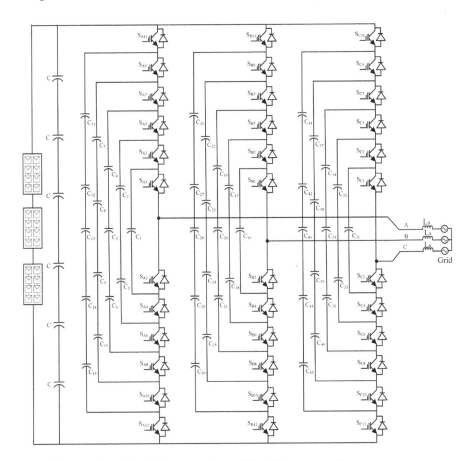

FIGURE 2.7   Seven-level flying capacitor multilevel inverter topology.

The advantages of the flying capacitor multilevel inverter are the availability of phase redundancies for capacitor voltage balancing, the control of real and reactive power flow, and the ability of the multilevel inverter to ride through short-duration outages and deep voltage sags due to having such a large number of capacitors.

The disadvantages of the flying capacitor multilevel inverter topology are: (i) complex control required to maintain the voltages across the capacitors, (ii) the complexity of pre-charging all of the capacitors to the same voltage level during the start-up process, (iii) poor switching utilization, (iv) poor efficiency for real power transmission, (v) bulky system due to a large number of capacitors, and (vi) difficulty in packaging for a high number of voltage levels.

Figure 2.8 shows the output phase voltage waveform, the frequency spectrum of the output phase voltage waveform, the line voltage waveform, and the frequency spectrum of the line voltage waveform for the H-bridge seven-level inverter. Here, the input DC voltage for H-bridge units is considered as 100 V. The switching technique is sinusoidal PWM with 1 kHz carrier frequency. The THD of the phase voltage waveform is 17.83% and that of the line voltage waveform is 10.64%.

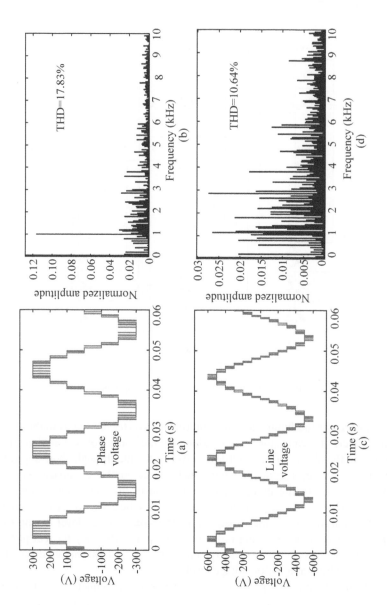

FIGURE 2.8 H-bridge seven-level inverter (a) phase voltage, (b) phase voltage frequency spectrum, (c) line voltage, and (d) line voltage frequency spectrum.

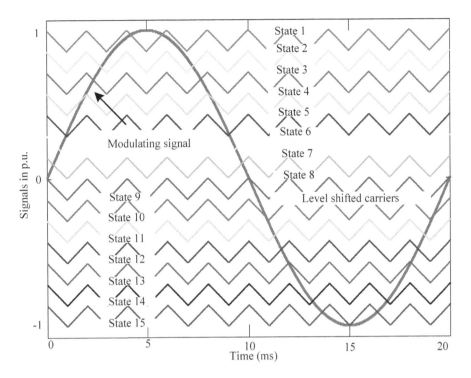

FIGURE 2.9   Level-shifted scheme for a 15-level inverter.

The NPC and the flying capacitor clamp multilevel inverters will show the same time domain and frequency domain response as those from the H-bridge multilevel inverter. Here, for the NPC and the flying capacitor multilevel inverters, the magnitude of the DC voltage source needs to be 600 V to generate the same output as the H-bridge multilevel inverter.

For multilevel inverters, there are two basic carrier-based PWM schemes: (a) the level-shifted carrier scheme and (b) the phase-shifted carrier scheme.

Figure 2.9 shows a level-shifted carrier scheme for a 15-level inverter and Figure 2.10 shows a phase-shifted carrier scheme for a 15-level inverter. For a 15-level inverter, the requirement is 14 carrier waves and one modulating signal. The modulating signal is compared with the 14 carrier waves and the corresponding gate pulses are produced. By comparing one carrier and the modulating signal, two gate pulses can be obtained. Thus, by comparing 14 carriers and the modulating signal, 28 gate pulses can be obtained. Figure 2.11 shows the gate pulse generation for the level-shifted PWM modeled in the MATLAB® software environment.

Figure 2.12 shows the gate pulse generation for the phase-shifted PWM simulated in the MATLAB software environment. Here, Fm is the fundamental frequency and, Fc is the carrier frequency. Figure 2.13 shows the output phase voltage and its frequency spectrum for the level-shifted carrier scheme in a 15-level cascaded H-bridge inverter.

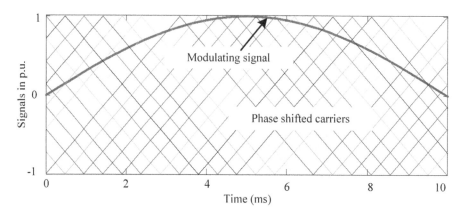

FIGURE 2.10   Phase-shifted scheme for a 15-level inverter.

Figure 2.14 shows the output phase voltage waveform and its frequency spectrum for the level-shifted carrier scheme in a 15-level CHB inverter. Here, for the same power converter topology and the same reference signal, the power converter output is different due to the different carrier scheme. The THDs of the output voltage waveforms of the level-shifted carrier scheme and the phase-shifted carrier scheme are 8.04% and 8.23%, respectively. Therefore, the level-shifted carrier-based PWM shows a better spectra property than the phase-shifted carrier-based PWM.

There are three alternative PWM strategies with different phase relationships for the level-shifted multicarrier modulation:

1. In-phase disposition (IPD), where all carrier waveforms are in phase
2. Phase opposition disposition (POD), where all carrier waveforms above zero reference are out of phase with those below zero by 180°
3. Alternate phase opposition disposition (APOD), where every carrier waveform is out of phase with its neighbor carrier by 180°

The performance of these carrier-based PWM techniques is analyzed in reference [8] and the IPD shows the best performance among the three carrier-based modulation schemes.

2.4   MAGNETIC LINKED POWER CONVERTERS FOR GRID INTEGRATION OF RENEWABLE ENERGY SOURCES

High-frequency magnetic linked power converters improve the power conversion efficiency as compared to the power frequency transformer-based power conversion systems. The magnetic link also provides galvanic isolation and can also generate multiple isolated balanced DC voltage sources. Compared with the conventional power frequency transformers, the high-frequency transformers have much smaller and lighter magnetic cores and windings, and thus much lower costs [9]. For fabrication of high-frequency transformers, the amorphous material has excellent magnetic characteristics, such as

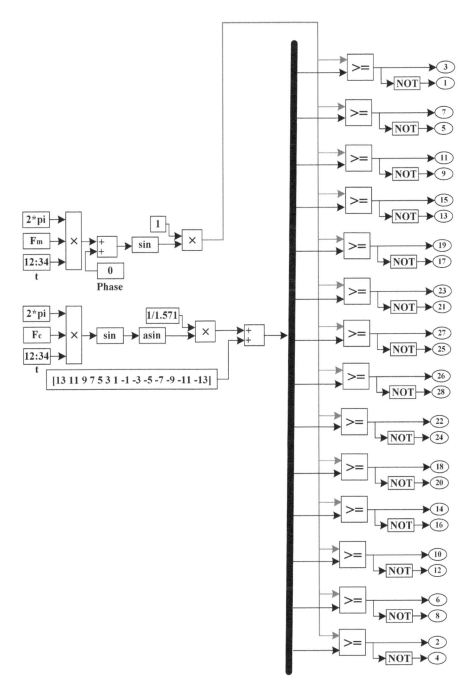

FIGURE 2.11   Gate pulse generation for level-shifted PWM in the MATLAB® software environment.

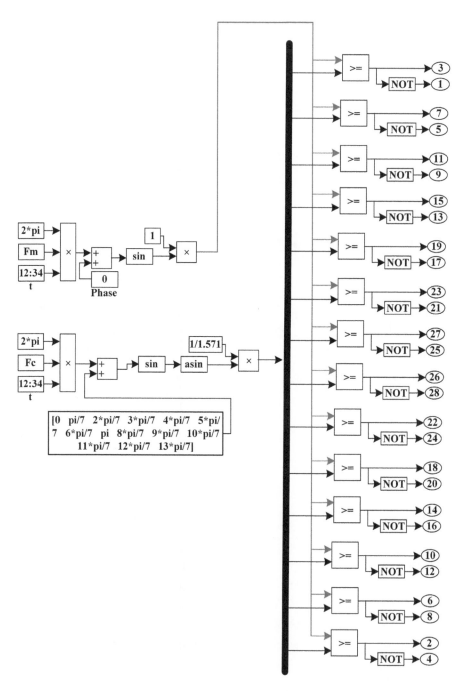

FIGURE 2.12 Gate pulse generation for the phase-shifted PWM in the MATLAB® software environment.

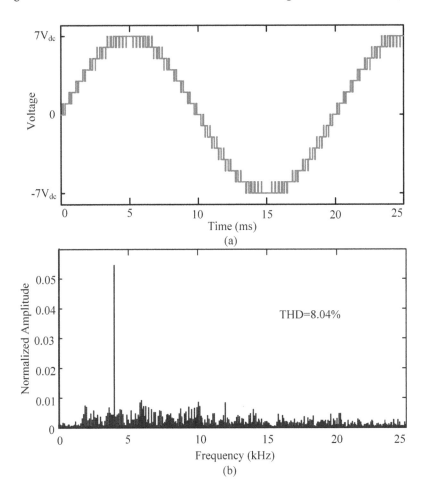

FIGURE 2.13 H-bridge 15-level inverter with level-shifted carrier scheme: (a) output phase voltage and (b) frequency spectrum of the output voltage.

high saturation flux density and relatively low specific core losses at medium to high frequencies [10]. The commercially available amorphous material is Metglas (e.g., the Metglas alloys 2605S3A and 2605SA1), which is manufactured by Hitachi Metals. The saturation flux density of the Metglas alloy 2605S3A is 1.41 T and the specific core loss at 10 kHz sinusoidal excitation of 0.5 T is 20 W/kg [11].

Due to system requirements and recent advances in power semiconductor devices and magnetic materials, the power electronic transformer (PET) has been receiving significant attention for the past two decades. Nowadays, MVA level PET has become a reality. In fact, there are currently various MVA level PETs that are used in practical applications. For example, a 1.2-MVA PET is currently in use by the Swiss Federal Railways [12]. As a result, the advanced magnetic material-based common high-frequency link may be the natural choice to generate multiple isolated and balanced DC supplies for the multilevel converters.

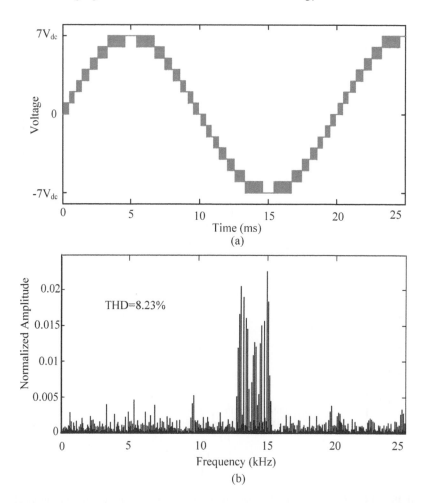

FIGURE 2.14 H-bridge 15-level inverter with phase-shifted carrier scheme: (a) output phase voltage and (b) frequency spectrum of the output voltage.

Figure 2.15 shows the circuit topology of a grid-connected medium-voltage converter with the CHB topology in conjunction with isolated DC-DC converters [13]. Here, the isolation is achieved using a high-frequency magnetic link which helps to supply the balanced DC sources and the galvanic isolation for the grid integration.

The direct AC–AC conversion system (with the help of a cyclo-converter) may decrease the energy conversion stage and boost the efficiency of the system. Figure 2.16 shows the circuit CHB converter topology based on a cyclo-converter for integrating the solar photovoltaic (PV) power plants with the medium-voltage grid [14]. Although the proposed topology decreases the energy conversion stages, the number of switching devices required is still high.

The operation of numerous PV arrays can cause a substantial voltage imbalance due to the partial shading, different ambient temperatures, nonuniformity of the solar irradiance, and inconsistency in the solar module degradation among the PV

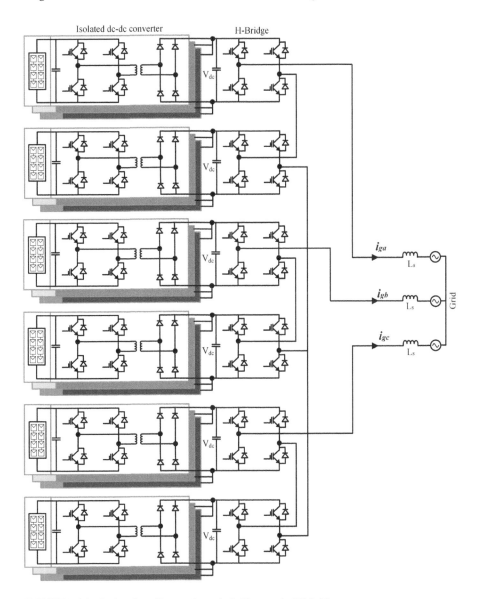

FIGURE 2.15 Isolated medium-voltage hybrid cascaded H-bridge converter.

arrays. Several research articles proposed the use of a common DC link, as depicted in Figure 2.17, to reduce the voltage imbalance issue, e.g., a common DC link for a medium-voltage solar PV power converter is reported in reference [15]. This power converter topology based on the common DC link for direct integration of multiple sources has some serious issues, such as the need for isolation and extra control strategy for solving the voltage imbalance and the common mode issue [16, 17].

The common DC link can be eliminated by using a common high-frequency magnetic link that eventually eliminates the use of the multiple isolated DC-DC power

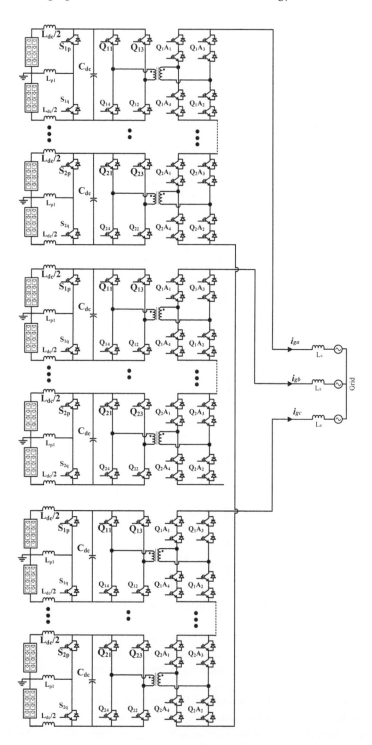

FIGURE 2.16   Cyclo-converter–based modified cascaded H-bridge converter.

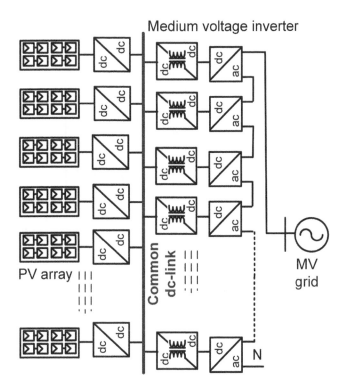

FIGURE 2.17 Single-phase medium-voltage converter based on a common DC link for direct grid integration of solar photovoltaic power plants.

converters. Recently, the common DC link has been replaced by a high-frequency (10 kHz) common magnetic link [11, 18] which inherently solves the need for multiple isolated DC-DC converters and a common DC link. In references [19, 20], a 1-kV inverter was tested experimentally with a magnetic link operating at 6 kHz and reasonable outcomes were obtained. Figure 2.18 shows the proposed single-phase medium-voltage converter based on the common magnetic link for direct grid integration of solar PV power plants.

However, the converter cost and reliability are affected due to the non-modularity of the common magnetic link power converters. Moreover, the power rating of a high-frequency magnetic link converter topology is limited by the power handling capacity of the electronic switching devices at high-frequency switching and the leakage inductances of the magnetic link [21].

On the other hand, several identical four windings (one primary winding and three secondary windings) high-frequency magnetic links can be used in parallel. Here, the primary windings of the magnetic links are excited using a single DC power source. This concept of parallel operation of multiple high-frequency magnetic links enables an option to use multiple low-power magnetic links as a replacement of a single high-power magnetic link. Therefore, the modularity is increased and the leakage inductance is reduced with this magnetic linked power converter topology.

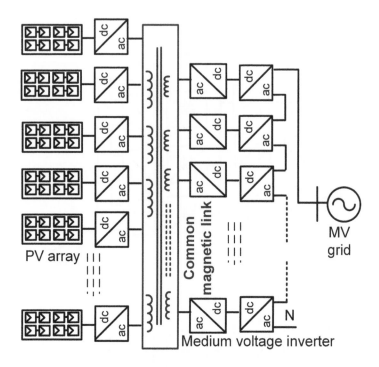

FIGURE 2.18 Single-phase medium-voltage converter based on a common magnetic link for direct grid integration of solar photovoltaic power plants.

A modular 5-kVA, 1.2-kV PV power converter was tested experimentally using two identical 2.5-kVA high-frequency magnetic links [1]. Figure 2.19 shows the circuit topology of the proposed medium-voltage PV power converter which is fully modular in construction.

Here, if a single module generates the AC output peak voltage of V_p, then by connecting N modules in series, the output peak voltage would be NV_p. Therefore, by connecting several modules, high voltages can be obtained to connect the multilevel inverter with the medium-voltage grid without stepping out with a power frequency transformer.

A new multilevel power converter topology known as a modular multilevel converter (MMC) was proposed in reference [22]. It has the property of scalability and modularity similar to the CHB power converter topology but has a single DC source like the NPC and the flying capacitor power converter topologies. The MMC topology is becoming popular due to its attractive features such as the need for only a single DC power source, internal power flow capability, high modularity, and scalability. The MMC is also suitable for industrial medium-voltage power converter applications, such as motor drives [23–25], high-voltage DC transmission systems [26], and static synchronous compensator (STATCOM) [27]. Although the MMC topology has been explored for different applications in recent years, only a few research papers have focused on its application in large-scale solar PV systems. A modular multilevel solar PV power converter with the MMC topology has been developed in reference [28], where a series of connected PV modules uses the common DC link.

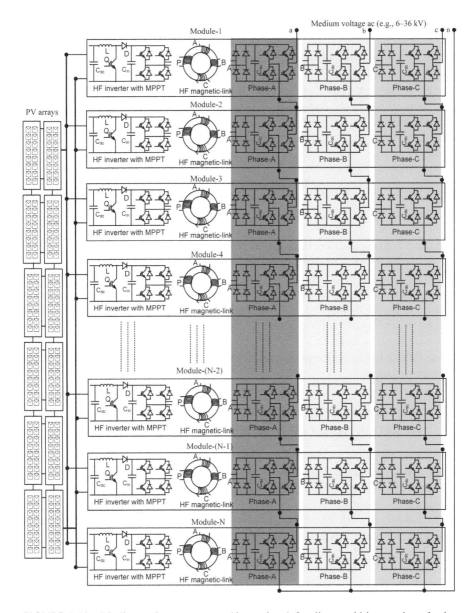

FIGURE 2.19 Medium-voltage converter (three-phase) for direct grid integration of solar PV power plants.

A medium-voltage solar PV power converter is presented in reference [29] using the H-bridge cell-based MMC power converter topology. In this proposed power converter topology, the fly-back converters are used to interconnect the PV modules with H-bridge cells. This topology provides the galvanic isolation to allow the grounding of the PV modules.

TABLE 2.1

Performance Matrix of the Magnetic Linked Power Converter Topologies

	Performance Indicator				
Converter Topology	Distributed MPPT	Large DC-link	Galvanic Isolation	Easily Scalable	Simple Control
Figure 2.15	Yes	No	Yes	Yes	No
Figure 2.17	Yes	No	Yes	Yes	No
Figure 2.18	Yes	No	Yes	Yes	Yes
Figure 2.19	Yes	No	Yes	Yes	Yes
Figure 2.20, Figure 2.21	Yes	No	Yes	Yes	No

Figure 2.20 depicts the circuit diagram of the proposed fly-back isolated DC-DC PV converter for medium-voltage applications. A dual active bridge (DAB) isolated DC-DC power converter was proposed in reference [30] to interconnect the PV modules with the half-bridge cells of the MMC topology. This isolated DAB power converter enables the distributed maximum power point tracking (MPPT) functionality and the grounding of the PV modules. Figure 2.21 depicts the circuit topology of the proposed DAB isolated DC-DC PV power converter for medium-voltage applications. Table 2.1 presents the performance metrics of different magnetic linked power converter topologies.

2.5   SWITCHING TECHNIQUES FOR THE MAGNETIC LINKED POWER CONVERTERS

The PWM switching strategy is the most commonly investigated and applied due to its better performance and easier implementation. Using the PWM technique, the gate pulses for the electronic switches can be easily generated by comparing the fundamental-frequency (50 Hz or 60 Hz) reference signal with the high-frequency carrier signals (usually triangular waveform). The sinusoidal PWM (SPWM) is the most commonly used switching technique, where a pure sinusoidal signal is used as a modulating signal. To reduce the third harmonic component and the multiples of the third harmonic component at the output voltage waveform, a triplen harmonic is injected into the pure sinusoidal modulating signal and this strategy is often referred to as the third harmonic injected PWM (THPWM). The space-vector-based hybrid PWM technique, often called the conventional space-vector PWM (CSVPWM), was proposed in reference [31] to reduce the harmonic distortion. The trapezoidal signal with long flattened top and bottom can also be used as the reference signal; such method is often referred to as the trapezoidal PWM (TRPWM) technique, where due to the flattening top and bottom portions, the switching loss of the electronic switches is reduced, but this switching technique produces a high THD compared to those from the other switching strategies. The flattened top and bottom portion in the reference signal can also be obtained by clamping the sinusoidal reference signal to a fixed value. In the sixty-degree PWM (SDPWM) technique, the sinusoidal reference

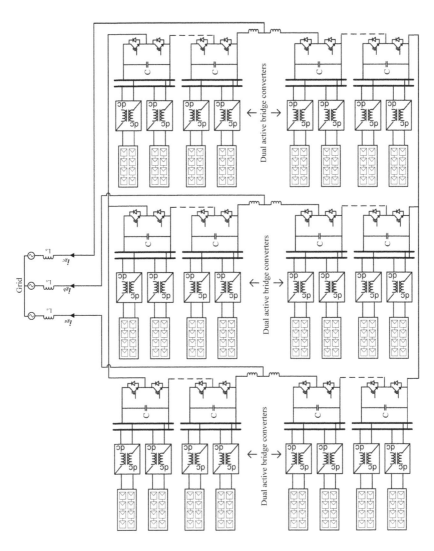

FIGURE 2.20 Fly-back isolated DC–DC medium-voltage photovoltaic converter.

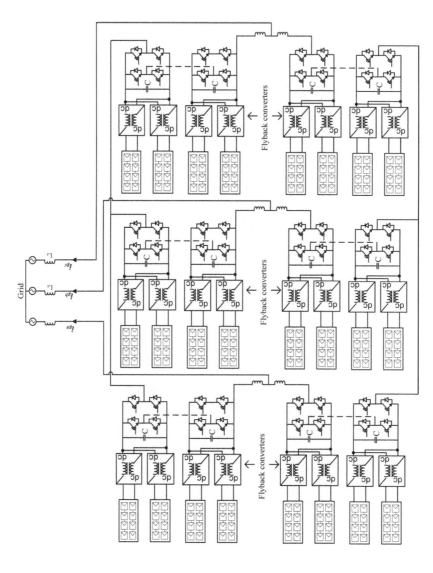

FIGURE 2.21   Dual active bridge isolated DC–DC medium-voltage photovoltaic converter.

signal is flattened for a period of 60°. The discontinuous reference signal-based PWM technique is often known as the bus clamping PWM (BCPWM). The BCPWM techniques were also well-studied for multilevel medium-voltage applications, especially to reduce the converter losses [32]. The sixty-degree BCPWM (SDBCPWM) and the thirty-degree BCPWM (TDBCPWM) are the most common forms of the BCPWM switching techniques. Recently, the third harmonic injected SDBCPWM (THSDBCPWM) and the third harmonic injected TDBCPWM (THTDBCPWM) techniques have been proposed to ensure the quality of the output power and reduce the converter losses [1].

Figure 2.22 shows different modulating signals for power converters. Figure 2.23 shows the output line voltages of an 11-kV 15-level H-bridge multilevel inverter for different switching schemes. Here, the carrier frequency used is 4 kHz and a level-shifted carrier scheme has been considered for the simulation. Figure 2.24 shows the frequency spectrum of the output line voltages of an 11-kV 15-level H-bridge inverter for different switching schemes. Here, the THD of output line voltages for CSVPWM, SPPWM, THPWM, SDBCPWM, TDBCPWM, TRPWM, THSDBCPWM, and THTDBCPWM switching techniques are 4.02, 4.63, 4.13, 4.64, 4.63, 4.8, 4.12, and 3.97%, respectively. The TRPWM produces the highest and the THTDBCPWM produces the lowest THD among the switching techniques. Figure 2.25 shows the THD for different modulation indices with different switching schemes.

The value of the THD reduces with the increase of the modulation index as depicted in Figure 2.15. Figure 2.26 shows the THDs for different modulation schemes with different multilevel inverters when the modulation index is 1. The THD reduces with the increase in the number of levels in the multilevel inverter.

2.6 MODELING OF INVERTER LOSSES

The inverter loss includes the conduction loss and the switching loss. For an IGBT module, the conduction loss occurs due to the conduction of the switch and the anti-parallel diode of the IGBT. The switching loss occurs due to the switch turn-on, the switch turn-off, the diode turn-on, and the diode turn-off operations. For the modern fast recovery diode, the turn-on loss is negligible. For the evaluation of inverter loss, the corresponding IGBTs data sheet is used. For the loss evaluation of a particular IGBT module, such as the 5SNA 1200G450300 from ABB HiPak, the characteristics curves from the datasheet [33] are used; a fifth-order polynomial equation is deduced using the curve fitting tool in the MATLAB environment.

2.6.1 CONDUCTION LOSS ANALYSIS

In the MATLAB/Simulink® model, the switch current due to switching only the switch and the switch current due to switching only the diode are obtained separately. The positive portion of the on-state device current represents the switch current due to switching only the switch, and the negative portion represents the switch current due to switching only the diode. Using the typical IGBT on-state characteristics curve and typical diode forward characteristics curve, fifth-order polynomial

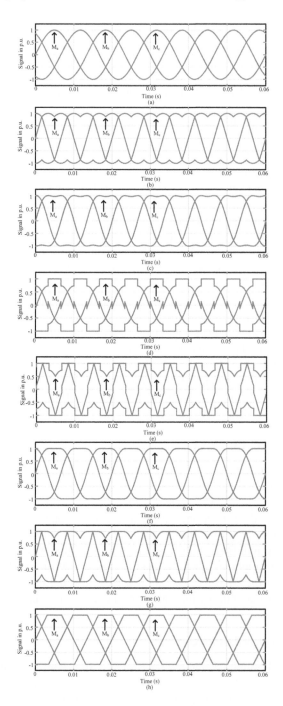

FIGURE 2.22 Modulating signals correspond to (a) SPWM, (b) CSVPWM, (c) THPWM, (d) SDBCPWM, (e) TDBCPWM, (g) THSDBCPWM, (h) THTDBCPWM, and (f) TRPWM.

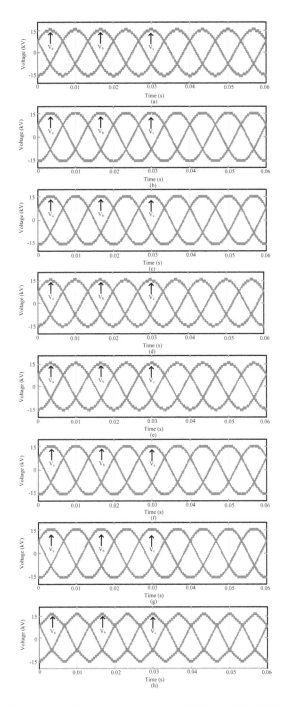

FIGURE 2.23 Output line voltage corresponds to (a) SPWM, (b) CSVPWM, (c) THPWM, (d) SDBCPWM, (e) TDBCPWM, (g) THSDBCPWM, (h) THTDBCPWM, and (f) TRPWM.

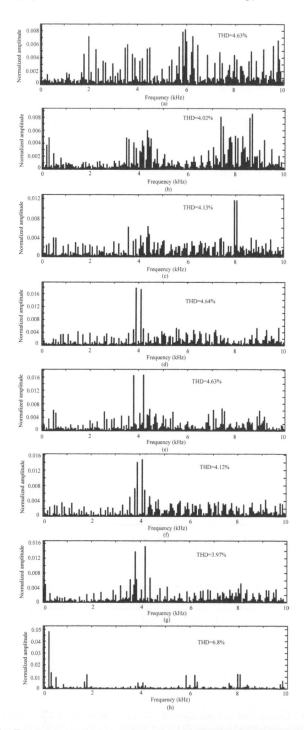

FIGURE 2.24 Frequency spectrum corresponds to (a) SPWM, (b) CSVPWM, (c) THPWM, (d) SDBCPWM, (e) TDBCPWM, (g) THSDBCPWM, (h) THTDBCPWM, and (f) TRPWM.

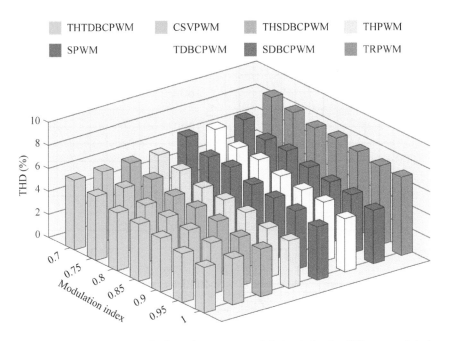

FIGURE 2.25   Total harmonic distortion versus modulation index for different modulation schemes.

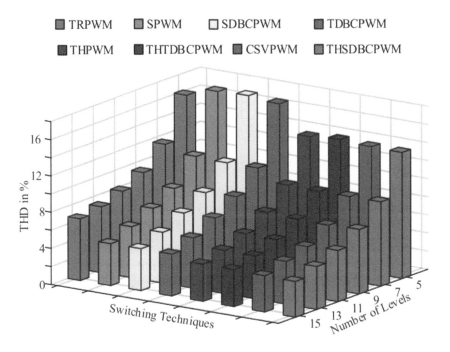

FIGURE 2.26   Total harmonic distortion for different modulation schemes with different multilevel inverters.

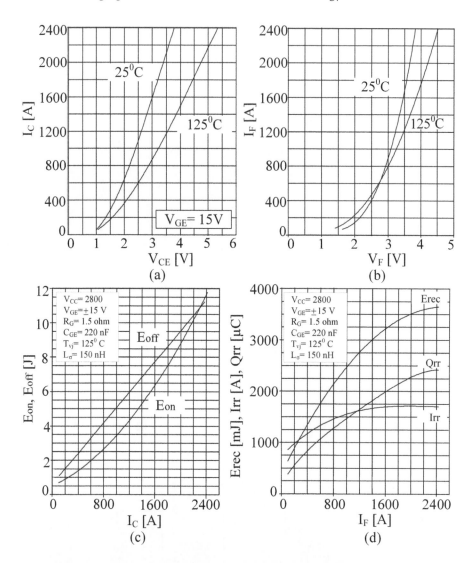

FIGURE 2.27 (a) Typical insulated gate bipolar transistor on-state characteristics, (b) typical diode forward characteristics, (c) typical energy loss per pulse vs. collector current, and (d) typical reverse recovery characteristics vs. the forward current.

equations can be used to fit the curves. The obtained fifth-order polynomial equations can then be used to derive the voltages across the switch and diode. Then using Equations (2.3) and (2.4) conduction losses of an IGBT are calculated. The fifth-order polynomial equations of Figure 2.27 (a) and (b) are:

$$v_{ce} = 1.157 \times 10^{-16} i_c^5 - 8.369 \times 10^{-13} i_c^4 + 2.382 \times 10^{-9} i_c^3 - 3.445 \times 10^{-6} i_c^2$$

$$+ 0.00417 i_c + 0.8435 \tag{2.3}$$

and

$$v_F = 2.272 \times 10^{-16} i_F^5 - 1.605 \times 10^{-12} i_F^4 + 4.356 \times 10^{-9} i_F^3 - 5.866 \times 10^{-6} i_F^2$$
$$+ 0.005096 i_F + 1.012 \tag{2.4}$$

where i_c and i_F are in (A) and V_{ce} & V_F are in (V).

The average conduction loss for the switch and diode can be calculated as [1, 34–36]:

$$P_{scl} = \frac{1}{2\pi} \int_0^{2\pi} [v_{ce}(t) \, i_c(t)] \, d(wt) \tag{2.5}$$

and

$$P_{dcl} = \frac{1}{2\pi} \int_0^{2\pi} [v_F(t) \, i_F(t)] \, d(wt) \tag{2.6}$$

If there are N numbers of IGBTs then the total conduction loss can be calculated as [1, 34–36]:

$$P_{cond} = \sum_{k=1}^{N} [P_{scl}(k) + P_{dcl}(k)] \tag{2.7}$$

2.6.2 SWITCHING LOSS ANALYSIS

The switching loss depends on the switching frequency and the current that is going to be switched. IGBT datasheet provides the losses for different switched currents. The fifth-order polynomial equation of switch turn-on energy and switch turn-off energy as a function of current is stated below:

$$E_{on} = 5.6 \times 10^{-31} i_c^5 - 6.248 \times 10^{-27} i_c^4 - 2.271 \times 10^{-12} i_c^3 + 1.246 \times 10^{-6} i_c^2$$
$$+ 0.00167 i_c + 0.5393 \tag{2.8}$$

and

$$E_{off} = 7.19 \times 10^{-32} i_c^5 + 1.599 \times 10^{-26} i_c^4 - 1.059 \times 10^{-12} i_c^3 + 3.129 \times 10^{-8} i_c^2$$
$$+ 0.004402 i_c + 0.6595 \tag{2.9}$$

where i_c is in (A) and E_{on} and E_{off} are in (J).

The fifth-order polynomial equation for the diode turn-off loss as a function of diode current can be expressed as follows:

$$E_{rec} = 1.059 \times 10^{-27} i_F^5 - 8.346 \times 10^{-24} i_F^4 + 5.942 \times 10^{-9} i_F^3 - 0.0006143 i_F^2$$
$$+ 2.989 i_F + 82.21 \tag{2.10}$$

where i_F is in (A) and E_{rec} is in (mJ).

The IGBT switching loss and diode reverse recovery loss for a fundamental period T_o can be expressed as:

$$P_{SL} = \frac{1}{T_0} \sum_{k=1}^{N} (E_{on_j}(i_c) + E_{off_j}(i_c))$$ (2.11)

and

$$P_{rrL} = \frac{1}{T_0} \sum_{k=1}^{N} E_{rec_j}(i_F)$$ (2.12)

Figure 2.28 shows the comparison of total power losses for different modulation schemes. Figure 2.28 shows that with the increase of the modulation index, the total power loss decreases. Figure 2.29 shows power losses for different THTDBCWM and THSDBCPWM switching techniques.

The advanced modulation technique THSDBCPWM scheme shows the lowest total power loss. However, the other THTDBCPWM scheme gives the best harmonic performance (THD = 3.97%) among the conventional modulation schemes.

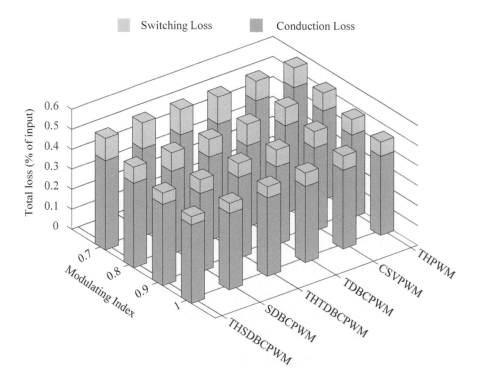

FIGURE 2.28 Comparison of total power losses with different modulation schemes.

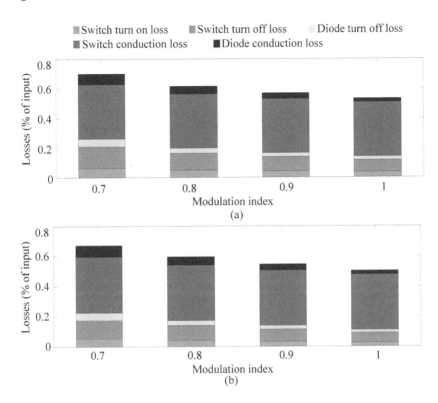

FIGURE **2.29** Losses of two advance modulations: (a) THTDBCPWM and (b) THSDBCPWM.

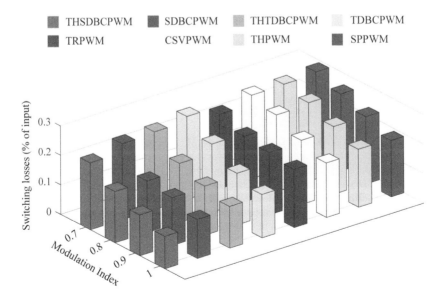

FIGURE **2.30** Switching losses of different modulation schemes as a percentage of input.

Here, the conduction losses increase during the flattened intervals of the advanced BCPWM technique, but the rate of increase of the conduction loss is comparatively lower than the rate of reduction of the switching loss. Therefore, total losses with the THSDBCPWM scheme are still lower than those from the other conventional schemes. Figure 2.30 shows the switching losses for different modulation schemes, where the SPPWM and the THSDBCPWM techniques show the highest and the lowest switching loss, respectively among the different modulation schemes.

2.7 CONCLUSION

In this chapter, different power converter topologies for applications in grid-connected renewable energy sources have been reviewed. Also, the performances of different switching algorithms are compared for different power converter topologies. The CHB and the modular multilevel inverters are the most promising topologies for the development of grid-connected power converters. The third harmonic injected bus clamping switching technique can reduce the harmonics at the output voltage and reduce the power converter losses. The high-frequency magnetic link can inherently solve the voltage balancing issues in the multilevel inverter topology. Thus, further research to design advanced high-frequency magnetic link and multilevel converter technologies will be a promising research area to make the grid-connected power conversion system more compact, reliable, and efficient.

REFERENCES

1. M. R. Islam, A. M. Mahfuz-Ur-Rahman, M. M. Islam, Y. G. Guo, and J. G. Zhu, "A modular medium-voltage grid-connected converter with improved switching techniques for solar photovoltaic systems," *IEEE Trans. Ind. Electron.*, vol. 64, no. 11, pp. 8887–8896, Nov. 2017.
2. A. M. Mahfuz-Ur-Rahman, M. R. Islam, K. M. Muttaqi, and D. Sutanto, "An advance modulation technique for single-phase voltage source inverter to integrate SMES into low-voltage distribution," *IEEE Trans. Appl. Supercond.*, vol. 29, no. 2, p. 5400305, Mar. 2019.
3. M. R. Islam, A. M. Mahfuz-Ur-Rahman, K. M. Muttaqi, and D. Sutanto, "State-of-the-art of the medium-voltage power converter technologies for grid integration of solar photovoltaic power plants," *IEEE Trans. Energy Conv.*, vol. 34, no. 1, pp. 372–384, Mar. 2019.
4. A. M. Mahfuz-Ur-Rahman, M. R. Islam, K. M. Muttaqi, and D. Sutanto, "An energy management strategy with advanced power electronic converter and control to achieve energy resiliency and minimum operating cost," In *Proceedings of IEEE IAS Annual Meeting*, Detroit, Michigan USA, Oct. 11–Oct. 15, 2020.
5. A. M. Mahfuz-Ur-Rahman, M. R. Islam, K. M. Muttaqi, and D. Sutanto, "A magnetic-linked multilevel active neutral point clamped converter with an advanced switching technique for grid integration of solar photovoltaic systems," *IEEE Trans. Ind. Appl.*, vol. 56, no. 2, pp. 1990–2000, Mar. 2020.
6. F. Z. Peng, J. S. Lai, J. W. McKeever, and J. VanCoevering, "A multilevel voltage-source inverter with separate DC sources for static VAR generation," *IEEE Trans. Ind. Appl.*, vol. 32, no. 5, pp. 1130–1138, Sept./Oct. 1996.
7. L. G. Franquelo, J. Rodriguez, J. I. Leon, S. Kouro, R. Portillo, and M. A. M. Prats, "The age of multilevel converters arrives," *IEEE Ind. Electron. Mag.*, vol. 2, no. 2, pp. 28–39, Jun. 2008.

8. A. Hassanpoor, S. Norrga, H. Nee, and L. Ängquist, "Evaluation of different carrier-based PWM methods for modular multilevel converters for HVDC application," In *Proceedings of 38th Annual Conference on IEEE Industrial Electronics Society*, Montreal, QC, pp. 388–393, 2012.

9. K. Basu and N. Mohan, "A high frequency link single stage PWM inverter with common-mode voltage suppression and source based commutation of leakage energy," *IEEE Trans. Power Electron.*, doi: 10.1109/TPEL.2013.2280600.

10. M. R. Islam, Y. G. Guo, and J. G. Zhu, "A medium-frequency transformer with multiple secondary windings for grid connection through H-bridge voltage source converters," In *Proceedings of International Conference On Power Electronics Machine Systems*, Sapporo, Japan, pp. 1–6, Oct. 21–24, 2012.

11. M. R. Islam, Y. G. Guo, and J. G. Zhu, "A high-frequency link multilevel cascaded medium-voltage converter for direct grid integration of renewable energy systems," *IEEE Trans. Power Electron.*, vol. 29, no. 8, pp. 4167–4182, Aug. 2014.

12. D. Dujic, C. Zhao, A. Mester, J. K. Steinke, M. Weiss, S. L. Schmid, T. Chaudhuri, and P. Stefanutti, "Power electronic traction transformer low voltage prototype," *IEEE Trans. Power Electron.*, vol. 28, no. 12, pp. 5522–5534, Dec. 2013.

13. Y. Yu, G. Konstantinou, B. Hredzak, and V. G. Agelidis, "Power balance of cascaded H-bridge multilevel converters for large-scale photovoltaic integration," *IEEE Trans. Power Electron.*, vol. 31, no. 1, pp. 292–303, Jan. 2016.

14. S. Essakiappan, H. S. Krishnamoorthy, P. Enjeti, R. S. Balog, and S. Ahmed, "Multilevel medium-frequency link inverter for utility scale photovoltaic integration," *IEEE Trans. Power Electron.*, vol. 30, no. 7, pp. 3674–3684, Jul. 2015.

15. S. Kouro, C. Fuentes, M. Perez and J. Rodriguez, "Single DC-link cascaded H-bridge multilevel multistring photovoltaic energy conversion system with inherent balanced operation," In *Proceedings of 38th Annual Conference of IEEE Industrial Electronics Society*, Montreal, QC, Canada, pp. 4998–5005, Oct. 25–28, 2012.

16. G. Buticchi, D. Barater, E. Lorenzani, and G. Franceschini, "Digital control of actual grid-connected converters for ground leakage current reduction in PV transformerless systems," *IEEE Trans. Ind. Informat.*,vol.8, no. 3, pp. 563–572, Aug. 2012.

17. H. A. Hamed, A. A. F. Abdou, S. S. Acharya, M. S. El Moursi, and E. E. El-Kholy, "A novel dynamic switching table based direct power control strategy for grid connected converters," *IEEE Trans. Energy Convers.*, vol. 33, no. 3, pp. 1086–1097, Jan. 2018.

18. M. R. Islam, Y. G. Guo, and J. G. Zhu, "A multilevel medium-voltage inverter for step-up-transformer-less grid connection of photovoltaic power plants," *IEEE J. Photovolt.*, vol. 4, no. 3, pp. 881–889, May 2014.

19. M. R. Islam, Y. G. Guo, and J. G. Zhu, "An amorphous alloy core medium frequency magnetic-link for medium voltage photovoltaic inverters," *J. Appl. Phys.*, vol. 115, no. 17, May 2014, Art. no. 17E710.

20. M. R. Islam, Y. G. Guo, and J. G. Zhu, "A medium-frequency transformer with multiple secondary windings for medium-voltage converter based wind turbine generating systems," *J. Appl. Phys.*, vol. 113, no. 17, p. 17A324, May 2013, Art. no. 17A324.

21. M. R. Islam, G. Lei, Y. G. Guo, and J. G. Zhu, "Optimal design of high frequency magnetic-links for power converters used in grid connected renewable energy systems," *IEEE Trans. Magn.*, vol. 50, no. 11, Nov. 2014, Art. no. 2006204.

22. A. Lesnicar and R. Marquardt, "An innovative modular multilevel con- verter topology suitable for a wide power range," In *Proceedings of IEEE Bologna. Power Technology Conference*, Bologna, Italy, Jun. 23–26, 2003.

23. M. Hagiwara, K. Nishimura, and H. Akagi, "A medium-voltage motor drive with a modular multilevel PWM inverter," *IEEE Trans. Power Electron.*, vol. 25, no. 7, pp. 1786–1799, Jul. 2010.

24. Y. Okazaki, et al., "Experimental comparisons between modular multilevel DSCC inverters and TSBC converters for medium-voltage motor drives," *IEEE Trans. Power Electron.*, vol. 32, no. 3, pp. 1805–1817, Mar. 2017.

25. Y. S. Kumar and G. Poddar, "Control of medium-voltage AC motor drive for wide speed range using modular multilevel converter," *IEEE Trans. Ind. Electron.*, vol. 64, no. 4, pp. 2742–2749, Apr. 2017.

26. M. Saeedifard and R. Iravani, "Dynamic performance of a modular multilevel back-to-back HVDC system," *IEEE Trans. Power Del.*, vol. 25, no. 4, pp. 2903–2912, Oct. 2010.

27. P. Sotoodeh and R. D. Miller, "Design and implementation of an 11-level inverter with FACTS capability for distributed energy systems," *IEEE J. Emerg. Sel. Topics Power Electron.*, vol. 2, no. 1, pp. 87–96, Mar. 2014.

28. J. Mei, B. Xiao, K. Shen, L. M. Tolbert, and J. Y. Zheng, "Modular multilevel inverter with new modulation method and its application to photovoltaic grid-connected generator," *IEEE Trans. Power Electron.*, vol. 28, no. 11, pp. 5063–5073, Nov. 2013.

29. S. Rivera, B. Wu, R. Lizana, S. Kouro, M. Perez and J. Rodriguez, "Modular multilevel converter for large-scale multistring photovoltaic energy conversion system," In *Proceedings of IEEE Energy Conversion Congress and Exposition,* Denver, CO, USA, pp. 1941–1946, Sep. 15–19, 2013.

30. H. Bayat and A. Yazdani, "A power mismatch elimination strategy for an MMC-based photovoltaic system," *IEEE Trans. Energy Convers.*, vol. 33, no. 3, pp. 1519–1528, Sep. 2018.

31. D. Zhao, V. S. S. P. K. Hari, G. Narayanan, and R. Ayyanar, "Space-vector-based hybrid pulse width modulation techniques for reduced harmonic distortion and switching loss," *IEEE Trans. Power Electron.*, vol. 25, no. 3, pp. 760–774, Mar. 2010.

32. A. M. Hava, R. J. Kerkman, and T. A. Lipo, "Simple analytical and graphical methods for carrier-based PWM-VSI drives," *IEEE Trans. Power Electron.*, vol. 14, no. 1, pp. 49–61, Jan. 1999.

33. ABB 5SNA 1200G450300 HiPak IGBT data sheet (online). Accessed: Jun. 23, 2020. Available at https://library.e.abb.com/public/801266f165d1481085834ceee4f74f91/5 SNA%201200G450300_5SYA%201401-05%2003-2016.pdf.

34. A. M. Mahfuz-Ur-Rahman, M. R. Islam, K. M. Muttaqi and D. Sutanto, "A novel active neutral point clamped multilevel converter with an advanced switching technique for grid integration of solar photovoltaic systems," In *Proceedings of IEEE IAS Annual Meeting,* Portland, OR, pp. 1–7, Sept. 23–27, 2018.

35. A. M. Mahfuz-Ur-Rahman, M. R. Islam, K. M. Muttaqi and D. Sutanto, "Performance analysis of switching techniques for asymmetric multilevel inverters," In *Proceedings of 5th IEEE Region 10 Humanitarian Technology Conference (R10-HTC),* Dhaka, Bangladesh, pp. 404–407, Dec. 21–23, 2017.

36. A. M. Mahfuz-Ur-Rahman, M. R. Islam, T. A. Fahim, M. M. Islam, K. M. Muttaqi and D. Sutanto, "Performance analysis of symmetric and asymmetric multilevel converters," In *Proceedings of 5th IEEE Region 10 Humanitarian Technology Conference (R10-HTC),* Dhaka, Bangladesh, pp. 383–386, Dec. 21–23, 2017.

3 Power Electronics for Wireless Charging of Future Electric Vehicles

Deepak Ronanki
Indian Institute of Technology Roorkee

Phuoc Sang Huynh and Sheldon S. Williamson
Institute of Technology University of Ontario

CONTENTS

3.1   INTRODUCTION

High cost, limited driving range, long charging time, and bulk energy storage systems are inherent concerns for growing widespread use of electric vehicles (EVs). The performance of battery systems not only depends on the type of battery, manufacturing process, and design but also on the charge and discharge profile [1]. Therefore, battery chargers play a crucial role in the promotion of EVs. They can be classified as on-board and off-board (standalone) chargers. Most widely used charging technology to charge the batteries is conductive/wired charging, in which a cable is plugged in from an AC utility to the onboard charger of an EV [2]. However, several key challenges are associated with the conductive/wired charging approaches, such as slow charging time, demand of substantial areas on the sides of highways or parking lots, potential traffic congestion, tripping hazards; EVs are also sometimes prone to vandalism [3]. Furthermore, the conductive charging does not facilitate an autonomous charging facility, thereby EVs must be equipped with heavy battery packs [4].

Nowadays, wireless power transfer (WPT)-based battery charging has become a prominent technology that mitigates the aforementioned concerns [4–8]. It enables WPT via an air gap/medium with several benefits such as safety, convenience, flexibility, weather immunity, and the possibility of range extension. Furthermore, wireless power electronic chargers dispense inherent galvanic isolation between the grid and EV [8]. Wireless power electronic chargers are deployed into three categories namely stationary, quasi-stationary, and dynamic. The stationary wireless charging is suitable for garages, homes, office parking lots, or any public/private charging lots [9]. In quasi-stationary charging, the charging pads can be located at bus stops, taxi ranks, layby's, and rest areas along the highways. The last type is dynamic wireless charging (DWC), where primary transmitters are installed on the dedicated charging lanes of highway roads [10]. Therefore, it offers automated charging opportunities that do not need any battery on the EV, and the power is transferred through a power rail. Consequently, an autonomous wireless charging facility overcomes the range-anxiety and battery overheads.

In WPT-based EV technology, the power transferred from the grid to the EV is typically utilizing magnetic coupling, resulting in an inductive power transfer (IPT). It consists of primary converters on the transmitter side to generate high-frequency inputs, compensation networks, an induction coupling pair, and a secondary rectifier on the receiver side to charge the battery on the EV [11]. Over the past years, significant research studies have been conducted by researchers to improve the performance of power transfer systems including the coupling links, power converter topologies, and related control schemes. In EV wireless charging systems, power electronic converters play a vital role to reduce the size, cost and maximize the overall system efficiency.

This chapter aims to provide an overview of the existing state-of-the-art coupling links, power converter topologies, and control schemes for IPT systems in EV charging applications. Initially, the fundamentals, industrial standards, types of coupler links, and charging forms of IPT systems for EV applications are extensively discussed. The main objective of this chapter is to provide a detailed overview of the architecture and control of various power converter topologies in dual-stage and single-stage conversion IPT systems. In this chapter, most widely adopted power conversion topologies for inductive-based WPT systems are selected and their performances are compared in terms of input power factor, input current distortion, current stress, voltage stress, power losses on the converters, and cost by considering the requirements of the Society of Automotive Engineers (SAE) J2954 standard [12]. The loss distribution includes a front-end rectifier, secondary rectifier, primary inverter, and passive components of dual-stage and single-phase power converter topologies. Finally, this chapter seeks to provide a discussion on the technical considerations and future opportunities of power converters in EV wireless charging applications.

This chapter is organized as follows:

- The principle of the operation, fundamentals, industrial standards, types of coupler links, and power conversion architectures of IPT systems are discussed in Section 3.2.
- Power electronic converter topologies for IPT wireless charging are presented in Section 3.3.
- The overview of the IPT control system and various control schemes are detailed in Section 3.4.
- Comparison studies of dual-stage and single-stage IPT systems are presented in Section 3.5.
- Future trends and opportunities in IPT systems are presented in Section 3.6.
- Finally, Section 3.7 provides the concluding remarks of this chapter.

3.2   FUNDAMENTALS OF INDUCTIVE POWER TRANSFER TECHNOLOGY

3.2.1   IPT-BASED EV CHARGING TECHNOLOGY

The architecture of an IPT-based EV charging system is shown in Figure 3.1. It essentially comprises an inductive coupling coil pair, compensation networks, primary converters to generate high-frequency inputs, and a secondary rectifier to convert AC current to DC to charge the battery. The primary side of the charging system is usually placed on the parking lot or dedicated lanes on the road. The secondary side is embedded in the vehicle and supplies the charging current to the battery [13]. Some of the commercially available WPT-based charging systems are listed in Table 3.1 [14–17].

However, the existence of multiple conversion stages and a bulk DC link capacitor increases the size, weight, and cost of the system. The power conversion systems on the primary side can be categorized as dual-stage and single-stage conversion systems. Typically, dual-stage conversion (AC–DC–AC) systems have been employed to excite the IPT systems as shown in Figure 3.2(a). These topologies are most widely

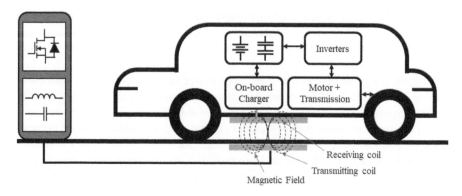

FIGURE 3.1 Architecture of an inductive power transfer-based electric vehicle charging system.

used in industry since each conversion stage can be separately designed and controlled to optimize specific performance indices [18]. Alternatively, matrix converters (MCs) for feeding the IPT systems have gained significant attention due to reduced conversion stage, thus forming a single-stage conversion system [19–22]. MCs allow direct conversion of low-frequency AC inputs (50–60 Hz) to high-frequency outputs (up to 85 kHz) without any intermediate conversion stage; therefore, they improve the IPT performance in terms of reliability, power density, and cost [23]. Also, these eliminate the DC link capacitor requirement on the primary side. However, a double line frequency ripple appears on the battery side. To overcome this issue, the sinusoidal ripple current (SRC) charging method [24] can be utilized which charges the batteries by double line frequency (100 or 120 Hz) current with negligible side effects. Thus, the single-phase MC combined with the SRC method of feeding the IPT systems would be advantageous with the elimination of life-limited capacitors. The single-stage EV-based IPT charging system using MCs is illustrated in Figure 3.2(b).

TABLE 3.1

Commercially Available Wireless Power Transfer-Based Charging Systems [14–17]

Manufacturer	Power (kW)	Frequency (kHz)	Air Gap (mm)	Efficiency (%)
Witricity	3.3	145	180	91.0
Conductix-Wampfler	60–180	20	40	91.7
Momentum Dynamics	3.3–10	–	610	93.6
HEVO Power	1–10	85	305	96.0
Bombardier	200	20	60	˜90.6

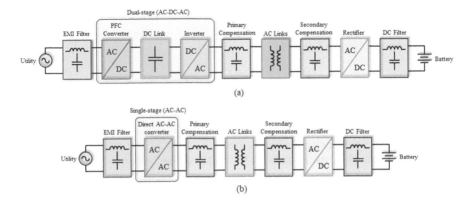

(a)

(b)

FIGURE 3.2 Configuration of inductive power transfer-based electric vehicle charging systems with (a) dual-stage power conversion and (b) single-stage power conversion.

3.2.2 ELECTRIC CIRCUIT AND ITS EQUIVALENT MODEL

Figure 3.3 illustrates the electric circuit of an inductive coupler and its equivalent models, which are T-circuit and mutual inductance models.

where, L_p and L_s are the self-inductances of the primary (transmitting) and secondary (receiving) coils, r_p and r_s are their parasitic resistances, M is the mutual inductance between two coils, and R_L is the AC load resistance.

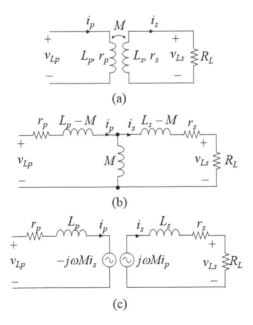

(a)

(b)

(c)

FIGURE 3.3 Inductive coupler: (a) electric circuit, (b) equivalent T-circuit model, and (c) equivalent mutual inductance model.

Assume that the primary coil is excited by a sinusoidal voltage V_{Lp} with angular frequency ω. The steady-state equations in the phasor domain can be obtained from the equivalent circuits as follows:

$$\dot{V}_{Lp} = r_p \dot{I}_p + j\omega L_p \dot{I}_p - j\omega M \dot{I}_s \tag{3.1}$$

$$j\omega M \dot{I}_p = j\omega L_s \dot{I}_s + r_s \dot{I}_s + R_L \dot{I}_s \tag{3.2}$$

where, $\dot{V}_{Lp}$ is the voltage across the primary coil, $\dot{I}_p$ and $\dot{I}_s$ are the currents flowing in the primary and secondary coils, respectively.

From Equation (3.2), the Equations (3.3) and (3.4) can be derived as follows:

$$\dot{I}_p = \frac{\dot{I}_s Z_s}{j\omega M} \tag{3.3}$$

$$\frac{\dot{I}_s}{\dot{I}_p} = \frac{j\omega M}{Z_s} \tag{3.4}$$

where, $Z_s = r_s + R_L + j\omega L_s$ is the secondary equivalent impedance.

Substitution of Equation (3.3) in Equation (3.1) results in:

$$\dot{I}_s = \frac{j\omega M}{Z_p Z_s + \omega^2 M^2} \dot{V}_{Lp} \tag{3.5}$$

where, $Z_p = r_p + j\omega L_p$ is the primary coil impedance.

Then, the output power is derived as follows:

$$P_L = \left| \dot{I}_s \right|^2 R_L = \frac{\omega^2 M^2 \left| \dot{V}_{Lp} \right|^2}{\left| Z_p Z_s + \omega^2 M^2 \right|^2} R_L \tag{3.6}$$

From Equation (3.3), the reflected impedance from the secondary side to the primary side can be derived as:

$$Z_{eq} = \frac{-j\omega M \dot{I}_s}{\dot{I}_p} = \frac{-j\omega M \dot{I}_s}{\dot{I}_s Z_s / j\omega M} = \frac{\omega^2 M^2}{Z_s} \tag{3.7}$$

The total impedance seen from the primary side end is given by:

$$Z_t = r_p + j\omega L_p + \frac{\omega^2 M^2}{Z_s} = \mathrm{Re}(Z_t) + j\mathrm{Im}(Z_t) \tag{3.8}$$

where,

$$\mathrm{Re}(Z_t) = r_p + \frac{(r_s + R_L)\omega^2 M^2}{(r_s + R_L)^2 + (\omega L_s)^2}, \text{ and } \mathrm{Im}(Z_t) = j\left[\omega L_p - \frac{\omega^3 L_s M^2}{(r_s + R_L)^2 + (\omega L_s)^2} \right]$$

are the real and imaginary parts of the total impedance Z_t, i.e., seen from the primary side end.

From Equations (3.4) and (3.8), the system efficiency is given by,

$$\eta = \frac{|i_s|^2 R_L}{|i_p|^2 \operatorname{Re}(Z_t)} = \frac{R_L}{(r_s + R_L)\left[\dfrac{r_p(r_s + R_L)}{\omega^2 M^2} + 1\right] + r_p \dfrac{L_s^2}{M^2}} \tag{3.9}$$

From Equations (3.6) and (3.9), for a given mutual inductance, frequency, and load resistance, the efficiency and power transfer capability can be enhanced by compensating secondary reactance. Additionally, a compensation network in the primary side is included to compensate for the reactance component of the total impedance Z_t (eq. 3.8); the voltage-ampere (VA) rating of the source is minimized. As a result, the size of the source is reduced.

3.2.3 Inductive Coupler Structures

The high-frequency AC link or coupler is the most crucial part of the IPT systems which comprises the transmitter and receiver coils separated by air as a medium. Its architecture deals with the design of the primary and secondary interface, which transfers power wirelessly through magnetic field coupling. The desired characteristics for coupling pads are the high coupling coefficient (k), maximum power transfer capability, high misalignment tolerance, and low losses [5, 25].

The AC link design can be classified into planar and nonplanar couplers based on the material used to fabricate the link as shown in Figure 3.4. Planar couplers are named after the geometry of the coil structure utilized. Similarly, an AC link is developed using different core shapes in non-planar couplers [26].

3.2.3.1 Non-Planar Couplers

The nonplanar couplers are usually made of the ferrite cores which improve the coupling coefficient between primary and secondary windings. Therefore, it offers higher efficiency with low Eddy current and hysteresis losses operating at a higher frequency. Furthermore, it reduces the unwanted stray magnetic radiation. However, the major drawback is its higher weight-to-power transfer ratio and sensitivity to horizontal misalignment. The different variants of ferrite cores are analyzed including *E–E*, *I–I*,

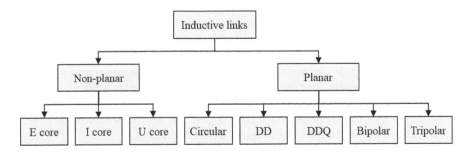

FIGURE 3.4 Classification of inductive link couplers.

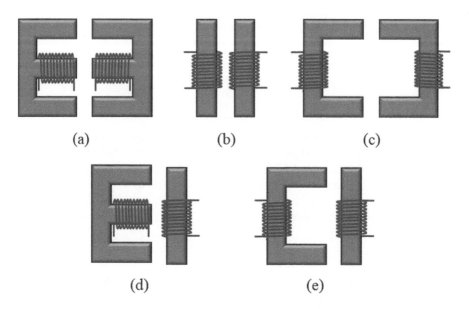

FIGURE 3.5   Nonplanar couplers: (a) E–E core, (b) I–I core, (c) U–U core, (d) E–I core, and (e) U–I core.

U–U, *E–I*, and *U–I* cores and are shown in Figure 3.5 [26]. These types of couplers are the most effective option for wireless charging of electric golf carts and mine carts as these provide bumper to wall charging option. A new ferrite core composed of paddle and disc type is presented to achieve better coupling efficiency with reduced weight even under the horizontal misalignment variation increased by 1.7 times [27].

3.2.3.2   Planar Coils

The Litz wire-based planar coil structures with ferrite spokes have minimized leakage inductance and improved misalignment tolerance. These structures are popularly used for the stationary and dynamic charging of EVs due to ease of manufacture for desired power transfer levels and have more visual appeal than bulky ferrite core configurations. The planar couplers can be further classified into nonpolarized and polarized charging pads based on the orientation of coupling (parallel and perpendicular). The most popular nonpolarized coupler structures are circular, rectangular, and square-shaped pads, as shown in Figure 3.6 [26]. Ferrite cores are added to these couplers with the optimal arrangement so that flux lines are guided from primary to the secondary side and increase coupling. Despite lower coupling coefficient, the circular charging pads employing symmetrical Archimedean spiral coil pairs are one of the widely adopted pad shapes in static EV battery charging applications. This is because of its identical misalignment tolerance in all directions that facilitates vehicle parking. The rectangle and square-shaped coil structures have better power transfer capability and higher flux areas with effective design.

The polarized charging pads utilize the horizontal component of flux and the most popular coil structures are circular pads, DD pads, DD-Q pads, and bipolar

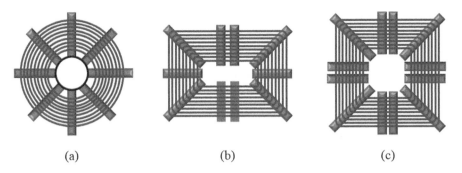

FIGURE 3.6 Nonpolarized planar couplers: (a) Circular, (b) rectangular, and (c) square.

pads as shown in Figure 3.7 [26, 28]. In the bipolar pad (BP) shown in Figure 3.7(c), two identical mutually decoupled coils are placed over the ferrite core that overlap at the center. It is also proven that BP exhibits identical performance in terms of coupling coefficient and misalignment tolerance to DDQ structure. However, it requires 25–30% less amount of copper and is simple to design.

A three-coil coupler, also named as tripolar pad [29], is mutually decoupled with an angle of 120° and can be driven independently to achieve the highest coupling factor. As a result of its assembly, it allows high rotational misalignment tolerance of a nonpolarized coil pad. However, it demands three separate inverters to drive three coils. Therefore, it increases the control complexity and system cost. The comparison of different coil structures in terms of magnetic flux flow, coupling coefficient, misalignment tolerance, and electric and magnetic fields (EMF) exposure are listed in Table 3.2.

3.2.4 Compensation Networks

The high magnetizing inductance and leakage inductance of the primary and secondary coils demand a higher value of input source VA rating to transfer active power to the load. Typically, capacitors are connected either in parallel or in series to eliminate the circulation of high reactive current. The primary requirement of a compensation network is to minimize the VA rating of the input power supply

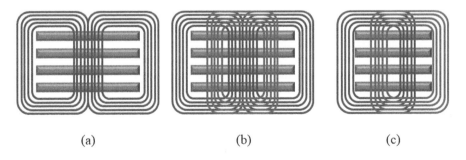

FIGURE 3.7 Types of planar coil charging pads: (a) DD, (b) DDQ, and (c) bipolar.

TABLE 3.2

Comparison of Different Charging Pad Structures [28]

Coil Structure	Magnetic Flux	Coefficient of Coupling	Misalignment Tolerance	EMF Exposure
Circular pad	Single-sided	Low	Low	High
DD coil	Double-sided	Medium	Low	Low
DDQ coil	Double-sided	High	High	Low
Bipolar pad	Double-sided	High	Medium	Low
Tripolar pad	Single-sided	High	High	Low

by producing the reactive power needed to form and sustain the magnetic field. To attain maximum power transfer, leakage inductance is cancelled on both primary and secondary sides. The key requirements of compensation networks include high efficiency, ability to minimize VA rating of input power supply, high misalignment tolerance, and high bifurcation tolerance. Typically, metalized film capacitors are utilized as compensation capacitors as they offer very low dissipation and negligible effect on system efficiency. The tuning of compensation circuits plays a vital role in system performance. Otherwise, it leads to rise in the VA rating of the power supply and stress on power semiconductor devices. To achieve unity power factor, the secondary coil is operated at or near the same resonant frequency, thereby the primary resonance cancels the primary leakage inductance. Generally, the compensation network is designed for a single zero phase angle, and it ensures system stability for variable frequency control to achieve bifurcation tolerance [28].

The compensation method at primary and secondary can be kept symmetric as in series-series (SS), series-parallel (SP), parallel-parallel (PP), parallel-series (SP). The simplest compensation techniques which use only a single capacitor on each side are presented in Figure 3.8.

The comparison summary of basic compensation topologies is listed in Table 3.3. These topologies are not so efficient for varying load conditions and coil position misalignments, which affects the maximum power transfer [30]. Moreover, high-order compensation circuits have more than one reactive component on each side such as double-sided inductor-capacitor-inductor (*LCL*) and double-sided inductor-capacitor-capacitor (*LCC*) topologies or asymmetric as in series-parallel, parallel-series, series-capacitor-inductor-capacitor (*CLC*), series-*LCL*, and series-*LCC* that are developed with aim to improve the system performance. However, a higher number of compensation elements increases power losses and the cost of the systems, which is not recommended for EV charging systems. The most common high-order compensation networks are the *LCL* [31], inductor-capacitor-capacitor-inductor (*LCCL*) [32], and parallel-series capacitor-capacitor (*CC*) [33].

Several studies are conducted to compare the basic compensation topologies with a constant voltage source and constant current source input in terms of coupling coefficient, load conditions, inverter switch voltage, and current ratings [34]. Optimal sizing of 200-kW IPT for four basic compensation circuits is investigated in reference [35]. The analysis reveals that SS compensation topology in Figure 3.8(a) demands the least

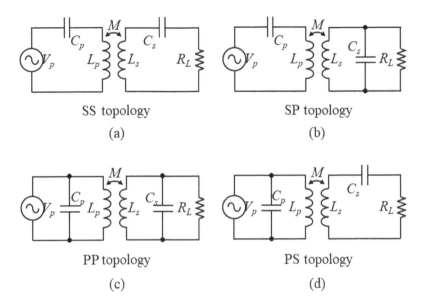

FIGURE 3.8 Basic compensation topologies for inductive power transfer: (a) SS, (b) SP, (c) PP, and (d) PS.

amount of copper among all compensation topologies, thereby reducing the system cost. Also, it offers better performance as the primary capacitance is independent of mutual inductance and load variation. The SS and SP compensation topologies are economically viable for high-power applications. On the flip side, PS and PP topologies are suitable for high-power current sources with long cables [36]. Typically, the performance of the

TABLE 3.3

Comparison Summary of Compensation Topologies

Topologies	Secondary Capacitance	Primary Capacitance	Theoretical Efficiency
SS		$C_p = \dfrac{1}{\omega_0^2 L_p}$	$\eta = \dfrac{\omega^2 M^2 R_L}{r_p(r_s + R_L)^2 + \omega^2 M^2(r_s + R_L)}$
SP		$C_p = \dfrac{1}{\left(L_p - \dfrac{M^2}{L_s}\right)\omega_0^2}$	$\eta = \dfrac{\omega^2 L_p^2 R_L}{\omega^2 R_L + r_s\omega^2 L_s^2 + r_s R_L^2 + \dfrac{r_p r_s^2 L_p^2}{M^2} + \dfrac{r_p L_s^4 \omega^2}{M^2}}$
PS	$C_s = \dfrac{1}{\omega_0^2 L_s}$	$C_p = \dfrac{L_p}{\left(\dfrac{\omega_0^2 M^2}{R_L}\right)^2 + \omega_0^2 L_p^2}$	$\eta = \dfrac{\omega^2 M^2 R_L}{r_p(r_s + R_L)^2 + \omega^2 M^2(r_s + R_L)}$
PP		$C_p = \dfrac{L_p - \dfrac{M^2}{L_s}}{\left(\dfrac{M^2 R_L}{L_s^2}\right)^2 + \left(L_p - \dfrac{M^2}{L_s}\right)^2 \omega_0^2}$	$\eta = \dfrac{\omega^2 L_p^2 R_L}{\omega^2 R_L + r_s\omega^2 L_s^2 + r_s R_L^2 + \dfrac{r_p r_s^2 L_p^2}{M^2} + \dfrac{r_p L_s^4 \omega^2}{M^2}}$

compensation networks varies with the alignment of coils and it requires perfect alignment to achieve maximum efficiency. To overcome the misalignment issues, a combination of series-parallel-series topology is presented in reference [37], which maintains constant output power under severe misalignment conditions.

The double-sided *LCC*-compensated topology with one external inductor and two external capacitors at each side to compensate the power transmitting coils is presented in reference [38]. In this configuration, the resonant tank acts as a current source to both the input and output, and the output currents are independent of the load conditions. Therefore, the resonant circuit is unaffected by the battery voltage variation and provides nearly unity power factor to the input side inverter. Consequently, the *LCC* topology is recommended for EV battery charging applications. The experimental prototype of 7.6 kW with this compensation topology achieved a DC–DC efficiency of 96% with a ground clearance of 200 mm [38].

3.2.5 EV CHARGING STANDARDS

Standardization of WPT-based EV charging in terms of the efficiency, safety criteria, electromagnetic interference (EMI) limits, and interoperability targets is essential for reliable commercialization of high-voltage and high-power systems. To have standardization in terms of design, testing, and safety of WPT-based EV charging systems, several standards and procedures are developed. The SAE J2954 defines specific criteria for interoperability, minimum performance, electromagnetic compatibility, safety, and testing for static WPT-based EV charging systems. Table 3.4 shows the SAE J2954 standard in terms of power level and operating frequency standards for static WPT charging. The manufacturers in different nations follow these common design guidelines to ensure the interoperability of various WPT-based charging infrastructure. The International Electrotechnical Commission (IEC)-61980-1 standard defines the total architecture of WPT systems from the supply network to the battery charging of an EV or any equipment of the same operating at a supply of 1000-V AC or 1500-V DC. Furthermore, the International Commission on Non-Ionizing Radiation Protection (ICNIRP) 2010 [39] has defined the limits of maximum permissible exposure which needs to be followed while designing and testing the WPT systems, as listed in Table 3.5.

The applicable standards and technical codes related to design, testing, maintenance, safety, and communication interfaces for WPT-based charging are tabulated in Table 3.6.

TABLE 3.4

Wireless Power Transfer Classification for Light-Duty Electric Vehicles—SAE J2954 [12]

WPT Levels	1	2	3	4
Maximum AC input power (kVA)	3.7	7.7	11	22
Minimum target efficiency at nominal alignment (%)	>85	>85	>85	To be defined (TBD)
Minimum target efficiency at offset position (%)	>80	>80	>80	TBD
Operating frequency (kHz)	81.38–90 (typical 85)			

TABLE 3.5

Electromagnetic Field Exposure Standard by ICNIRP, 2010 [39]

Parameters	RMS Level	Peak Level
Magnetic field	27 µT or 21.4 A/m	38.2 µT or 30.4 A/m
Electric field	83 V/m	117 V/m
Contact current	17 mA @ 85 kHz	24 mA @ 85 kHz

TABLE 3.6

Industrial Standards for Wireless Power Transfer Charging [4, 26, 28]

Standard	Description
SAE J2954/1, /2	WPT for light-duty plug-in/electric vehicles (EV) and alignment methodology
ICNIRP 2010	Guidelines for limiting exposure to time-varying electric and magnetic fields
IEEE C95-1234	IEEE standard for safety levels concerning human exposure to radiofrequency electromagnetic fields, 3 kHz to 300 GHz
IEEE standard C93.1	Restrict the frequency electric field and magnetic field exposure to outside humans
ISO 19363	Electrically propelled road vehicles – magnetic field WPT – safety and interoperability requirements
ISO 15118-1	Road vehicles – vehicle to grid communication interface– Part 1: General information and use-case definition
ISO 15118-2	Road vehicles – vehicle to grid communication interface– Part 2: Network and application protocol requirements
ISO 15118-8	Road vehicles – vehicle to grid communication interface– Part 8: Physical layer and data link layer requirements for wireless communication
SAE J2954	Wireless charging of electric and plug-in hybrid vehicles
SAE J2954_201605	WPT for light-duty plug-in/EV and alignment methodology
SAE J2954_201606	SAE EV inductively coupled charging (Jun. 2014)
SAE J2847-6	Communication between wirelessly charged vehicles and wireless EV chargers
SAE J2931-6	Signalling communication wirelessly charged EVs
IEC 61980-1	EV WPT systems, Part 1: General requirements
IEC 61980-2	EV WPT systems, Part 2: Specific requirements for communication between EV and infrastructure with respect to WPT systems
IEC 61980-2	EV WPT systems, Part 3: Specific requirements for the magnetic field power transfer systems
ISO/AWI PAS 19363	Electrically propelled road vehicles – magnetic field WPT – safety and interoperability requirements
UL 2750	Wireless charging equipment for EVs

3.3   POWER ELECTRONIC CONVERTER TOPOLOGIES FOR INDUCTIVE WIRELESS CHARGING

Power electronic converters play a vital role in the IPT-based EV charging systems in terms of maximizing the system efficiency, power density, and cost. Figure 3.2(a) and (b) illustrate the two different variants of power conversion types and the corresponding conversion stages. The first stage of the IPT system is a front-end converter which has a basic function to correct the power factor and maintain low current total harmonic distortion (THD) (<5%) as specified by the Institute of Electrical and Electronics Engineers (IEEE)-1547, and IEC1000-3-2. This section is mainly focused on a review of frond-end power electronic converter topologies for IPT-based EV charging systems. They are broadly classified as dual-stage and single-stage based on the power conversion stages, as shown in Figure 3.9.

3.3.1   DUAL-STAGE POWER CONVERSION

In the dual-stage power conversion system shown in Figure 3.2(a), an active front-end AC–DC converter is required to convert the grid voltage to an intermediate DC link voltage. Furthermore, it has to shape the input current for harmonic reduction and maintain power near unity. Therefore, these converters should act as power factor correction (PFC) converters. Then, a high-frequency inverter connected to the front-end rectifier via a DC link filter generates high-frequency current/voltage feeding the primary coil. The main advantage of the IPT systems using dual-stage AC-DC-AC conversion is that the PFC rectifier and the inverter are decoupled through DC link capacitors or inductors; therefore, they can be separately controlled to optimize specific performance indices. For the inversion stage, a current source inverter (CSI) or a voltage source inverter (VSI) can be employed to excite the IPT systems depending on the primary compensation network configuration. Figure 3.10 shows the compatibility of inverter types and primary compensation topologies of the IPT systems. Figure 3.11(a)–(c) shows three CSI topologies namely push-pull, half-bridge, and full-bridge inverters which are commonly used in the IPT systems. Generally, a

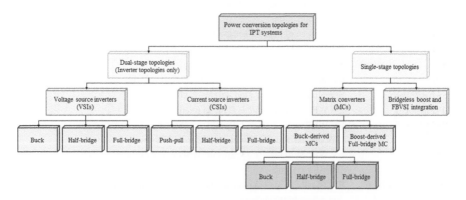

FIGURE 3.9   Classification of front-end converter topologies for inductive power transfer systems.

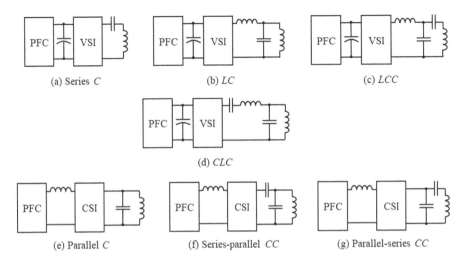

FIGURE 3.10 Compatibility between inverter types and primary compensation circuits in inductive power transfer systems. (a) Series-*C*, (b) *LCL*, (c) *LCCL*, (d) parallel-*C*, and (e) parallel-series *CC*.

large inductor is connected between DC voltage and the inverter to form a quasi-current source at the high frequency in the CSI topologies.

In the push-pull inverter (Figure 3.11(a)), two additional splitting inductors are used to introduce the continuous path for current flow. The benefits of the push-pull inverter are the reduction of switching device number and simple

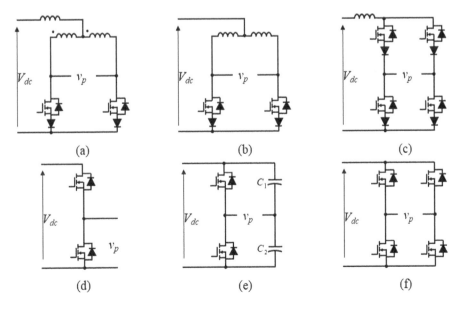

FIGURE 3.11 Inverter topologies: (a) Push-pull, (b) half-bridge current source inverter (CSI), (c) full-bridge CSI, (d) buck, (e) half-bridge voltage source inverter (VSI), and (f) full-bridge VSI.

gating circuit because the switches are referenced to the ground. Compared to a full-bridge inverter, the push-pull inverters can generate a higher resonant tank voltage due to the action of the two splitting inductors. However, adding the splitting inductors increases the size, cost, and losses of the overall system. The main drawback of CSI topologies is the demand for bulky inductors and blocking diodes that increase the size and cost of the system. Normally, a single parallel compensating capacitor is added to the primary circuit. However, active switches in the CSIs for high-power applications experience high voltage stress when the power level rises [40], which is not suitable for EV charging applications. To conquer this demerit, a parallel-series *CC* compensation circuit is incorporated with the CSIs which mitigates device stresses and harmonic distortion in primary coil current [33].

Buck, half-bridge, and full-bridge topologies shown in Figure 3.11(d)–(f) are the most commonly used VSI topologies in IPT systems. These topologies are compatible with the single capacitor series, *LCL*, and *LCCL* compensation networks [11]. Among them, the series compensation network is simple to implement and cost-effective. However, the IPT system experiences severe instability under light load conditions or the absence of the receiver [28, 41]. To overcome these concerns, LCL or LCCL tanks are employed which have a high tolerance to coil misalignments [28]. However, the connection of these circuits with VSIs increases low-order harmonics which deviates the zero-phase-angle operation of the inverters [42]. Furthermore, these circuits demand precise design of the inductors as its inductance value has a high impact on power transfer capability [43]. Some of the developed IPT systems utilizing the dual-stage power conversion systems in terms of power electronic converter topology, compensation topology, and power transistor technology are presented in Table 3.7.

3.3.2   SINGLE-STAGE AC–AC CONVERSION

MCs provide direct AC–AC power conversion with only one stage for powering the IPT systems which include buck (Figure 3.12(a)), half-bridge (Figure 3.12(b)), and

---

TABLE 3.7
Comparison of Various Developed Inductive Power Transfer Systems

Power Level (kW)	Frequency (kHz)	Power Converter	Compensation Topology	Power Transistor Technology	Maximum DC/DC Efficiency (%)	Ref.
3.6	40	FBVSI	SS	Si MOSFET	91	[44]
3.3	23	FBVSI	SP	SiC MOSFET	91.77	[45]
7.7	79	FBVSI	LCC	Si MOSFET	96	[38]
3.6	95	FBVSI	LCC	SiC MOSFET	93.36	[46]
50	85	FBVSI	SS	SiC MOSFET	93.8	[47]
5	140	FBVSI	SS	Hybrid SiC MOSFET & Si IGBT	98	[48]
2.1	100	FBVSI	SS	GaN GIT	93.6	[49]

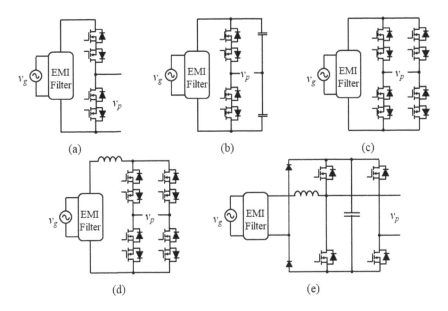

FIGURE 3.12   Single-stage conversion topologies: (a) buck matrix converter, (b) half-bridge matrix converter, (c) full-bridge matrix converter, (d) boost-derived matrix converter, and (e) bridgeless boost converter.

full-bridge (Figure 3.12(c)). Line current distortion and power factor deterioration arise if the diode rectifier is employed on the secondary side. To overcome this issue, an active full-bridge rectifier is utilized instead of a diode rectifier on the battery side, whose phase-shift angle tracks the line-voltage waveform and is utilized to shape the line current [19]. In this configuration, the synchronization control should be achieved between the primary and secondary converters in every switching cycle. As a result, the implementation and control complexity increases.

To overcome the above concern, a boost-derived full-bridge MC (FBMC) with a parallel-series *CC* compensation circuit in the primary side is presented in reference [22]. It can regulate power flow and shape the line current through two control loops. A single-stage topology operating in discontinuous conduction mode (DCM) that integrates bridgeless boost PFC converter and full-bridge VSI was introduced in reference [50] for IPT applications. However, this topology operates in DCM results in high losses, device stress, and EMI. Figure 3.12(d) and (e) illustrate the boost-derived FBMC and bridgeless boost PFC converter topologies.

Various three-phase MC-based IPT systems have been introduced in the literature. In reference [51], a three-phase MC was proposed to convert low-frequency three-phase utility voltage to a single-phase high-frequency voltage that can feed the primary series resonant circuit. The switching strategy introduced in this topology minimizes commutation steps and is independent of the primary current sign. The configuration of the system is illustrated in Figure 3.13(a). In reference [52], a silicon carbide (SiC)-based three-phase MC, as shown in Figure 3.13(b), was introduced to improve the efficiency of the IPT systems. A three-phase MC with switching

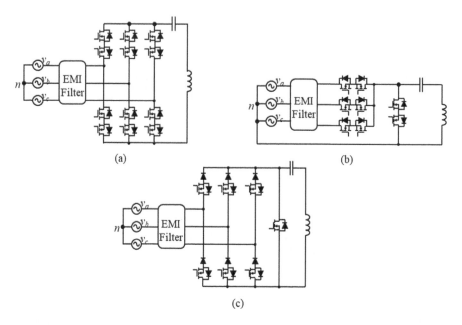

(a)   (b)

(c)

FIGURE 3.13   Three-phase matrix converter configuration for IPT systems. (a) [51], (b) [52], and (c) [23].

frequency reduction and soft-switching operation for the IPT system supply [23] is shown in Figure 3.13(c). This converter composed of seven switches with reverse blocking capability can regulate the transferred power by using the discrete energy injection control scheme with eight operating modes.

3.4   POWER CONTROL SCHEMES

Several methods have been reported to control the power flow in IPT systems. It can be performed on the primary side, secondary side, or both sides. The taxonomy of IPT control systems is presented in Figure 3.13. The primary side control is suitable for static EV charging applications since the onboard configuration remains simple. In the primary-side control, the status of the battery SOC on the secondary side is transmitted to the primary side through a communication link, and then complete control is performed on the primary side [53] by fixed frequency, variable frequency, and discrete energy injection techniques. In fixed frequency control, the switching frequency of the inverter is kept at a constant value, and the duty cycle or the phase (phase shift control) of inverter switches is varied to regulate the power flow [55]. To achieve soft switching operation, the frequency is selected in such a way that its value is slightly different from the primary resonant frequency. Another possible way to control the power flow with constant switching frequency is the employment of a front-end DC–DC converter before the inverter to regulate input DC voltage [56]. A constant duty cycle of 50% with varied switching frequency is used to regulate the output power in the variable switching frequency control schemes [57]. However, it causes a large circulating current, large device losses, and coil losses since the

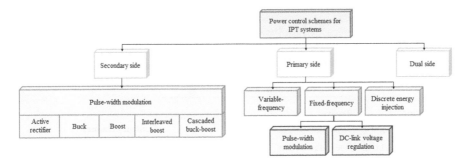

FIGURE 3.14   Classification of power control schemes for inductive power transfer systems.

operating frequency is largely different from the resonant frequency. Furthermore, it requires severe attention with respect to the bifurcation phenomenon [58]. A discrete energy injection control for the matrix buck converter is presented in reference [20] to control the magnitude of the primary current. This control scheme operates at the reduced switching frequency and offers soft switching. However, the main concern is the detection of zero-crossings of primary current for zero-current switching (ZCS) operation. Furthermore, the current sag arises during the zero-crossings of the input grid voltage. As a result, it diminishes the average power that is transferred.

The secondary side control system is widely adopted in the IPT systems where the multiple pick-up coils are coupled to a single primary coil. In these systems, the magnitude and the frequency of primary current are fixed. The power flow is controlled on the secondary side through an active rectifier or a back-end DC–DC converter such as buck, boost, interleaved boost, and cascaded buck-boost for each secondary coil as shown in Figure 3.14 [11]. The dual-side control is preferred in bidirectional IPT systems. In such systems, both the duty cycle and phase-shift angle of the primary and secondary converters are usually manipulated to regulate the power flow and maintain soft-switching.

Dual- and single-stage inverter topologies are summarized with regard to component requirement and control schemes and are listed in Table 3.8 and Table 3.9, respectively.

3.5   PERFORMANCE COMPARISON AND DISCUSSION

In this section, the performance of a conventional dual-stage topology and two potential single-stage topologies including buck and boost-derived FBMCs are compared regarding input power quality, current stress, voltage stress, power losses, and cost.

3.5.1   DESIGN CONSIDERATIONS

The complete schematic of a dual-stage IPT-based charging system is shown in Figure 3.15(a) [59]. It consists of a boost rectifier at the front end which is utilized to regulate a constant DC voltage (V_{dc}) across the DC link capacitor (C_i). Also, it acts as a PFC converter and can shape the grid current. A full-bridge VSI is most widely used topology at the primary side to produce a high-frequency voltage (v_p) feeding

TABLE 3.8

Dual-Stage Power Conversion Topologies (Excluding Front-End PFC Stage) and Control Schemes of the IPT Systems

Power Conversion Topologies	Figures	Component Requirement		Control Schemes
		Passive Components	Switches	
Push-pull	3.11(a)	One inductor, one phase-splitting transformer	Two reverse blocking	• Variable switching frequency • DC link voltage control • Secondary-side control
HB CSI	3.11(b)	Two inductors	Two reverse blocking	• Dual-side control
FB CSI	3.11(c)	One inductor	Four reverse blocking	• Variable switching frequency • DC link voltage control • Pulse width modulation (duty cycle control) • Secondary-side control • Dual-side control
Buck	3.11(d)	None	Two reverse conducting	• Discrete energy injection
HB VSI	3.11(e)	Two capacitors	Two reverse conducting	• Variable switching frequency • DC link voltage control • Pulse width modulation (duty cycle control) • Secondary-side control • Dual-side control
FB VSI	3.11(f)	None	Four reverse conducting	• Discrete energy injection • Variable switching frequency • DC link voltage control • Pulse width modulation (phase-shift control) • Secondary-side control • Dual-side control

the primary coil as a bulky and costly inductor required in CSIs. SS compensation network is adopted for this configuration since it is cost-effective, simple, highly efficient at a low coupling coefficient, and independent of load conditions [28]. To minimize the VA rating of the primary side and maximize the power transfer, the resonant circuits at primary and secondary sides are tuned to the same resonant frequency (ω_0), which is equal to the inverter switching frequency.

$$\omega_0 = \frac{1}{\sqrt{L_p C_p}} = \frac{1}{\sqrt{L_s C_s}} \qquad (3.10)$$

where, L_p and C_p are self-inductance and tuning capacitors of the primary circuit, respectively. Similarly, L_s and C_s denote self-inductance and tuning capacitors of the secondary circuit, respectively.

TABLE 3.9

Single-Stage Power Conversion Topologies and Control Schemes of the IPT Systems

Power Conversion Topologies	Figures	Component Requirement Passive Components	Switches	Control Schemes
Single-phase buck MC	3.12(a)	None	Two bidirectional	• Discrete energy injection • Pulse width modulation (duty cycle control)
Single-phase HBMC	3.12(b)	Two capacitors	Two bidirectional	• Secondary-side control • Dual-side control
Single-phase FBMC	3.12(c)	None	Four bidirectional	• Discrete energy injection • Pulse width modulation (phase-shift control) • Secondary-side control • Dual-side control
Single-phase boost-derived FBMC	3.12(e)	One inductor	Four bidirectional	• Pulse width modulation • Secondary-side control • Dual-side control
Single-phase bridgeless boost	3.12(f)	One inductor, one capacitor	Two diodes, four reverse conducting	• Pulse width modulation • Secondary-side control • Dual-side control
Three-phase FBMC	3.13(a)	None	Six bidirectional	• Pulse width modulation
Three-phase buck MC	3.13(b)	None	Four bidirectional	• Secondary-side control • Dual-side control
Three-phase unidirectional FBMC	3.13(c)	None	Six reverse blocking + one reverse conducting	• Discrete energy injection

Abbreviations: FBMC, full-bridge matrix converter; MC, matrix converter; HBMC, half-bridge matrix converter.

The power flow regulation can be performed by employing the phase-shift control for inverter at the primary side. The power transferred from primary to secondary side operating at the resonance frequency (ω_0) can be given by:

$$P_o = \frac{8V_{dc}V_b}{\pi^2\omega_0 M}\sin \pi D_p \tag{3.11}$$

where, D_p represents the duty cycle of the primary voltage (V_p) and M denotes the mutual inductance, which can be given by:

$$M = k\sqrt{L_p L_s} \tag{3.12}$$

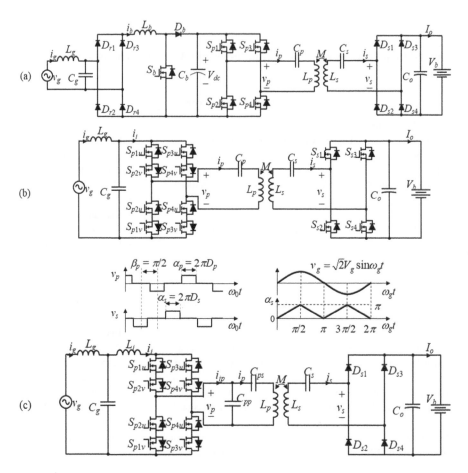

FIGURE 3.15 Inductive power transfer charging system fed by (a) dual-stage power converter (PFC and FB VSI), (b) buck-derived FBMC, and (c) boost-derived FBMC. FBMC, full-bridge matrix converter; PFC, power factor correction; VSI, voltage source inverter.

where, k is the coupling coefficient that varies from 0.1 to 0.3 in EV charging applications. The main concern in a dual-stage topology is low power density due to the existence of a bulky DC link capacitor and multiple conversion stages. The MCs can be alternative topology with the reduced conversion stages. Figure 3.15(b) illustrates a buck-derived FBMC [60] feeding the IPT charging system. The FBMC comprises four bidirectional switches which can directly convert low-frequency (50–60 Hz) grid voltage to resonant frequency (typical 85 kHz as per SAE J2954) voltage feeding the inductive AC link. The phase-shift pulse-width modulation (PWM) strategy is utilized to control switches (S_{pnu} and S_{pnv}) ($n = 1, 2, 3, 4$). In contrast, in dual-stage topology, an active rectifier is employed instead of a diode rectifier in the secondary side for shaping the input current. The primary voltage V_p is 90° lagging with secondary voltage V_s to synchronize the primary and secondary converters in every switching cycle. The duty cycle of V_s is controlled in such a way that it follows the primary grid voltage waveform to shape the input current as shown in Figure 3.15(b).

The power transferred by adjusting the duty cycle of the primary voltage is given by:

$$P_o = \frac{4\sqrt{2}V_g V_b}{\pi^2 \omega_0 M} \sin \pi D_p \tag{3.13}$$

where, V_g represents the root mean squares (RMS) value of the grid voltage.

Although the buck-derived FBMC-based power conversion system eliminates the intermediate conversion stage, it demands high-frequency communication to synchronize the PWM patterns of the primary and secondary converters in every switching cycle. Moreover, precise synchronization is difficult to achieve. As a result, control complexity increases. The boost-derived MC [22] can overcome the above concern and its topology is shown in Figure 3.15(c). Like a boost converter, it shapes the grid current and regulates the power flow through two control loops. Tables 3.10 and 3.11 show the passive component design and switching stresses of the converter components.

TABLE 3.10
Design of Passive Components and Compensation Networks

Topologies	Components	Designed Values	Current/Voltage Peak
Dual-stage [59]	L_b	$L_b = \dfrac{V_g^2}{P_o \% \Delta I_b f_s}\left(1 - \dfrac{\sqrt{2}V_g}{V_{dc}}\right)$	$\hat{I}_{Lb} = \sqrt{2}\dfrac{P_o}{V_g} + \dfrac{\Delta I_b}{2}$
	C_b	$C_b = \dfrac{1}{\% \Delta V_{dc}}\dfrac{P_o}{\omega_g V_{dc}^2}$	$\hat{V}_{Cb} = V_{dc} + \dfrac{\Delta V_{dc}}{2}$
	C_p and C_s	$C_p = \dfrac{1}{\omega_0^2 L_p}$	$\hat{V}_{Cp} = \dfrac{4V_b}{\pi\omega_0^2 C_p M}$
		$C_s = \dfrac{1}{\omega_0^2 L_s}$	$\hat{V}_{Cs} = \dfrac{4V_{dc}}{\pi\omega_0^2 C_s M}$
Buck-derived FBMC [60]	C_p and C_s	$C_p = \dfrac{1}{\omega_0^2 L_p}$	$\hat{V}_{Cp} = \dfrac{4V_b}{\pi\omega_0^2 C_p M}$
		$C_s = \dfrac{1}{\omega_0^2 L_s}$	$\hat{V}_{Cs} = \dfrac{4\sqrt{2}V_g}{\pi\omega_0^2 C_s M}$
Boost-derived FBMC [22]	L_i	$L_i = \dfrac{\sqrt{2}V_g}{2\Delta I_i f_s}$	$\hat{I}_{Li} = \dfrac{\sqrt{2}P_o}{V_g} + \dfrac{\Delta I_i}{2}$
	C_{pp}, C_{ps}, and C_s	$C_s = \dfrac{1}{\omega_0^2 L_s}$	$\hat{V}_{Cs} = \dfrac{\pi P_o}{\omega_0 C_s V_b}$
		$C_{pp} = \dfrac{L_p - \dfrac{1}{\omega_0^2 C_{ps}}}{\dfrac{\omega_0^4 M^4}{R_{oeq}^2} + \omega_0^2\left(L_p - \dfrac{1}{\omega_0^2 C_{ps}}\right)^2}$	$\hat{V}_{Cpp} = \hat{V}_p = \sqrt{\left(\dfrac{4V_b}{\pi M}\right)^2\left(L_p - \dfrac{1}{\omega_0^2 C_{ps}}\right)^2 + \left(\dfrac{\pi\omega_0 M P_o}{V_b}\right)^2}$
		Note: C_{ps} is designed to limit the peak of primary voltage v_p, which is the voltage stress on MC switches	$\hat{V}_{Cps} = \dfrac{4V_b}{\pi\omega_0^2 C_{ps} M}$

TABLE 3.11

Device Stress of Different Power Converter Topologies

Topologies	Components	Current Stress	Voltage Stress
Dual-stage	S_b and D_b	$\hat{I}_{Sb} = \hat{I}_{Db} = \dfrac{\sqrt{2}P_o}{V_g} + \dfrac{\Delta I_b}{2}$	$\hat{V}_{Sb} = \hat{V}_{Db} = V_{dc} + \dfrac{\Delta V_{dc}}{2}$
	S_{pn}	$\hat{I}_{Spn} = \dfrac{4V_b}{\pi \omega_0 M}$	$\hat{V}_{Spn} = V_{dc} + \dfrac{\Delta V_{dc}}{2}$
	D_{sn}	$\hat{I}_{Dsn} = \dfrac{4V_{dc}}{\pi \omega_0 M}$	$\hat{V}_{Dsn} = V_b$
Buck-derived FB MC	S_{pnx}	$\hat{I}_{Spnx} = \dfrac{4V_b}{\pi \omega_0 M}$	$\hat{V}_{Spnx} = \sqrt{2}V_g$
	S_{sn}	$\hat{I}_{Ssn} = \dfrac{4\sqrt{2}V_g}{\pi \omega_0 M}$	$\hat{V}_{Ssn} = V_b$
Boost-derived FB MC	S_{pnx}	$\hat{I}_{Spnx} > \dfrac{\sqrt{2}P_o}{V_g} + \dfrac{\Delta I_i}{2}$	$\hat{V}_{Spnx} = \hat{V}_p$
	D_{sn}	$\hat{I}_{Dsn} > \dfrac{\pi P_o}{V_b}$	$\hat{V}_{Dsn} = V_b$

(n = 1, 2, 3, 4 and $x = u, v$)

3.5.2 PERFORMANCE COMPARISON

In this section, the performance of selected architectures is evaluated in terms of input current waveform quality, input power factor, device stress, power loss, and the cost. The performance of selected power conversion systems is conducted on designed IPT-based charging systems in compliance with level 1 of static wireless charging standard for light-duty vehicles provided in SAE J2954 technical information report [19]. The specifications of designed IPT systems are power rating P_o = 3.3 kW, operating frequency f_s = 85 kHz, battery voltage V_b = 300–400 V, and grid voltage V_g = 208 V. The components are selected and designed based on characteristics presented in Tables 3.10 and 3.11. The parameters of each type of power conversion architecture are shown in Table 3.12. In all power conversion topologies, SiC-MOSFETs and Schottky diodes manufactured by Rohm are considered. To limit current harmonic injection caused by the switching power converters, LC filters are employed between the grid and the charging systems. By performing the spectrum analysis of the input current waveforms (i_i), LC filters can be designed. The details of each component of selected topologies are listed in Table 3.13.

The results of the selected three topologies are shown in Figure 3.16. It shows a double line frequency fluctuation in transferred power due to the absence of a bulky capacitor in MC based topologies. Consequently, fluctuating charging current is observed in Figure 3.16 (b) and (c).

TABLE 3.12

Specifications of Selected Dual- and Single-Stage Inductive Power Transfer Charging Systems

Topologies	Parameter	Symbol	Value	Unit
Dual-stage	Primary, secondary, mutual inductance	L_p, L_s, M	356, 328, 65	μH
	Compensation capacitors	C_p, C_s	10, 11	nF
	Boost inductor	L_b	0.215	mH
	DC-bus capacitor	C_b	1540	μF
	DC-bus voltage	V_{dc}	400	V
	Grid inductor	L_g	0.215	mH
	Grid capacitor	C_g	0.78	μF
	Output capacitor	C_o	500	μF
Buck-derived FBMC	Primary, secondary, mutual inductance	L_p, L_s, M	111, 111, 24	μH
	Compensation capacitors	C_p, C_s	32, 32	nF
	Grid inductor	L_g	0.215	mH
	Grid capacitor	C_g	0.78	μF
	Output capacitor	C_o	500	μF
Boost-derived FBMC	Primary, secondary, mutual inductance	L_p, L_s, M	111, 111, 24	μH
	Compensation capacitors	C_{ps}, C_{pp}, C_s	43, 115, 32	nF
	Boost inductor	L_i	0.215	mH
	Grid inductor	L_g	0.036	mH
	Grid capacitor	C_g	0.136	μF
	Output capacitor	C_o	500	μF

3.5.2.1 Input Power Factor and Input Current Distortion

Ensuring high power quality, high power factor, and low current distortion are essential features of an EV charger. Therefore, harmonic and power factor analyses are carried out for three selected topologies at different loadings (20, 50, and 100% of load). It is noticed from Figure 3.16 that all three power topologies generate sinusoidal grid currents with a power factor of 0.99.

The THDs of grid current in all three topologies are summarized in Figure 3.17(a). The continuous input current with ripple at a twice-switching frequency is observed in the boost-derived FBMC. Subsequently, a significant harmonic reduction of grid current is achieved with a smaller input filter. The buck-derived FBMC injects higher current harmonics to the grid due to discontinuity in input current. Therefore, the grid current quality is highest in boost-derived FBMC and is followed by dual-stage converter and buck-derived FBMC.

3.5.2.2 Switching Stress

Figure 3.17(b) and (c) illustrate the maximum device current and voltage stress of the selected topologies. The dual-stage converter topology displays the lowest switch current stress in both primary and secondary converters. Despite using the parallel-series CC compensation, the switches of boost-derived FBMC experience high

TABLE 3.13

Details of Main Components of Selected Power Converter Topologies

Topologies	Components	Symbol	Part Number	Quantity
Dual-stage	Front-end rectifier diodes	D_{gn}	SCS240AE2C-ND	4
	Boost diode	D_b	SCS240AE2C-ND	1
	Boost switch	S_b	SCT3060ALGC11-ND	1
	Primary inverter switches	S_{pn}	SCT3120ALHRC11-ND	4
	Secondary rectifier diodes	D_{sn}	SCS230AE2HRC-ND	4
	Boost inductor	L_b	HF5712-561M-25AH	2 parallel
	DC-bus capacitor	C_b	LGN2X221MELC50	7 parallel
	Grid inductor	L_g	HF5712-561M-25AH	2 parallel
	Grid capacitor	C_g	B32656T7394K000	2 parallel
Buck-derived FBMC	Primary MC switches	S_{pnx}, S_{pnx}	SCT3030ALGC11-ND	8
	Secondary rectifier diodes	S_{sn}	SCT3060ALGC11-ND	4
	Grid inductor	L_g	HF5712-561M-25AH	2 parallel
	Grid capacitor	C_g	B32656T7394K000	2 parallel
Boost-derived FBMC	Primary MC switches	S_{pnx}, S_{pnx}	SCT2080KEC-ND	8
	Secondary rectifier diodes	$D_{sn}{}^*$	SCS240AE2C-ND	4
	Boost inductor	L_i	HF5712-561M-25AH	2 parallel
	Grid inductor	L_g	HF467-980M-25AV	2 parallel
	Grid capacitor	C_g	B32654A1683K000	2 parallel

(n = 1, 2, 3, 4 and $x = u, v$)

voltage stress. Although the buck-derived FBMC experiences low switch voltage stress (grid voltage peak) it suffers from high switch current stress.

3.5.2.3   Loss Distribution and Efficiency

The power losses of the selected three power electronic converter topologies are simulated using the electrothermal modules in the PSIM software platform. The total power conversion efficiency of the three topologies under different loading conditions (20, 50, and 100%) is shown in Figure 3.17(d). It can be seen that the buck-derived FBMC system exhibits the highest efficiency (almost 98%) at full load conditions and gradually decreases to 93% at light load conditions. On the flip side, the efficiency of the boost-derived FBMC system steadily increases from 92.5 to 96% with a decrease in load (100 to 20%). The efficiency of the dual-stage IPT system fairly remains the same in the range of 94–96.5% over a wide load range. The detailed loss distribution of the three IPT systems is illustrated in Figure 3.18. It is noticed that conduction losses of switches contribute a major portion of the total losses.

3.5.2.4   Cost Analysis

To evaluate the economic viability of the three IPT systems, cost analysis is performed with the component data listed in Table 3.14.

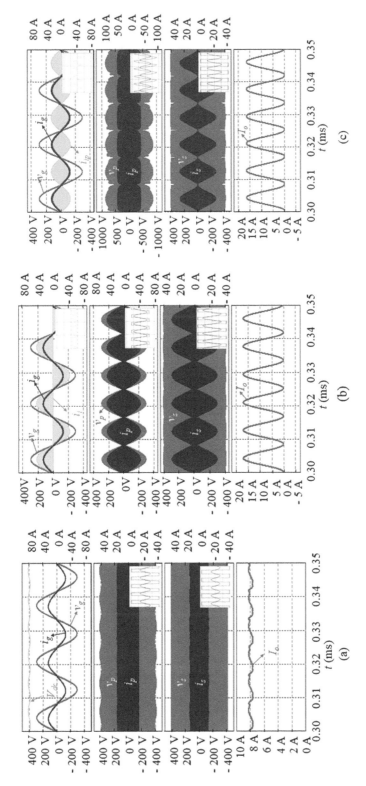

FIGURE 3.16 Results of IPT charging systems fed by (a) dual-stage power converter (PFC and full-bridge VSI), (b) buck-derived FBMC, and (c) boost-derived FBMC.

TABLE 3.14

Cost Details of the Active and Passive Components Utilized in This Study

Components	Manufacturer Part Number	Rating	Unit Cost ($)
Diode	SCS230AE2HRC-ND	650V/30A	8.97
	SCS240AE2C-ND	650V/40A	12.75
MOSFET	SCT3120ALHRC11-ND	650V/21A	7.02
	SCT3060ALGC11-ND	650V/39A	8.74
	SCT3030ALGC11-ND	650V/70A	19.46
	SCT2080KEC-ND	1200V/40A	17.77
Gate driver IC	UCC5390SCD	N/A	2.16
Gate driver supply	R12P21503D	+15V/-3V/2W	7.11
Inductor	HF467-980M-25AV	25A/72μH	21.15
	HF5712-561M-25AH	25A/430μH	29.25
Capacitor	LGN2X221MELC50 (Electrolytic)	600 V/220μF	7.78
	B32656T7394K000 (Film)	500 V/0.39μF	4.23
	B32654A1683K000	500 V/0.068μF	1.01

Source: Digikey and Mouser.

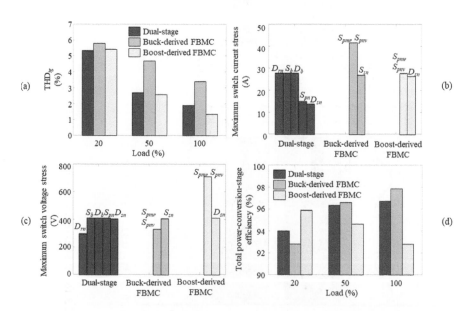

FIGURE 3.17   Performance comparison of three inductive power transfer charging systems: (a) Grid current total harmonic distortion, (b) device current stress, (c) device voltage stress, and (d) power-conversion-stage efficiency.

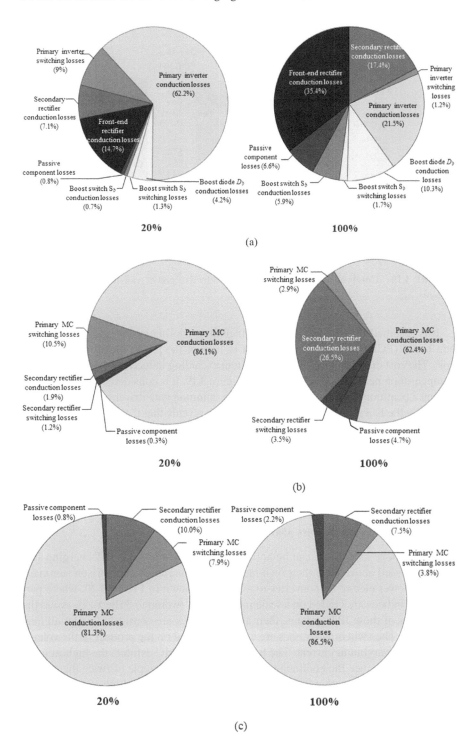

FIGURE 3.18 Power loss distribution: (a) dual-stage topology, (b) buck-derived full-bridge matrix converter (FBMC), and (c) boost-derived FBMC.

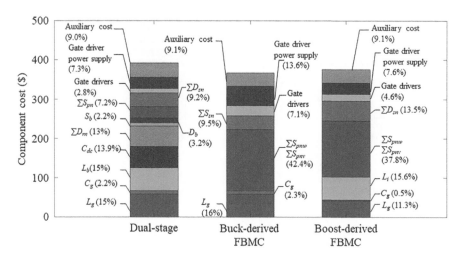

FIGURE 3.19 Component cost structure of the inductive power transfer power conversion systems.

It is assumed that the auxiliary cost including cooling system cost, printed circuit board (PCB) cost, and housing cost is 10% of the power electronic converter cost. The cost distribution of each IPT system excluding inductive coupling coils and compensation networks are illustrated in Figure 3.19. It is considered that MOSFETs having a common source connection utilize a common gate driver power supply and driven by the isolated gate drivers. The analysis reveals that the cost of single-stage systems is lower than that of the dual-stage IPT systems.

The buck-derived FBMC system contributes 6.3% less cost than the dual-stage system. Therefore, it is the most cost-effective solution. It is found that power devices of MCs dominate a major part of the total cost of the single-stage systems, whereas the costs of the passive components occupy large portions in the dual-stage systems.

3.5.3 Comparison Studies and Their Discussion

From the above analysis, it can be noticed that the most popular IPT charging systems discussed above have their own merits and demerits. A comparison summary of these power electronic converter topologies is shown in Figure 3.20, where performance indices are graded on a scale range from 1 (worst) to 3 (best). To assess the efficiencies of these IPT systems, their average values are considered under all load conditions. The switching stresses are evaluated based on the product of the voltage stresses and maximum current. The boost-derived FBMC exhibits the highest input current quality since the continuous input current with a ripple of twice the switching frequency. On the other hand, buck-derived FBMC displays high efficiency, high power density, and cost reduction due to less component count (Figure 3.20). The dual-stage power conversion systems depict the lowest device stress and maintain a comparable efficiency over a wide load range. Furthermore, it is more matured in terms of control, easy design, and manufacturability. Thus, this configuration

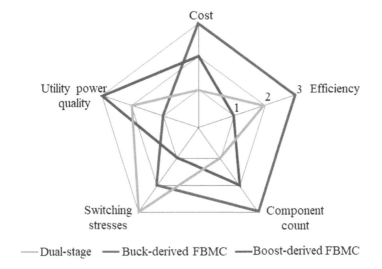

FIGURE 3.20 Comparison summary of the dual-stage and single-stage inductive power transfer charging systems.

is widely adopted in academic research, industries, and power electronic charger manufacturers.

3.6 FUTURE TRENDS AND OPPORTUNITIES

Over the past few decades, significant improvements in power conversion stages and control strategies have been carried out for IPT-based EV charging systems. One of the main concerns is the design and development of high-frequency power electronic converters to meet the current standards and future requirements. With the current state–of-the-art in power semiconductor devices, advanced control techniques, and digital controllers, further improvements are needed to improve the performance of power conversion systems in terms of power density, scalability, efficiency, and reliability to stimulate the IPT-based EV charging infrastructure. One of the prime factors to enhance efficiency is the reduction in power conversion losses from the source to the input of the coil. The employment of advanced soft-switching control schemes for the existing power electronic converters and development of new reduced-switch-count power converters can reduce switching losses to a greater extent. Consequently, it improves the thermal design and power density of overall IPT systems. Numerous soft-switching control techniques have been reported in the literature and are categorized into three main groups: variable resonant networks, with auxiliary DC–DC converter, and with active inverter/rectifier control. However, these control strategies demand resonant components and extra DC–DC converters. Furthermore, it depicts high control complexity and has an operation range limitation.

Single-stage power conversion topologies like MCs can be one of the attainable contenders with the excretion of life-limited aluminum electrolytic capacitors and employing enhanced charging techniques such as sinusoidal ripple current charging.

The existing or advanced power electronic converter topologies employing wide bandgap (WBG) devices help to improve efficiency and power density as it allows operation at higher switching frequencies with low switching losses [61].

The utilization of gallium nitride (GaN) devices in IPT systems has opened up a new research area as it offers a low voltage drop and ability to operate at the higher switching frequency. Furthermore, it generates comparatively lower thermal losses during operation which facilitates the passive cooling in order to improve the power density and cost-effectiveness of the power electronic converter. However, some serious concerns still exist with regard to the manufacturing process, the gate driver design, device characterization, packaging with higher current ratings, thermal management, busbar layout, and reliability. Hence, further research is needed to resolve the aforementioned issues. Moreover, operating the GaN devices can enhance the performance of IPT systems in various ways, such as passive component size reduction, higher tolerance to coil misalignment, and transfer distance extension. To further reduce the size and loss of IPT systems, the integration of magnetic components of power electronic converters, compensation networks, and coupling coils is another stimulating solution, and is an active area of research. Also, the utilization of advanced materials and nanotechnology can reduce the volume and weight of passive components.

Last decade, industrial manufacturers demand power electronic converters that have modularity, scalability, flexibility in assembly, simple manufacturing process, and reduced mean time to failure (MTTF). The prominent candidates to enhance the fault tolerance and output power capability for IPT systems are multiphase parallel inverter [62], modular multilevel converter [63], and parallel IPT topologies [64], which can open up more attention in developing fault-tolerant control schemes and advanced power topologies. The development of advanced high-performance control techniques, such as model predictive control, disturbance-observer-based control, and sliding mode control, can be employed to improve the performance and controllability of power converters in IPT systems. Another important concern regarding control and monitoring is the existing control algorithms utilize a large number of sensors and communication channels. Thus, it is necessary to develop sensorless control methods and parameter estimation approaches to reduce the cost and complexity of the system. The sensorless control scheme presented in [65] allows the reduction of sensors and the removal of a communication link between the ground (primary side) and the vehicle (secondary side), which reduces component count, cost, and enhance the reliability of the system.

Another stimulating research area is bidirectional power flow, in integration with multiple energy sources such as solar, wind, fuel, and supercapacitors. Vehicle-to-grid (V2G) takes EVs as a movable backup energy source, which can effectively improve the quality and reliability of the grid by providing grid support services, such as peak shaving, and voltage control. In combination with the bidirectional wireless power transfer systems, solar-based EV wireless vehicle charging (S2V) systems is a bridge to further integrate this type of renewable energy into the power system. However, these areas are still under research, which further attention and investigation for developing advanced power electronic converter topologies and control schemes must comply with future IPT charging standards. Developing converter topologies, energy management systems, and control schemes, which can work in multiple operating modes, is a challenging task and has a more significant trend in the near future.

3.7   CONCLUDING REMARKS

This chapter exhaustively discussed the fundamentals, coupling links, industrial standards, power electronic converter topologies, and control schemes in dual-stage and single-stage conversion IPT systems for EV charging applications. Most widely adopted dual- and single-stage power conversion topologies for IPT systems are selected and their important concepts are illustrated through case studies. The key performance of each power topology is holistically evaluated in terms of input power factor, input current distortion, current stress, voltage stress, power losses on the converter, and the cost. It is concluded that buck-derived FBMC exhibits high efficiency, low component count, and low cost over the other topologies. In contrast, boost-derived FBMC provides the greatest input current quality and conventional dual-stage topology has the lowest stress on switching devices. Finally, the future trends and opportunities of IPT-based EV charging systems are briefly summarized showing the importance and definitive increase the availability of the EVs in near future.

REFERENCES

1. A. Khaligh and S. Dusmez, "Comprehensive topological analysis of conductive and inductive charging solutions for plug-in electric vehicles," *IEEE Trans. Veh. Technol.*, vol. 61, no. 8, pp. 3475–3489, Oct. 2012.
2. D. Ronanki, A. Kelkar, and S.S. Williamson, "Extreme fast charging technology-prospects to enhance sustainable electric transportation," *Energies*, vol. 12, no. 19, pp. 3721–3737, Sep. 2019.
3. A. Khaligh and M. D'Antonio, "Global trends in high-power on-board chargers for electric vehicles," *IEEE Trans. Vehi. Tech.*, vol. 68, no. 4, pp. 3306–3324, April 2019.
4. A. Ahmad, M. S. Alam, and R. Chabaan, "A comprehensive review of wireless charging technologies for electric vehicles," *IEEE Trans. Transport. Electrific.*, vol. 4, no. 1, pp. 38–63, Mar. 2018.
5. C. S. Wang, O. H. Stielau, and G. A. Covic, "Design considerations for a contactless electric vehicle battery charger," *IEEE Trans. Ind. Electron.*, vol. 52, no. 5, pp. 1308–1314, Oct. 2003.
6. R. Bosshard and J. W. Kolar, "Inductive power transfer for electric vehicle charging: Technical challenges and tradeoffs," *IEEE Power Electron. Mag.*, vol. 3, no. 3, pp. 22–30, Sept. 2016.
7. S. Y. Choi, B. W. Gu, S. Y. Jeong, and C. T. Rim, "Advances in wireless power transfer systems for roadway-powered electric vehicles," *IEEE J. Emerg. Sel. Top. Power Electron.*, vol. 3, no. 1, pp. 18–36, Mar. 2013.
8. C. C. Mi, G. Buja, S. Y. Choi, and C. T. Rim, "Modern advances in wireless power transfer systems for roadway powered electric vehicles," *IEEE Trans. Ind. Electron.*, vol. 63, no. 10, pp. 6533–6545, Oct. 2016.
9. J1772: SAE electric vehicle and plug in hybrid electric vehicle conductive charge coupler - SAE International [Online], 2017. Available at https://www.sae.org/standards/content/j1772_201710/.

10. S. Lukic and Z. Pantic, "Cutting the cord: Static and dynamic inductive wireless charging of electric vehicles," *IEEE Electrific. Mag.*, vol. 1, no. 1, pp. 57–64, Sep. 2013.
11. P. S. Huynh, D. Ronanki, D. Vincent, and S. S. Williamson, "Overview and comparative assessment of single-phase power converter topologies of inductive wireless charging systems," *Energies*, vol. 13, no. 9, pp. 2150–2172, May 2020.
12. J2954: Wireless power transfer for light-duty plug-in/electric vehicles and alignment methodology—SAE International [Online], 2019. Available at http://standards.sae.org/wip/j2954/.
13. V. Cirimele, M. Diana, F. Freschi, and M. Mitolo, "Inductive power transfer for automotive applications: State-of-the-art and future trends," *IEEE Trans. Ind. Appl.*, vol. 54, no. 5, pp. 4069–4079, Sept.–Oct. 2018.
14. Charging electric buses quickly and efficiently: bus stops fitted with modular components make 'charge & go' simple to implement [Online], Jul. 2013. Accessed: Feb. 28, 2016. Available at http://www.conductix.us/en/news/2013-05-29/chargingelectric-busesquickly-and-efficiently-busstops-fittedmodular-components-make-charge-go.
15. T. M. Fisher, K. B. Farley, Y. Gao, H. Bai, and Z. T. H. Tse, "Electric vehicle wireless charging technology: A state-of-the-art review of magnetic coupling systems," *Wireless Power Transf.*, vol. 1, no. 2, pp. 87–96, 2014.
16. DRIVE: Electric Vehicles. [Online]. Accessed: May 2, 2016. Available at http://witricity.com/products/automotive/.
17. Review and evaluation of wireless power transfer (wpt) for electric transit applications. [Online]. Accessed: Jun. 5, 2016. Available at https://www.transit.dot.gov/sites/fta.dot.gov/files/FTA_Report_No._0060.pdf.
18. R. Bosshard and J. W. Kolar, "All-SiC 9.5 kW/dm3 on-board power electronics for 50 kW/85 kHz automotive IPT system," *IEEE J. Emerg. Sel. Top. Power Electron.*, vol. 5, no. 1, pp. 419–431, Mar. 2017.
19. S. Weerasinghe, U. K. Madawala, and D. J. Thrimawithana, "A matrix converter-based bidirectional contactless grid interface," *IEEE Trans. Power Electron.*, vol. 32, no. 3, pp. 1755–1766, Mar. 2017.
20. H. L. Li, A. P. Hu, and G. A. Covic, "A direct AC–AC converter for inductive power-transfer systems," *IEEE Trans. Power Electron.*, vol. 27, no. 2, pp. 661–668, Feb. 2012.
21. W. Sulistyono and P. Enjeti, "A series resonant AC-to-DC rectifier with high-frequency isolation," *IEEE Trans. Power Electron.*, vol. 10, no. 6, pp. 784–790, Nov. 1993.
22. S. Samanta and A. K. Rathore, "A new inductive power transfer topology using direct AC–AC converter with active source current waveshaping," *IEEE Trans. Power Electron.*, vol. 33, no. 7, pp. 5565–5577, July 2018.
23. M. Moghaddami, A. Anzalchi, and A. I. Sarwat, "Single-stage three-phase AC–AC matrix converter for inductive power transfer systems," *IEEE Trans. Ind. Electron.*, vol. 63, no. 10, pp. 6613–6622, Oct. 2016.
24. L. Chen, S. Wu, D. Shieh, and T. Chen, "Sinusoidal-ripple-current charging strategy and optimal charging frequency study for Li-ion batteries," *IEEE Trans. Ind. Electron.*, vol. 60, no. 1, pp. 88–97, Jan. 2013.
25. J. M. Miller, O. C. Onar, and M. Chinthavali, "Primary-side power flow control of wireless power transfer for electric vehicle charging," *IEEE J. Emerg. Sel. Topics Power Electron.*, vol. 3, no. 1, pp. 147–162, Mar. 2013.
26. D. Vincent, P. S. Huynh, N. A. Azeez, L. Patnaik, and S. S. Williamson, "Evolution of hybrid inductive and capacitive AC links for wireless EV charging—a comparative overview," *IEEE Trans. Transport. Electrific.*, vol. 5, no. 4, pp. 1060–1077, Dec. 2019.
27. F. Nakao, Y. Matsuo, M. Kitaoka and H. Sakamoto, "Ferrite core couplers for inductive chargers," In *Proceedings of Power Conversion Conference*, Osaka, Japan, pp. 850–854, Apr. 2002.

28. D. Patil, M. K. McDonough, J. M. Miller, B. Fahimi, and P. T. Balsara, "Wireless power transfer for vehicular applications: Overview and challenges," *IEEE Trans. Transport. Electrific.*, vol. 4, no. 1, pp. 3–37, March 2018.

29. S. Kim, A. Zaheer, G. Covic and J. Boys, "Tripolar pad for inductive power transfer systems," In *Proceedings of* 40th Annu*al Conference of* IEEE Ind*ustrial* Electronic *Society* (IECON), Dallas, TX, USA, pp. 3066–3072, Oct./Nov. 2014.

30. J. Kim, D. Kim, and Y. Park, "Analysis of capacitive impedance matching network for simultaneous wireless power transfer to multiple devices," *IEEE Trans. Ind. Electron.*, vol. 62, no. 5, pp. 2807–2813, May 2013.

31. F. Lu, H. Zhang, H. Hofmann, and C. C. Mi, "An inductive and capacitive combined wireless power transfer system with LC-compensated topology," *IEEE Trans. Power Electron.*, vol. 31, no. 12, pp. 8471–8482, Dec. 2016.

32. W. Li, H. Zhao, S. Li, J. Deng, and T. Kan, "Integrated LCC compensation topology for wireless charger in electric and plug-in electric vehicles," *IEEE Trans. Ind. Electron.*, vol. 62, no. 7, pp. 4215–4225, July 2013.

33. S. Samanta and A. K. Rathore, "A new current-fed CLC transmitter and LC receiver topology for inductive wireless power transfer application: Analysis, design, and experimental results," *IEEE Trans. Transport. Electrific.*, vol. 1, no. 4, pp. 357–368, Dec. 2013.

34. Y. H. Sohn, B. H. Choi, E. S. Lee, G. C. Lim, G. H. Cho, and C. T. Rim, "General unified analyses of two-capacitor inductive power transfer systems: Equivalence of current-source SS and SP compensations," *IEEE Trans. Power Electron.*, vol. 30, no. 11, pp. 6030–6045, Nov. 2013.

35. J. Sallan, J. L. Villa, A. Llombart, and J. F. Sanz, "Optimal design of ICPT systems applied to electric vehicle battery charge," *IEEE Trans. Ind. Electron.*, vol. 56, no. 6, pp. 2140–2149, Jun. 2009.

36. J. M. Miller and B. R. Long, "What all technology adopters should know about WPT for high power charging," In IEEE PELS Workshop *of Emerging* Techno*logy*, Wireless Power (WOW), Knoxville, TN, USA, Oct. 2016.

37. J. L. Villa, J. Sallan, J. F. S. Osorio, and A. Llombart, "High misalignment tolerant compensation topology for ICPT systems," *IEEE Trans. Ind. Electron.*, vol. 59, no. 2, pp. 945–951, Feb. 2012.

38. S. Li et al., "A double-sided LCC compensation network and its tuning method for wireless power transfer," *IEEE Trans. Veh. Technol.*, vol. 64, no. 6, pp. 2261–2273, Jun. 2013.

39. ICNIRP guidelines for limiting exposure to time-varying electric, magnetic and electromagnetic fields (1 Hz to 100 kHz), 2010 [Online]. Available at https://www.icnirp.org/cms/upload/publications/ICNIRPLFgdl.pdf.

40. A. Kamineni, G. A. Covic, and J. T. Boys, "Self-tuning power supply for inductive charging," *IEEE Trans. Power Electron.*, vol. 32, no. 5, pp. 3467–3479, May 2017.

41. K. Aditya and S. S. Williamson, "A review of optimal conditions for achieving maximum power output and maximum efficiency for a series–series resonant inductive link," *IEEE Trans. Transport. Electrific.*, vol. 3, no. 2, pp. 303–311, June 2017.

42. Y. Yao, Y. Wang, X. Liu, F. Lin, and D. G. Xu, "A novel parameter tuning method for double-sided LCL compensated WPT system with better comprehensive performance," *IEEE Trans. Power Electron.*, vol. 33, no. 10, pp. 8525–8536, 2018.

43. U. K. Madawala and D. J. Thrimawithana, "A bidirectional inductive power interface for electric vehicles in V2G systems," *IEEE Trans. Ind. Electron.*, vol. 58, no. 10, pp. 4789–4796, Oct. 2011.

44. K. Aditya and S. S. Williamson, "Design guidelines to avoid bifurcation in a series–series compensated inductive power transfer system," *IEEE Trans. Ind. Electron.*, vol. 66, no. 5, pp. 3973–3982, May 2019.

45. O. C. Onar, M. Chinthavali, S. Campbell, P. Ning, C. P. White and J. M. Miller, "A SiC MOSFET based inverter for wireless power transfer applications," In Proc*eedings of* IEEE Appl*ied* Power Electro*nics* Conf*erence* Expo*sition* (APEC), Fort Worth, TX, USA, pp. 1690–1696, Mar. 2014.

46. J. Deng, W. Li, T. D. Nguyen, S. Li, and C. C. Mi, "Compact and efficient bipolar coupler for wireless power chargers: Design and analysis," *IEEE Trans. Power Electron.*, vol. 30, no. 11, pp. 6130–6140, Nov. 2013.

47. R. Bosshard and J. W. Kolar, "Multi-objective optimization of 50 kW/85 kHz IPT system for public transport," *IEEE J. Emerg. Sel. Topics Power Electron.*, vol. 4, no. 4, pp. 1370–1382, Dec. 2016.

48. S. G. Rosu, M. Khalilian, V. Cirimele and P. Guglielmi, "A dynamic wireless charging system for electric vehicles based on DC/AC converters with SiC MOSFET-IGBT switches and resonant gate-drive," In Proc*eedings of* 42nd Annu*al Conference of* IEEE Ind*ustrial* Electroni*c Society* (IECON), Florence, Italy, pp. 4465–4470, Oct. 2016.

49. A. Q. Cai and L. Siek, "A 2-kW, 95% efficiency inductive power transfer system using gallium nitride gate injection transistors," *IEEE J. Emerg. Sel. Topics Power Electron.*, vol. 5, no. 1, pp. 458–468, Mar. 2017.

50. J. Liu, K. W. Chan, C. Y. Chung, N. H. L. Chan, M. Liu, and W. Xu, "Single-stage wireless-power-transfer resonant converter with boost bridgeless power-factor-correction rectifier," *IEEE Trans. Ind. Electron.*, vol. 65, no. 3, pp. 2145–2155, March 2018.

51. A. Ecklebe, A. Lindemann, and S. Schulz, "Bidirectional switch commutation for a matrix converter supplying a series resonant load," *IEEE Trans. Power Electron.*, vol. 24, no. 5, pp. 1173–1181, May 2009.

52. N. X. Bac, D. M. Vilathgamuwa, and U. K. Madawala, "A SiC-based matrix converter topology for inductive power transfer system," *IEEE Trans. Power Electron.*, vol. 29, no. 8, pp. 4029–4038, Aug. 2014.

53. J. M. Miller, O. C. Onar, and M. Chinthavali, "Primary-side power flow control of wireless power transfer for electric vehicle charging," *IEEE J. Emerg. Sel. Topics Power Electron.*, vol. 3, no. 1, pp. 147–162, Mar. 2013.

54. K. Aditya and S. S. Williamson, "Comparative study on primary side control strategies for series-series compensated inductive power transfer system," In Proc*eedings of* IEEE 25th Int*ernational* Symp*osium of* Ind*ustrial* Electroni*cs* (ISIE), Santa Clara, CA, pp. 811–816, 2016.

55. A. Kamineni, M. J. Neath, G. A. Covic, and J. T. Boys, "A mistuning-tolerant and controllable power supply for roadway wireless power systems," *IEEE Trans. Power Electron.*, vol. 32, no. 9, pp. 6689–6699, Sept. 2017.

56. P. Si, A. P. Hu, S. Malpas, and D. Budgett, "A frequency control method for regulating wireless power to implantable devices," *IEEE Trans. Biomed. Circuits Syst.*, vol. 2, no. 1, pp. 22–29, Mar. 2008.

57. C. S. Wang, G. A. Covic, and O. H. Stielau, "Power transfer capability and bifurcation phenomena of loosely coupled inductive power transfer systems," *IEEE Trans. Ind. Electron.*, vol. 51, no. 1, pp. 148–157, Feb. 2004.

58. K. Aditya and S. S. Williamson, "Design guidelines to avoid bifurcation in a series-series compensated inductive power transfer system," *IEEE Trans. Ind. Electron.*, vol. 66, no. 5, pp. 3973–3982, May 2019.

59. A. A. S. Mohamed, A. Berzoy, F. G. N. d. Almeida, and O. Mohammed, "Modeling and assessment analysis of various compensation topologies in bidirectional IWPT system for EV applications," *IEEE Trans. Ind. Appl.*, vol. 53, no. 5, pp. 4973–4984, Oct. 2017.

60. E. A. Jones, F. F. Wang, and D. Costinett, "Review of commercial GaN power devices and GaN-based converter design challenges," *IEEE J. Emerg. Sel. Topics Power Electron.*, vol. 4, no. 3, pp. 707–719, Sept. 2016.

61. Q. Deng, J. Liu, D. Czarkowski, W. Hu, and H. Zhou, "An inductive power transfer system supplied by a multiphase parallel inverter," *IEEE Trans. Ind. Electron.*, vol. 64, no. 9, pp. 7039–7048, Sept. 2017.

62. D. Ronanki and S. S. Williamson, "Modular multilevel converters for transportation electrification: Challenges and opportunities," *IEEE Trans. Transport. Electrific.*, vol. 4, no. 2, pp. 399–407, June 2018.

63. H. Hao, G. A. Covic, and J. T. Boys, "A parallel topology for inductive power transfer power supplies," *IEEE Trans. Power Electron.*, vol. 29, no. 3, pp. 1140–1151, 2014.

64. R. Ruffo, V. Cirimele, M. Diana, M. Khalilia, A. L. Ganga, and P. Guglielmi, "Sensorless control of the charging process of a dynamic inductive power transfer system with an interleaved nine-phase boost converter," *IEEE Trans. Ind. Electron.*, vol. 65, no. 10, pp. 7630–7639, Oct. 2018.

4 Power Electronics-Based Solutions for Supercapacitors in Electric Vehicles

Deepak Ronanki
Indian Institute of Technology Roorkee

Yashwanth Dasari and Sheldon S. Williamson
University of Ontario Institute of Technology

CONTENTS

4.1   INTRODUCTION

In recent times, a paradigm shift towards the lowering exhaust gas emissions, energy security, and higher fuel efficiency has led to the electrification of the conventional vehicles through a wide spread of electric vehicle (EV) technology. In view of this current trend, the deployment of battery-powered electric vehicles (BEVs), hybrid electric vehicles (HEVs), and plug-in hybrid electric vehicles (PHEVs) has helped to stimulate greener transportation. The commercially available EVs, HEVs, and PHEVs employ lithium-ion (Li-ion) battery as a primary energy source to propel its traction system on account of high energy density, low self-discharge, and longer cycle life [1]. However, rapid accelerations of an EV in urban driving schedules results in higher current demands from the energy storage system such as batteries. Subsequently, the long-term performance of the Li-ion battery pack is affected. Moreover, Li-ion batteries have short cycle life and cannot be charged from the regenerative braking power during instant higher decelerations.

Hence, a standalone battery system may not be ample to fulfill the peaky demand periods and transient load variations on a single charge in the all-electric mode. In such cases, the batteries are enforced to handle high discharge rates in shorter intervals. Therefore, the battery systems need to be oversized to meet the load demands. Consequently, the volume, weight, and cost of the energy storage system (ESS) are increased [2]. Furthermore, their aging is expedited owing to periodic charge and discharge, which hassles its replacement before the scheduled vehicle maintenance. Indeed, all these concerns in battery systems are hindering the rapid commercialization of EVs.

Supercapacitors (SCs) that have high specific power density, smaller equivalent series resistance (ESR), faster dynamic response, longer lifecycle, and ability to operate in extremely low or high temperatures can be the solution to abovementioned problems. The characteristics of Li-ion batteries and SCs are summarized in Table 4.1. Typically, SCs offer a quick charge/discharge capability by resulting in wide voltage variations at their terminals. They are primarily used as a backup power source and deliver pulsed power to the loads. Therefore, these are mostly utilized in pulsed power applications such as electric railway traction systems [4], grid-connected systems [5], electric transit buses [6], more electric aircraft [7], electric ship power systems [8, 9], and solid-state transformer-based systems [10, 11]. The SCs are used in conjunction with the batteries as secondary storage devices that fulfill the aforementioned concerns of an EV, thus forming a hybrid energy storage system (HESS). Furthermore, the batteries in HESS are downsized and designed to serve only the smaller transients for a longer duration. As a result, such configuration enhances the power management, control flexibility, and driving range of an EV [12].

TABLE 4.1
Characteristics of Li-Ion Batteries and Supercapacitors [3]

Function	Units	Li-Ion Battery	Supercapacitor
Specific energy	Wh/kg	230	1–10
Specific power	kW/kg	2.5	Up to 10
Cycle life	–	<5000	>500,000
Cost	$/Wh	2	20

However, the specific energy density of SCs is low compared to Li-ion batteries. Subsequently, an SC cell cannot deliver its full energy to the load. Generally, SCs are rated at lower voltages ($\cong 2.5$ V) offering a higher capacitance due to their internal chemical characteristics. To meet the load voltage requirements, these are connected in series to build up the overall bank voltage. However, the nonlinear characteristics of individual SCs in a bank prompt a voltage imbalance during rapid charge/discharge operations. As a result, the cells undergo an uneven charge and discharge behavior. Therefore, voltage equalization techniques are recommended to overcome these issues. A simplest and low-cost approach is the use of a resistor in parallel with individual SC in a bank. However, this passive solution dissipates the excess energy of SCs as heat. Also, the series connection of SCs results in a reduction of the overall capacitance that causes the SCs to store less energy. In such cases, the power delivery duration of an SC bank is significantly reduced which lacks in serving a continuous series of transient power demands on a single charge. Hence, they are required to get recharged multiple times in HESS either from a battery or the regenerative braking power. Furthermore, it increases the terminal voltage variations during charge/discharge modes and results in higher current stress on the active and passive components of the subsequent power converter. Therefore, the energy extraction of SCs is crucial for determining the performance of the ESSs.

The aforementioned issues are addressed in this chapter by looking in depth into the power electronics (PE) intensive active SC cell voltage equalizers and bank switching architectures. First, the fundamentals of SCs and their charge/discharge behavior under various conditions are discussed. Also, the electric equivalent circuits of the SCs to evaluate their performance and future predictions are deliberated. The SCs demand voltage equalization and power management approaches for safe and efficient operation. The design, sizing, and control of various voltage equalizers and bank switching techniques are presented. The performance of each bank switching architecture is analyzed through simulation studies in the PLECS software platform. The results are monitored in terms of individual cell voltages, SC cell energy utilization, and power delivery duration. Also, the design and control of existing PE interfacing topologies including power allocation techniques and converter control strategies for HESS are discussed in detail. Finally, a case study on an EV with battery-SC HESS is discussed by employing PE enabled power management schemes under two different driving patterns in the MATLAB®/Simulink® environment. Overall, the fundamentals and PE intensive solutions for SCs in this chapter will refresh the reader's knowledge on this specific topic.

This chapter is organized as follows:

- The fundamental details of SCs such as construction, classification, electrical characteristics, charge/discharge behavior for various types of loads, and modeling approaches are discussed in Section 4.2.
- The past and recent developments of passive and active (PE enabled) SC cell voltage equalization circuits are given in Section 4.3.
- The implementation of PE intensive SC bank switching approaches and their performance evaluation studies are presented in Section 4.4.

- The overview of PE interfaces and their control algorithms for battery/SC-based HESSs are discussed in Section 4.5.
- A detailed case study on an EV with HESS to analyze the performance of the PE enabled converters under standard driving schedules is presented in Section 4.6.
- Section 4.7 provides the concluding remarks of this chapter.

4.2 FUNDAMENTAL OF SUPERCAPACITORS

4.2.1 CONSTRUCTION AND CLASSIFICATION OF SCs

The SCs are developed based on the principles of conventional capacitors by employing two porous carbon electrodes separated by a dielectric medium. Additionally, the SCs use an electrolyte as a dielectric medium and a separator to split the charges between the electrodes as shown in Figure 4.1. When the electrodes are excited, the diameter of the pores is increased. The carbon electrodes absorb those pores and offer high capacitance per unit. As the surface area of the carbon electrodes is high, the nearly ideal charge and discharge characteristics of an SC can be obtained [13].

The electrolyte used in SCs can be either aqueous or nonaqueous that flows through the insulator during the charge and discharge cycles. Most of the SCs are equipped with an aqueous electrolyte which offers low ESR. However, it provides lower decomposition voltages. For instance, the SC with sulfuric acid as an electrolyte offers an operating voltage of 1.2 V. On the other hand, the nonaqueous electrolyte can offer higher decomposition voltages yet results in higher ESR values, which are not recommended for the construction of SCs.

Moreover, the high voltage design of an SC results in dissociation of carbon electrodes and weakens the SC in case of prolonged usage. During the chemical process, the separator forms a double layer on its surface. Thus, the SCs are also termed as electric double-layer capacitors (EDLC) or ultracapacitors.

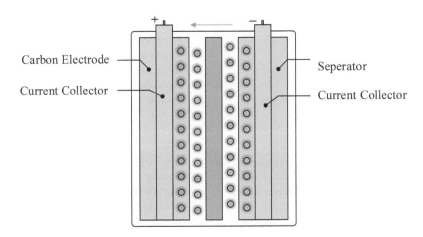

FIGURE 4.1   Construction of a supercapacitor.

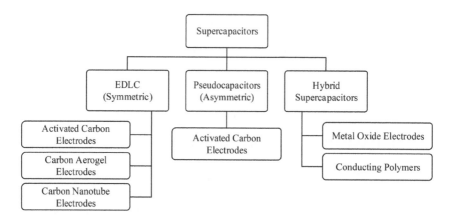

FIGURE 4.2 Classification of SCs.

Depending on the chemical properties and the charge storage capabilities [14], SCs are categorized into conventional EDLCs, pseudocapacitors, and hybrid SCs as shown in Figure 4.2. These SCs store the charge in either faradaic, non-faradaic, or hybrid combinations. The characteristics of each SC type are summarized in Table 4.2.

The characteristics of some of the commercially available SCs are presented in Table 4.3.

4.2.2 PARAMETERS AFFECTING THE PERFORMANCE OF AN SC

The ESR, capacitance, and voltage ratings define the capability of an SC for any pulsed power application. However, these are highly dependent on the operating

TABLE 4.2
Characteristics of Supercapacitor Technologies

Parameters	Units	Symmetric	Asymmetric	Hybrid
Storage	–	Double layer	Double layer + pseudocapacitance	Double layer + faradaic
Electrodes	–	Carbon	Carbon, metal oxides, conducting polymers	Carbon, intercalation materials
Electrolyte	–	Organic	Aqueous	Organic
Power density	kW/kg	9	5	4
Energy density	Wh/kg	5	30	100
Operating temperature	°C	−40/80	−25/60	−40/60
Advantages	–	Power density, maturity	Energy and power trade-off	Energy density
Disadvantages	–	Energy density	Price, efficiency, lifetime	Power density
Applicability	–	Commercial	Material research	Manufacturing research

TABLE 4.3

Characteristics of Commercially Available Supercapacitors [15–17]

Manufacturer	Classification	Power Density (kW/kg)	Energy Density (Wh/kg)	Temperature (°C)	Application
Maxwell	Symmetric	2.4–14	0.7–7.4	−40 to 65	Transport
Ioxus Inc.	Symmetric	23–34	4.5–6.3	−40 to 85	Transport, renewables
Panasonic Corp.	Symmetric	0.29–3.65	1.4–4.1	−40 to 70	Electronic devices
Nesscap Co. Ltd.	Asymmetric,	4.9-6.2	4.8–8.8	−25 to 60	Heavy vehicles
	Symmetric	6–17	2–5.7	−40 to 85	
Vinatech	Hybrid,	0.4–0.8	4–12	−25 to 60	Transport, wind
	Symmetric	1–11.3	0.7–6.5	−40 to 70	turbines
LS Mtron	Symmetric	0.9–2.4	3.3–7.5	−40 to 65	Electronics, devices

temperature, frequency of the charge currents, and the applied voltages. The operating temperature is an essential part of an SC energy storage system which influences the internal resistances and overall capacitance that changes the behavior of an SC bank [18]. The capacitance variation of a 50-F SC with respect to the temperature is shown in Figure 4.3. At extremely lower temperatures, the SCs offer high internal resistance and less capacitance due to the variation in the viscosity of an electrolyte. Moreover, the ions are accelerated toward the electrodes which influence the self-discharge rate of SCs.

On the other hand, the ESR and capacitance of an SC are highly affected by the frequency of the applied currents and voltages. The electrochemical impedance spectroscopy (EIS) method is used to determine the chemical characteristics of an SC. Also, it helps in analyzing the impact of the frequency on the ohmic resistance and the equivalent capacitance. The real and imaginary part of the impedance is determined using the EIS approach by measuring the current amplitude and phase with respect to the injected voltage. The SC capacitance also varies with the applied voltage and its dependent part accounted by the differential capacitance. The frequency response of capacitance of an SC from Maxwell technologies "BCAP0350" at different DC voltages applied by a potentiostat is shown in Figure 4.4. It is observed that the SC capacitance at low frequencies is determined by the overvoltage of the kinetically inhibited charge transfer and mass transport processes. However, the SC capacitance at higher frequencies (> 1 Hz) does not depend upon the voltage [19].

4.2.3 Parameter Definitions

The electrical behavior and performance of an SC are measured using some important parameters such as the capacitance, state of charge (SoC), overall energy, and the cell energy utilization efficiency of an SC.

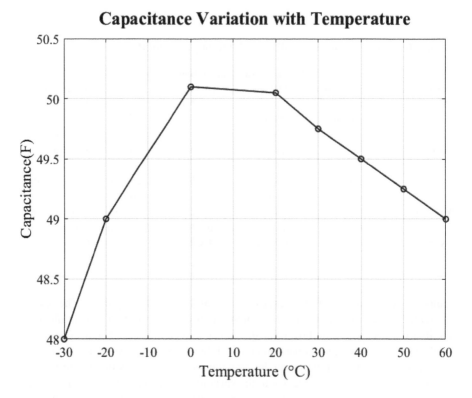

FIGURE 4.3   Variation of supercapacitor capacitance with temperature.

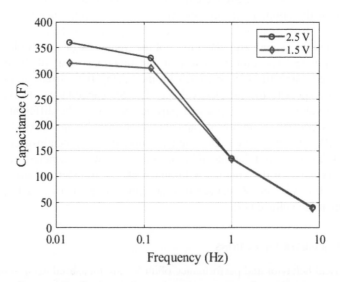

FIGURE 4.4   Frequency response of the capacitance of a supercapacitor from Maxwell (BCAP0350).

4.2.3.1 Capacitance

The capacitance of an SC is defined by the parameters of its internal construction that include relative permittivity (ε_r), the distance between electrodes (d), and the surface area (A) and is given as:

$$C = \frac{\varepsilon_r \times A}{d} \tag{4.1}$$

4.2.3.2 Energy

The energy stored in an SC is directly related to its rated operating voltage (V_R) and the electric charge (Q) available in it. The formula to obtain the charge and energy from the overall capacitance (C) of an SC is given as:

$$Q = CV_R \tag{4.2}$$

$$E_{SC} = \frac{1}{2}CV_R^2 \tag{4.3}$$

4.2.3.3 Total Energy Utilization

Typically, SCs cannot deliver all the energy onto the load due to its chemical limitations. The initial (E_i) and final energies (E_f) of an SC are used to determine its total cell energy utilization efficiency that is given by:

$$\eta_{SC} = \frac{(E_i - E_f)}{E_i} \tag{4.4}$$

where, $E_i = \frac{1}{2}CV_i^2$, and $E_f = \frac{1}{2}CV_f^2$

4.2.3.4 State of Charge

As the SC is charged and discharged multiple times, it is essential to monitor the SoC of that SC which depends upon the terminal voltage (V_T), capacitance (C), total charge (Q), and the SC current (i_{sc}). The initial charge (Q_i) value is used to define the SoC which is typically "1" for a fully charged SC and "0" for a discharged SC. The SoC of an SC is given as:

$$SoC = \left[\frac{(Q_i - \int_0^t i_{sc}(\tau)d\tau)}{Q} \right] \times 100 \tag{4.5}$$

4.2.4 Equivalent Electrical Model Representation of SCs

The modeling of SCs is an important approach to exploit the electrochemical principles, optimal utilization, integration, and management of the SCs for a given application. Furthermore, it is beneficial to determine the SC performance, state of health (SoH), lifetime estimation, cost calculations, control strategy, and future predictions. The variation of chemical characteristics such as self-discharge rates, leakage currents, losses, and internal resistances are essential to be considered for mathematical modeling of SCs. Generally, these are modeled based on the core principles of double-layer behavior, molecular theory, transmission line properties, and thermal behavior. However, these mathematical modeling approaches have too many variables and parameters to represent the exact behavior of an SC. Furthermore, it requires complex computations that are difficult to implement in real-time studies. To replicate the aforementioned characteristics, the simplified analytical models are mostly utilized in PE applications, which represent the complete electrical performance of an SC. These electrical models include parasitic inductances, self-discharge, and faradaic efficiency. Thus, it is an effective tool to estimate the performance of the SCs. The mathematical complexity and accuracy of the estimation process depend upon the type of model chosen for the application. Therefore, it is essential to explore the different electrical equivalent circuits which are explained as follows.

4.2.4.1   Single Branch *R-C* Model

The simple SC model is developed using the series connection of a resistor and a capacitor, and is shown in Figure 4.5 [20].

The resistor (R) acts as the internal ESR and the capacitor (C) is the overall capacitance of an SC. The main advantage of this model is fast computer simulation due to its simplicity in the fitting process of its two parameters. Nevertheless, dynamic performance, losses, and other characteristics are not considered in this model.

4.2.4.2   First-Order Equivalent Model

To address the dynamic characteristics of an SC, the simple RC model is further improved with two resistors, one inductor, and one capacitor as shown in Figure 4.6 [21]. The resistor (R_{ESR}) is assumed as the internal ESR, and a capacitor (C) is the overall capacitance of an SC. Besides, a resistor (R_p) is connected in parallel with the capacitor to replicate the leakage current behavior. Moreover, an inductor (L_C) is connected in series accounts for the behavior due to a variation in loads.

4.2.4.3   Multi Branch *R-C* Model

The single branch model provides the exact behavior of an SC. However, the precise results are obtained by increasing the level of branches. An example of a multi-branch

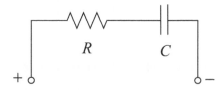

FIGURE 4.5   *R-C* model of a supercapacitor.

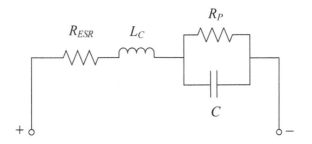

FIGURE 4.6 First-order R-C model of a supercapacitor.

R-C model is shown in Figure 4.7 [22], which consists of three R-C circuits connected in parallel combination. These three branches have different time constants
that represent slow, medium, and fast response of an SC. Therefore, this circuit provides the feasibility of modeling the charge and discharge characteristics of an SC in
terms of hours, minutes, and seconds, respectively.

4.2.4.4 Five-Stage R-C Ladder Model

To obtain the characteristics of an SC for the frequencies of 10 kHz, a five-stage ladder model is implemented as shown in Figure 4.8 [23]. The main advantage of this
model is that the circuit can be reduced to a second-order or fourth-order RC model
by simply modifying the resistor and capacitor values.

4.2.4.5 Zubeita and Bonert Model

The three-branch R-C model is further modified in reference [24] to analyze the
accurate dynamic behavior of an SC for the desired time. A differential capacitor is
included in the three-stage R-C model which comprises a voltage-dependent capacitor and a constant capacitor, as shown in Figure 4.9. For instance, to obtain the accurate response of an SC for 30 minutes, three R-C branches are considered. Each
branch has its time constants which provide the behavior of an SC for different periods. The first branch is incorporated with a differential capacitor while the second

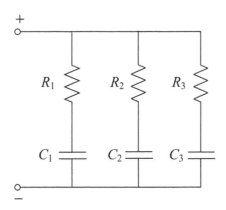

FIGURE 4.7 Three-branch R-C model of a supercapacitor.

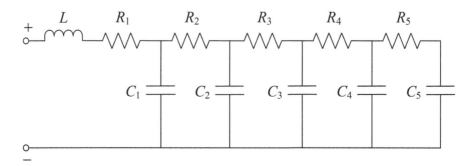

FIGURE 4.8 Five-stage ladder model of a supercapacitor.

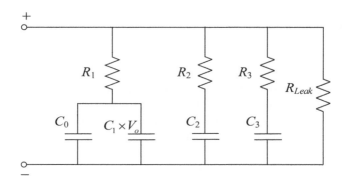

FIGURE 4.9 Zubeita and Bonert model.

and third branches have different time constants that account for the behavior of an SC in minutes and hours, respectively.

4.2.4.6   Franda Model

A two-branch R-C model is proposed in reference [25] to analyze the behavior of an SC in PE applications as shown in Figure 4.10.

This model comprises a two-stage R-C circuit which provides the feasibility to analyze the behavior from seconds to minutes.

4.2.4.7   Dynamic SC Model for EV Applications

The SCs for EV applications are typically subjected to operate in numerous charges and discharge cycles in shorter periods. Consequently, the capacity and voltage ratings are dynamically altered in accordance with the time. To approximate the behavior of an SC in such pulsed power applications, a dynamic model of an SC is demanded as shown in Figure 4.11 [26].

This model helps to analyze the characteristics of an SC for a specific amount of time with different voltages and capacitances. The model comprises two individual R-C branches, the first branch is developed with a resistor (R_1) and two parallelly

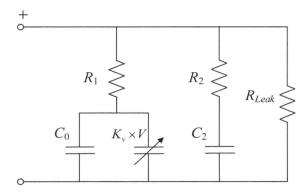

FIGURE 4.10 Franda model.

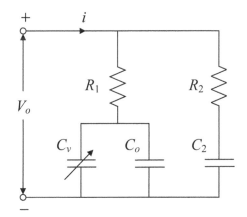

FIGURE 4.11   *R-C* equivalent dynamic model of a supercapacitor.

connected capacitors (C_v and C_o). The capacitance C_v is a variable capacitor which accounts for the varying capacitance with regard to the voltage. Moreover, this branch in the circuit offers the instantaneous behavior of an SC in seconds during charge and discharge operations. On the other hand, the second branch comprises a simple series-connected resistor (R_2) and capacitor (C_2) which model the ion redistribution at the end of charge and discharge operations.

A detailed analysis is performed on a 470-F (C_{tot}) SC to estimate the values of resistors and capacitors in the dynamic model. The resistor R_1 accounts for the ESR of an SC. The parameters of C_o, C_v, and R_2 of an SC are varied and can be obtained in real-time simulations. The charge and discharge tests are performed on SCs for different ratings of C_o. It is noticed that the C_o is varied and nearly equals to 57.2% of the total capacitance [26].

$$C_o = 57.2\% \times (C_{tot}) \tag{4.6}$$

TABLE 4.4

Parameters of a Supercapacitor Model

Variable	Units	Values @470 F
R_1	Ω	0.0025
C_0	F	270
C_v	F/V	190
R_2	Ω	0.9
C_2	F	100

The resistor R_2 from the second branch is estimated after the charge operation. The characteristic equation to estimate the resistor R_2 is given as:

$$R_2 = m \times (1.32 + (0.0003 \times (t_{on})^2) + (6.12e^{(-9)} \times (t_{on})^4) + (3.9e^{(-15)} \times (t_{on})^6) - (8.32e^{(-12)} \times (t_{on})^5) - (1.98e^{(-6)} \times (t_{on})^3)) \tag{4.7}$$

where,

$$m = 16.5 + (9.38e^{(-13)} \times (C_{tot})^6) - (0.028 \times C_{tot}) \tag{4.8}$$

The variable t_{on} is the total charging time, and C_{tot} is the total capacitance of an SC. The model parameters of an SC rated 470 F are given in Table 4.4.

4.2.5 CHARGE AND DISCHARGE CHARACTERISTICS

Compared to the conventional capacitors, the SCs consume higher currents and store maximum charge with less ESR. Consequently, they deliver a high amount of power in shorter periods. The charging and discharging procedure of SCs are performed using various approaches as discussed below.

4.2.5.1 Constant Voltage Charge

The SC with the rated voltage (V_R) is connected to a constant voltage source that results in higher in-rush currents. Thus, it is recommended to employ a protective resistor (R_p) in series with the SC bank. The value of R_p is selected based on the maximum discharge current of an SC, and is given as:

$$R_p = \frac{V_c}{I_{max}} \times ESR \tag{4.9}$$

where V_c is the constant charging voltage and I_{max} is the maximum discharge current. The constant voltage charging circuit is shown in Figure 4.12.

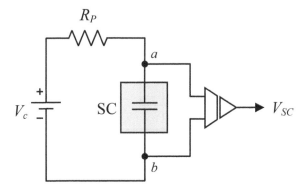

FIGURE 4.12 Constant voltage charging of a supercapacitor.

The total time required to charge (t_c) an SC bank from minimum to maximum rated voltage (V_R) depends upon the total capacitance and ESR of an SC and is given by:

$$t_c = \ln\left(\frac{V_c}{V_c - V_R}\right) \times (R_p + ESR) \times C_{tot} \qquad (4.10)$$

A detailed simulation for charging is conducted on an SC rated 50 F, 2.7 V with an ESR of 20 mΩ. A constant voltage charging approach is considered with a charge voltage of 3 V. The protective resistor rated 0.05 Ω is employed. The initial voltage of an SC is set to 0.5 V. The simulation results in terms of the terminal voltage during charging are shown in Figure 4.13. The results show that a rise in the SC voltage is observed from 0.5 V to its maximum rated voltage of 2.7 V in 8 seconds.

4.2.5.2 Constant Current Charge

A constant current (I_c) is supplied to charge an SC from the minimum voltage (V_{min}) to its maximum rated voltage (V_{max}). The circuit to charge an SC using a constant

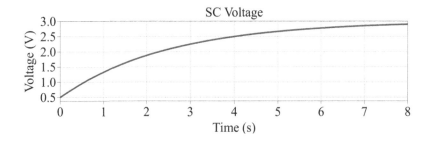

FIGURE 4.13 Results of constant voltage charging of a supercapacitor.

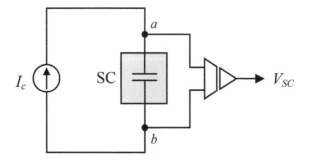

FIGURE 4.14 Constant current charging of a supercapacitor.

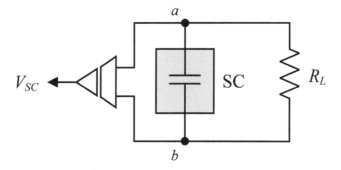

FIGURE 4.15 Constant resistive discharge circuit.

current is shown in Figure 4.14. The charging time of an SC with this approach depends upon the total capacitance and the charge current.

$$t_c = (V_R - V_{min}) \times \frac{C_{tot}}{I_C} \tag{4.11}$$

4.2.5.3 Constant Resistive Discharge

The discharge behavior of an SC is studied by incorporating a constant resistor as a load (R_L). The SCs typically undergo a voltage drop from the rated voltage (V_R) to its maximum voltage (V_{max}) due to its internal ESR. Therefore, the SC bank voltage is varied from V_{max} to V_{min}. The discharging circuit of an SC onto a resistive load is shown in Figure 4.15.

The total time required to discharge an SC with a capacitance of C_{tot} from maximum voltage to its minimum voltage is given as:

$$t_d = \ln\left(\frac{V_{max}}{V_{min}}\right) \times (ESR + R_L) \times C_{tot} \tag{4.12}$$

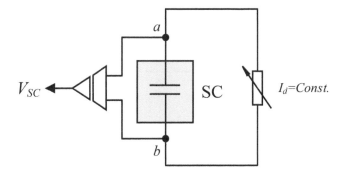

FIGURE 4.16   Constant current discharge circuit.

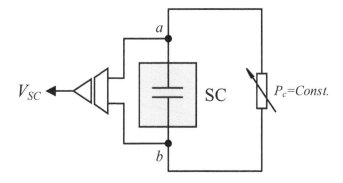

FIGURE 4.17   Constant power discharge circuit.

4.2.5.4   Constant Current Discharge

The discharge of SC can be performed by drawing a constant current (I_d) as shown in Figure 4.16. The total time required to discharge an SC using a constant current approach is obtained as:

$$t_d = (V_{max} - V_{min}) \times \frac{C_{tot}}{I_d} \tag{4.13}$$

4.2.5.5   Constant Power Discharge

Several pulsed power applications require the SCs to be incorporated with a PE converter to stabilize their output voltage. To understand the discharge behavior of an SC with a power converter, a constant power load is connected as shown in Figure 4.17.

The total time required to discharge an SC with a capacitance of C_{tot} for a constant power (P_c) discharge is given as:

$$t_d = (V_{max}^2 - V_{min}^2) \times \left(\frac{C_{tot}}{2} \times P_c \right) \tag{4.14}$$

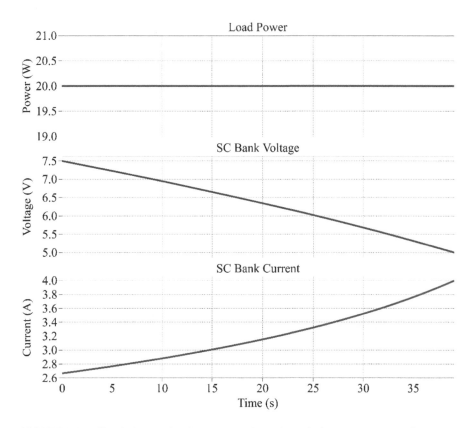

FIGURE 4.18   Simulation results of a supercapacitor voltage during constant power discharge.

The simulation studies are performed to verify the discharge characteristics of an SC bank rated 50 F, 7.5 V. A constant power load of 20 W is drawn from the bank. The maximum and minimum voltages of an SC are 7.5 V and 5 V, respectively. The simulation results of an SC terminal voltage and current during discharge is shown in Figure 4.18. The results indicate that the SC voltage is dropped from 7.5 V to 0.5 V. The SC current is raised from 2.7 A to 4 A as the load requires a constant power of 20 W.

4.3   SC CELL VOLTAGE EQUALIZATION

SCs are chemically constructed by employing an aqueous electrolyte which has the lower decomposition voltages. Thus, the rated voltage of the commercially available SCs typically ranges from 1.5 V to 3 V. To develop an SC bank with maximum terminal voltage, the SCs are stacked up in series combinations. However, each SC in a bank has individual tolerance level and other discrepancies in physical characteristics. Consequently, each SC in a series-connected bank is subjected to discharge and charge with differences in voltage levels. This affects their energy utilization efficiencies and may result in overcharging of individual cells.

Addressing the aforementioned concerns, a voltage balancing circuit is implemented which equalizes the individual voltages of SCs during charge and discharge operations. There are quite a few equalization circuits available in the literature and these are categorized into passive and active balancing circuits.

4.3.1 PASSIVE BALANCING CIRCUITS

A passive balancing circuit uses passive components such as resistors and diodes for each cell, which dissipates the excess voltage of an SC in a bank as heat. This circuit is less complicated and best suited for balancing the smaller currents. Thus, these circuits are highly implemented in low-power applications. Furthermore, it offers low cost with compact sizing. However, these circuits are not recommended for high-power applications where SCs are required to offer high energies to the load. The passive voltage balancers using resistors and diodes are discussed.

4.3.1.1 Resistive Balancer

The voltage levels of individual SCs are balanced by employing a leakage resistor in parallel with each SC in the bank as shown in Figure 4.19. The main function of this resistor is to dissipate the excess energy of an SC as heat. The balancing resistors connected to the individual SCs are designed based on its corresponding leakage currents, which is a function of an SC voltage.

The parallel resistor for an SC with a rated voltage V_R and leakage current I_{leak} is given by:

$$R_{leak} = \frac{V_R}{I_{leak}} \tag{4.15}$$

Typically, the leakage resistance for an SC is varied from 10 kΩ to 1 MΩ. To balance the SC bank voltages in small charge and discharge cycles, the leakage current of SCs is very high. Thus, this circuit is suitable for low cost and low-power applications.

4.3.1.2 Zener Diode Balancer

Zener diode-based balancing approach is implemented to protect the SC bank from over-voltages of individual SC cells. The balancing circuit using the Zener diodes is

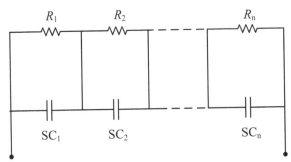

FIGURE 4.19   Resistive balancing circuit.

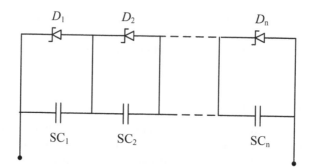

FIGURE 4.20   Zener diode-based voltage balancer.

shown in Figure 4.20. The resistors are replaced by Zener diodes that have a clamping voltage similar to the SC rated voltages. Therefore, these diodes circumvent the individual voltages of overcharged SCs during charge and discharge operations. This circuit has low power losses compared to the resistive balancing approach. However, these are implemented in low power applications, as the voltage rating of Zener diodes varies between 2.4 V and 2.6 V.

4.3.2   ACTIVE BALANCING CIRCUITS

An active balancing approach uses a PE-based active circuit with proper voltage management control. Unlike passive circuits, an active balancing circuit transfers the excess charge from a high-energy SC cell to the low-voltage SC in the bank and offers a high energy utilization efficiency of SCs. Thus, these are mostly implemented in high-power applications. An active balancing of an SC bank is performed based on three different practices: (a) the timely dissipation of excess energy from an overcharged SC cell, (b) storing the excess charge and releasing it onto the weak SC cell, (c) converting the energy from DC to AC and balancing each SC in the bank. Based on these procedures, various active voltage balancers are distinguished and discussed.

4.3.2.1   Switched Resistors

The SCs in a bank are actively balanced by using the timely dissipation approach. A resistive balancing circuit with switches is implemented across the SCs as shown in Figure 4.21 [27].

The switches are operated with a simple control algorithm that monitors the individual voltage of SCs in the bank. If the voltage of an SC reaches to a maximum voltage, the corresponding switch is activated and the energy is dissipated as heat. The switch is further deactivated when the SC voltage reaches to a nominal level. Though this circuit offers a better voltage balancing of SCs, it results in higher power losses.

4.3.2.2   Shuttling Capacitor

Another stimulating approach to balance the voltages of SCs is the utilization of conventional capacitors. The balancing circuit comprises a shuttling capacitor with six

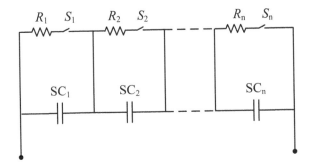

FIGURE 4.21   Switched resistors-based supercapacitor balancing circuit.

switches to balance three SCs in a bank as shown in Figure 4.22. The excess energy of an overcharged SC is stored in the shuttling capacitor and is redirected to the weak SC cell. Advanced switching control is employed to operate the switches between the weak and strong SC cells by following their corresponding voltages, thereby offering a quick voltage-balancing feature.

4.3.2.3   Buck-Boost–Based Equalization

An active balancing circuit based on a simplified buck-boost converter associated with SCs in a bank is introduced which transfers the excess energy to its adjacent cells. The buck-boost–based equalization circuit is shown in Figure 4.23.

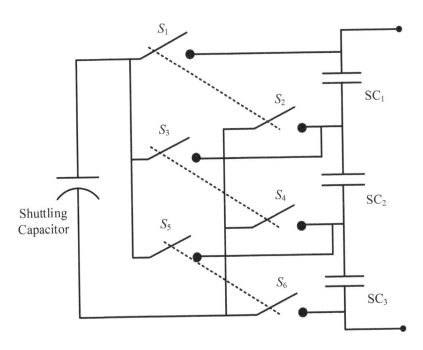

FIGURE 4.22   Shuttling capacitor-based supercapacitor voltage balancing circuit.

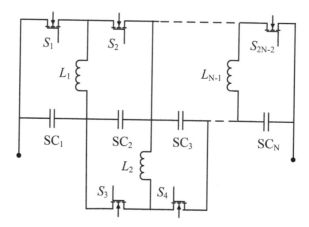

FIGURE 4.23 Buck-boost converter-based supercapacitor voltage equalization circuit.

The control monitors the individual SC voltages and operates the corresponding metal–oxide oxide–semiconductor field effect transistors (MOSFETs) by altering the duty ratio during charge and discharge operations. As a result, the overvoltage SCs provides excess charge to the weak SCs. However, these circuits necessitate the over-sized inductors and an accurate voltage sensing equipment for the operation.

4.3.2.4 Flyback-Based Equalization

The flyback converter-based operation is employed in the bank to balance the individual SC cell voltages. This circuit comprises diodes and inductors with a single MOSFET as shown in Figure 4.24.

A simple control scheme is implemented which creates a path for high currents to the low voltage SC cells by operating only one MOSFET. This approach does not require any extra voltage sensing circuit.

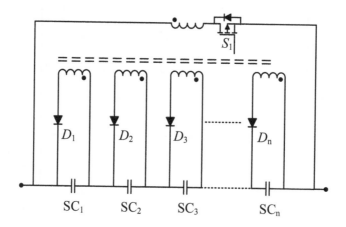

FIGURE 4.24 Flyback converter-based supercapacitor voltage equalization circuit.

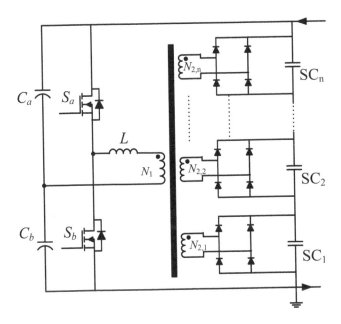

FIGURE 4.25 Multi-winding transformer-based supercapacitor voltage equalization.

4.3.2.5 Multi-Winding Transformer

The transformer-based balancing circuit undergoes DC to AC conversion of the SC energy to equalize the corresponding cell voltages. The equalization circuit comprises the capacitors, a half-bridge DC–DC converter, and the full-bridge rectifiers for each SC in a bank as shown in Figure 4.25. The control transfers the output of the half-bridge converter to the multi-output transformer which has the same coupling ratio. An individual rectifier converts the energy from AC to DC and fed to the SCs. Thus, all voltages of SCs in the bank are equalized. This circuit does not require any voltage sensing equipment and the voltage comparison circuits. Furthermore, the transformers in this circuit isolate the input and output sides, thereby enabling the galvanic isolation.

Simple control is implemented which creates a path for high currents to the low-voltage SC cells by switching of the MOSFETs. A detailed comparison of SC voltage balancing circuits in terms of efficiency, circuit complexity, control, and the cost are summarized in Table 4.5.

4.4 BANK SWITCHING CIRCUITS

Commercially available SCs are typically rated at lower voltages due to the electrolyte used in their chemical construction. However, these are connected in series combination to form a bank to match the load desired voltages. As a consequence, the total capacitance of an SC bank is significantly reduced and it provides a reduced storage capacity. In contrast, the SCs connected in parallel combinations result in a higher capacitance which allows the bank to store and deliver larger currents in

TABLE 4.5

Comparison of Supercapacitor Cell Voltage Balancing Techniques [28–30]

Balancers	Efficiency	Control	Circuit Complexity	Cost	Modularity
Resistor	Poor	Excellent	Low	Low	Yes
Zener diode	Poor	Excellent	Low	Low	Yes
Switched resistors	Moderate	Excellent	Low	Low	Yes
Shuttling capacitor	Good	Excellent	High	Medium	No
Buck-boost	Good	Good	Medium	Medium	Yes
Flyback	Good	Good	High	High	Yes
Multi-transformer	Moderate	Good	High	High	No

shorter durations. This distinctive feature of SCs enables the research to focus on employing bank-switching circuits. A switching circuit with a predefined control is employed in an SC bank which actively makes the SCs transition into the series and parallel combinations.

The SCs are well known for their high specific power density with fast charging capabilities. Nevertheless, they offer a low specific energy density which affects their utilization of total energy. The energy utilization of an SC bank without any switching circuit is nearly varied from 49 to 72% [31]. Consequently, the power delivery duration of an SC bank is significantly reduced. Furthermore, the total energy available in the bank is directly proportional to its operating voltage. Thus, the bank voltage oscillates between the maximum and minimum voltage points during the charge and discharge operations. To deliver the pulsed power to the load, a bank needs to be recharged simultaneously.

The terminal voltage of an SC bank during discharge operation typically drops from maximum to a minimum point (not 0 V). The minimum discharge voltage of an SC bank is a function of the load power and the maximum discharge current of an SC (manufacturer specific) which helps in identifying the total energy utilization efficiency of an SC. The minimum discharge voltage of an SC is calculated as:

$$V_{\min} = \frac{P_{\text{Load}}}{I_{\max}} \tag{4.16}$$

A detailed analysis is carried out to identify the minimum discharge voltage for different SCs under constant power load varying from 0.5 to 30 W. Two SCs rated 25 F and 50 F are considered with their maximum voltage of 2.6 V and charged to their maximum rated voltages (2.7 V). A constant power load is connected to each SC and allowed to discharge completely to its minimum points. The variation of minimum discharge voltage points of each SC is shown in Figure 4.26.

To deliver a constant 30 W power onto the load, the minimum discharge voltage of a 25-F SC is 1.44 V offering 66% of energy utilization efficiency. Whereas, a 50-F SC is deep discharged from 2.6 V to 0.8 V resulting in 89% of utilization efficiency. Thus, it is clear that the energy utilization efficiency of an SC is limited and varies with the load power demand.

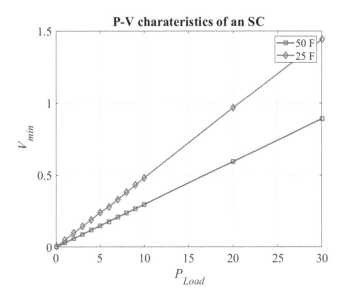

FIGURE 4.26   Minimum discharge voltage of a supercapacitor.

To address these issues, an active switching circuit with a proper bank management scheme is implemented in SC banks which enhances the energy utilization and power delivery duration in pulsed power applications. The timely switching of the SC bank outcomes a change in the overall capacity and alters the maximum discharge current of SCs in a bank. Thus, the minimum discharge voltage of an SC is varied and delivers maximum energy to the load. Furthermore, the power delivery duration of the bank is increased which helps in delivering power for a longer duration on a single charge. Several bank switching techniques and control schemes are proposed [32–36] to deal with this issue, which is deliberated in detail as follows:

4.4.1   SERIES-PARALLEL SWITCHING

An SC bank with a simple series-parallel change-over circuit is shown in Figure 4.27 [32]. The SCs in the bank are initially connected in parallel combination to meet the load voltage requirements. At this stage, the SC bank has high capacitance and can store a high amount of energy. When the terminal voltage drops to a preset threshold, the bank switches from parallel to series combination and adds up the terminal voltage. Likewise, the charge mode of operation starts with the series combination and switches to the parallel combination. The switches Q_1, Q_2, Q_3, and Q_4 are used to perform the bank transition. However, the bank transition of the series-parallel circuit results in higher voltage variations which affects the efficiency of the subsequent power converter. Thus, it is not recommended for high-power applications.

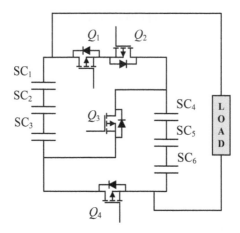

FIGURE 4.27 Series-parallel supercapacitor bank switching circuit.

4.4.2 MULTI-SHIFT SERIES-PARALLEL SWITCHING

The multi-shift series-parallel circuit presented in Figure 4.28 comprises the "$N+1$" number of switching levels [34]. Where "N" is the total number of legs for each branch. For instance, the multi-shift series-parallel circuit can be operated in "3" switching levels by simply substituting the value as "$N = 2$" in the circuit. The bank transition in this circuit is performed by adding only one SC from each branch and the process is continued until all the SCs are connected in series combination.

The simulation results of multi-shift series-parallel SC bank (7.5 V, 400 F) during discharge are shown in Figure 4.29. It is noticed that the SC cell voltage is dropped from 2.5 V to 0.85 V and has 88.4% of cell energy utilization efficiency.

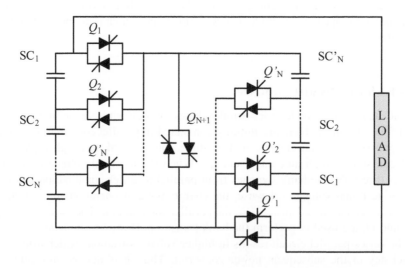

FIGURE 4.28 Multi-shift series-parallel supercapacitor bank switching circuit.

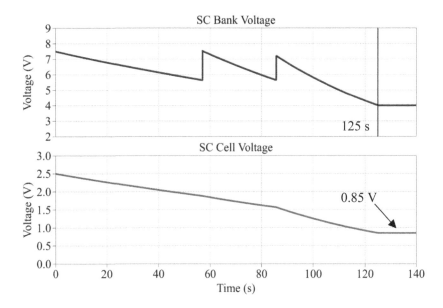

FIGURE 4.29   Results of a multi-shift series-parallel supercapacitor bank switching circuit.

4.4.3   THREE-LEVEL TRANSITION CIRCUIT

The three-level switching of SC bank shown in Figure 4.30 is to enhance the energy utilization of SCs [35].

The simulation results of an SC bank rated 7.5 V, 400 F is shown in Figure 4.31. The results show that SC cell voltage is declined from 2.5 to 0.67 V offering 92.8% of

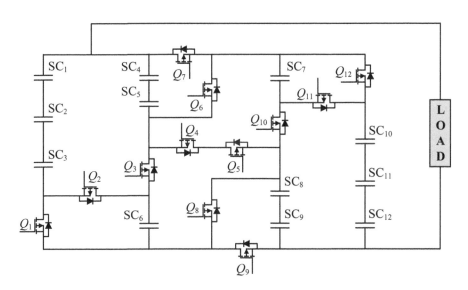

FIGURE 4.30   Three-level bank transition circuit.

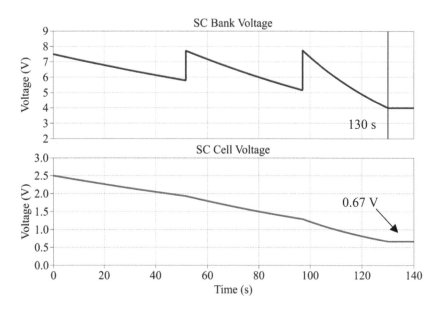

FIGURE 4.31   Results of a three-level bank transition circuit.

SC energy utilization efficiency. The power delivery duration of the bank observed is 130 second which is slightly higher than the multi-shift series-parallel circuit.

4.4.4   FOUR-BRANCH THREE-LEVEL SWITCHING

To enhance the power delivery duration of an SC bank, the aforementioned three-level switching circuit is modified as shown in Figure 4.32 [36]. The configuration comprises 12 active switches and 3 diodes which makes the three levels of bank transition. The switching modes of this configuration are shown in Figure 4.33. Initially, all the SCs in the bank are connected as a four-branch parallel mode with each branch having three SCs connected in series as shown in Figure 4.33(a). In the second level, the SCs are reconfigured and form a three-branch parallel configuration with each branch having four SCs in series as shown in Figure 4.33(b). Accordingly, the third level changeover has only two branches with six SCs shown in Figure 4.33(c).

The simulation results of the four-branch three-stage switched SC bank are shown in Figure 4.34. The results show that the total SC bank voltage is dropped from 7.5 to 4 V in 133.3 s. Moreover, the deep discharge of SCs is noticed where, each SC cell in the bank is discharged till 0.66 V. Thus, the energy utilization efficiency of each SC in the bank is 93.1%.

The overall performance of these circuits is evaluated and summarized in Table 4.6. To reduce the voltage fluctuations across the SC bank, the three-level circuits are highly preferred for high-power applications. Among them, the four-branch three-stage circuit offers the maximum power delivery duration with increased SC cell energy utilization efficiency.

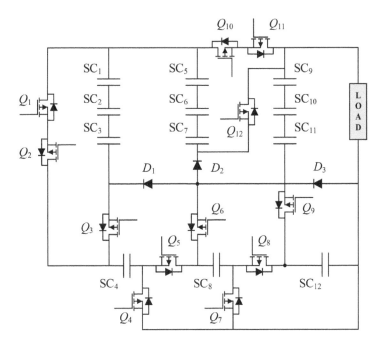

FIGURE 4.32 Four-branch three-level bank switching circuit.

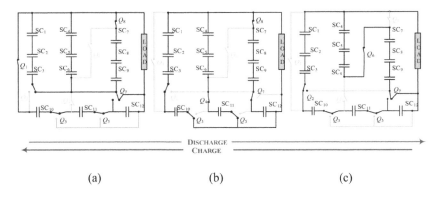

FIGURE 4.33 Operating modes of the four-branch three-level bank switching circuit.

4.5 SUPERCAPACITORS FOR HYBRID ENERGY STORAGE SYSTEMS

An ideal ESS should offer high specific energy density, high specific power density, less maintenance, and compact sizing at a lower cost. Although the batteries and fuel cells (FC) as an energy source offer high specific energy densities, they lack in transient power delivering capabilities. On the other hand, the SCs comprise high specific power densities, which cannot store a high amount of energy. A Ragone plot for different ESSs in terms of specific energy and power densities are shown in Figure 4.35 [37].

Due to these drawbacks, the SCs are combined with batteries or fuel cells to form a HESS that offers both energy and power simultaneously to the load. Consequently,

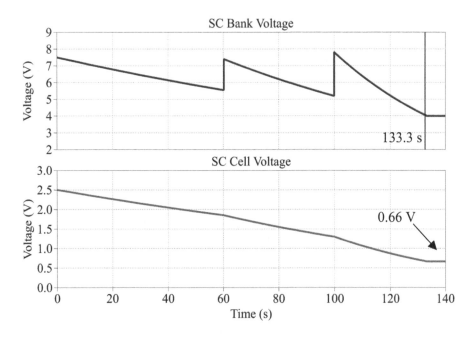

FIGURE 4.34   Results of the four-branch three-level switching circuit.

TABLE 4.6
Comparison of Bank Switching Techniques

Bank Switching Techniques	Switching Levels	Control Complexity	Power Delivery Time (s)	Energy Utilization (%)	Cost
Non-switched	0	–	120.1	71.2	–
Multi-shift	3	Medium	125.0	88.4	Medium
Three-level	3	High	130.0	92.8	High
Four-branch	3	High	133.3	93.1	High

the sizing and weight of the overall ESS are significantly reduced. A two or more energy storage devices connected in parallel using a power electronic interface forms a HESS. There are several forms of HESSs available such as battery-SC hybrid, battery-FC-SC hybrid, etc., which are mostly employed in electric vehicular applications, microgrids, and renewable energy systems.

4.5.1   HYBRIDIZATION OF BATTERY AND SUPERCAPACITORS FOR EVs

At present, most of the EVs are equipped with Li-ion batteries to drive their electric propulsion system. However, the practical driving pattern of an EV includes rapid accelerations and braking, especially in the urban driving scenarios. Consequently, the battery pack has to release a large amount of energy in shorter durations. In such conditions, high currents are drawn from the battery, which affects their calendar life. This issue can be handled by employing an oversized battery pack, thereby increasing

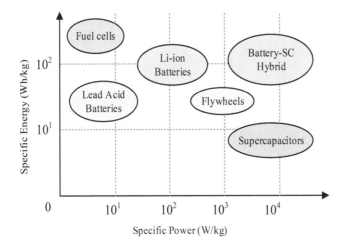

FIGURE 4.35   Ragone plots for various energy storage systems.

the number of the depth of discharge/charge cycles, weight, and the cost of the ESS [38]. Alternatively, SCs are the most widely used storage devices for pulsed power applications that offer high power densities with a faster dynamic response. They also have a quick charging capability, which absorbs and delivers higher bursts of currents in a shorter span. Therefore, these are connected with batteries to form a HESS. In such configuration, the batteries are sized to supply the constant power loads, and SCs to deal with transient power demands. Also, SCs are capable to recuperate the energy from regenerative braking, which enhances the driving range of an EV.

Nevertheless, a PE interface with a proper power management control is always required to implement such HESS for EVs. Furthermore, the load power allocation scheme that efficiently allots the transient power to SCs and constant power demands to the battery pack. The generic block diagram of the battery-SC HESS driving an electric propulsion system using the PE interfaces is shown in Figure 4.36.

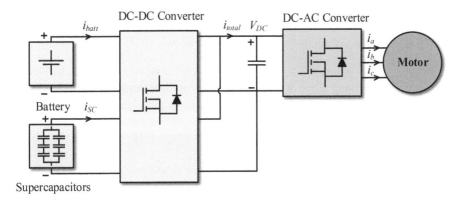

FIGURE 4.36   Block diagram of a battery-supercapacitor hybrid energy storage system driving an electric powertrain.

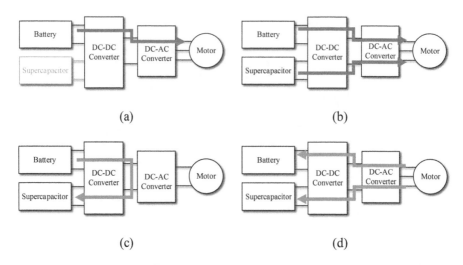

FIGURE 4.37 HESS modes of operation during: (a) battery driving an electric motor, (b) battery and SC bank driving an electric motor, (c) battery charging an SC bank, and (d) SC charging from regenerative braking power.

The battery and SCs are connected to a PE interface to step up their voltage and smoothen the currents. The power control allocates the load power between battery and SCs for the whole energy utilization of HESS. To deliver the maximum energy to the motor effectively, the battery-SC hybrid system is designed to operate in four different modes as shown in Figure 4.37.

If the EV is operated in highway driving patterns, then the load demands constant power from the HESS. Thus, only a battery pack is used to serve the load as shown in Figure 4.37(a). As the EV undergoes unanticipated power demands such as rapid accelerations during off-roading conditions, both the battery and SC banks deliver their energy to drive the load as shown in Figure 4.37(b). Typically, an SC bank for EV applications has less energy storage capability. As a result, these are required to be recharged simultaneously to serve another set of transient power demand. In such cases, batteries are used to recharge the SC bank while serving the load. The power flow during this mode is shown in Figure 4.37(c). The EV is subjected to sudden braking operations in urban driving schedules. Thus, kinetic energy at wheels is absorbed and converted to electric energy, which can be stored in batteries or SCs to enhance the driving range of an EV. The power flow of the energy storage system charging from the regenerative braking power is shown in Figure 4.37(d).

4.5.2 Power Electronic Interfaces

The SoC of an energy storage element displays a direct relation with its terminal voltage. A typical ESS results in wide terminal voltage variation during charge and discharge operations. Therefore, a PE converter is employed with an ESS that accepts wide terminal voltage variations at the input and delivers a stabilized voltage at the output. Consequently, the battery pack and SC bank associated with a power

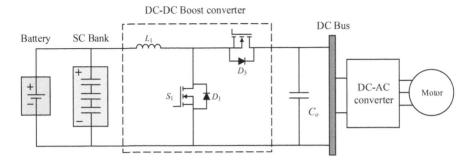

FIGURE 4.38   Passive-cascaded HESS interface.

converter to step-up their corresponding voltages and generate a constant DC link voltage. However, several interfacing topologies are available to hybridize the batteries and SCs based on their characteristics which are discussed as follows.

4.5.2.1   Passive Cascaded Configuration

The battery and SCs are connected in parallel combination with a bidirectional DC–DC PE converter associated with an SC bank as shown in Figure 4.38. The voltage rating of the battery is equal to the voltage rating of an SC bank. The battery recharges the SC bank simultaneously whenever it has been used to deliver the pulsed power. On the other hand, the bidirectional power flow enables the SC bank to discharge and recover the regenerative braking power from the load. However, this configuration requires an oversized battery that discharges onto the SCs and serving the load demand simultaneously without any control.

4.5.2.2   Active Cascaded Configuration

The aforementioned power interface is upgraded by employing a unidirectional power converter to the battery pack, which steps up the corresponding voltages as shown in Figure 4.39 [39]. This topology reduces the sizing of batteries and efficiently controls their discharge current. On the other hand, the SC bank (when detected) absorbs the transient currents from the load with the help of a bidirectional power converter.

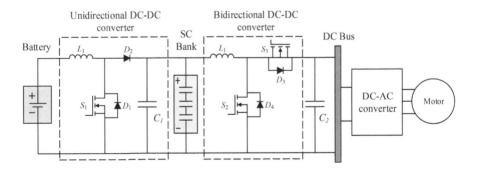

FIGURE 4.39   Active-cascaded HESS interface.

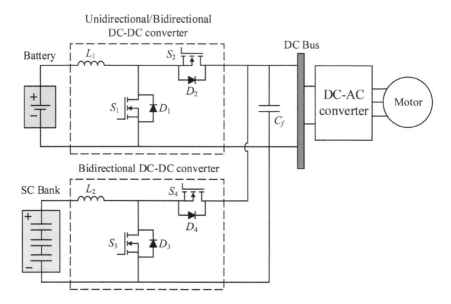

FIGURE 4.40   Parallel-active interface.

4.5.2.3   Parallel Active Configuration

To achieve high performance and better control over battery and SC bank currents, a parallel active power interface is employed. This approach uses two bidirectional converters to the battery pack and the SC bank as shown in Figure 4.40 [40].

The HESS with this configuration facilitates lower sizing of both the battery pack and the SC banks. The power management with this topology allows the regenerative braking power to be stored in either batteries or SCs as per the requirement. However, employing two individual bidirectional DC–DC converters results in addition of the complexity of control and the overall system cost.

4.5.2.4   Multiple-Input Converter Topology

The multiple-input converter configuration uses a single DC–DC converter with dual inputs for each source. A separate switch and diode are used for each input as shown in Figure 4.41 [41]. An additional switch and diode can be included if a bidirectional operation is required. Furthermore, the converter comprises a single inductor which is shared by the battery and SC bank. As the SC bank voltage varies widely, the voltage rating of the battery is selected higher than the SC bank.

The converter is able to operate in buck, boost, and buck-boost modes as per the power flow. Moreover, the weight and volume of the converter are significantly reduced by incorporating only a single inductor for multiple inputs. However, this topology requires a complex control scheme to operate effectively.

4.5.2.5   Dual-Source Bidirectional Converters

The parallel-active configuration is modified as a dual-source converter by reducing a switch from one of the bidirectional DC–DC converters. In this topology, the

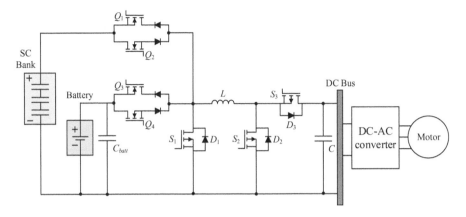

FIGURE 4.41   Multiple-input DC–DC converter interface.

battery pack and SC bank are combined at the input stage with a single converter as shown in Figure 4.42 [42]. The total cost of this configuration is slightly lesser than the parallel-active topology. However, this circuit requires an intricate control system.

4.5.2.6 Multiple Dual-Active Bridge Converters

The hybridization of battery pack and SC bank can also be realized by employing the transformers which can be operated high frequencies. The multiple dual active bridge converter as a power interface is shown in Figure 4.43 [43].

Typically, the transformers are beneficial where they provide isolation between the input sources and the DC link. However, the transformers increase the volume and weight of the power converter. Moreover, the converter uses multiple switches, which accounts for the higher system cost.

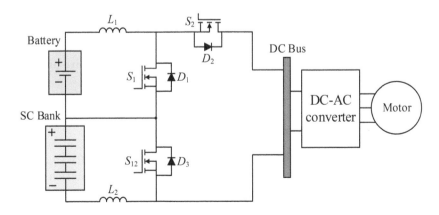

FIGURE 4.42   Dual-source bidirectional converter interface.

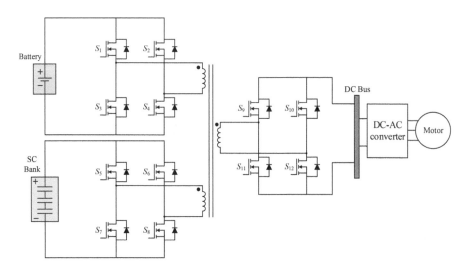

FIGURE 4.43 Dual-active bridge converter interface.

4.5.3 Sizing of HESS for an EV

Typically, the sizing of HESS for an EV involves in determining the power and energy requirements of the propulsion system, identification of the power allocation control, and sizing of an individual ESS (battery and SCs). The steps to identify the efficient sizing parameters for HESS are discussed.

4.5.3.1 Power and Energy Ratings of EV powertrain

In practice, the EV is operated in a dynamic environment with varied driving patterns. Thus, the propulsion system of an EV is selected based on the load power requirement, which is defined from the tractive force (F_t) acting on an EV for a specific driving cycle.

The overall power rating of an EV propulsion system is given by:

$$P_{\text{load}} = \frac{(F_t \times V_s)}{\eta_{\text{tr}}} \tag{4.17}$$

The total power required by the ESS (battery-SC hybrid) to drive an electric motor is obtained as:

$$P_{\text{ESS}} = \frac{(F_t \times V_s)}{\eta_{\text{EV}}} \tag{4.18}$$

where, V_s is the velocity of the vehicle and the total efficiency of an EV (η_{EV}) which depends upon the transmission (η_{tr}), motor (η_{mo}), and the power conversion efficiencies (η_{pc}) which is calculated from:

$$\eta_{\text{EV}} = \eta_{\text{tr}} \times \eta_{\text{mo}} \times \eta_{\text{pc}} \tag{4.19}$$

The energy storage of HESS is sized based on the total energy to be delivered by the source. This is obtained by integrating P_{ESS} with respect to time. However, the sizing of ESS is varied with the consideration of the regenerative braking power.

4.5.3.2   Battery and SC Sizing

The aforementioned parallel-active approach is considered for the sizing of an ESS in this study. The load power (P_{load}) is typically allocated between the battery pack and the SC bank. Thus, these are sized to meet the dynamic response of an EV. The following assumptions are made for the sizing: a() the battery is only used for a constant power load; (b) SC bank is used to deal with peak power demands; and (c) regenerative braking is considered which recharges only SC bank.

A power profile for a specific driving pattern is considered for sizing. The battery pack must contain the total energy that can cover a range of 100 km [44]. The recommended voltage rating of the battery is lower than the DC link voltage.

$$V_{batt} < V_{DC\_link} \tag{4.20}$$

The SC bank is designed to serve only the peak power demands. As the bidirectional boost converter is associated with an SC bank, the maximum voltage rating of the bank must be lesser than the DC link voltage. Thus, the converter steps-up the corresponding voltage and matches the output voltage of the boost converter, which is connected to a battery pack.

$$V_{SC} < V_{DC\_link} \tag{4.21}$$

The overall energy consideration for battery and SC bank using the parallel-active approach is given in (4.22). Therefore, the sum of energies of the battery pack (E_{batt}) and SC bank (E_{sc}) should be greater than the energy required by the load. Where E_{load} is the energy required by the load.

$$E_{batt} + E_{SC} > E_{load} \tag{4.22}$$

The energy of an SC bank in HESS should match with the initial energy of the bank switched SCs which is given as:

$$E_{SC} = E_{SC(Switching)} \tag{4.23}$$

4.5.4   LOAD POWER ALLOCATION

To avoid the battery pack from serving the pulsed power demands, a load power allocation algorithm is implemented. This scheme allocates the constant power to the battery pack and peaky power demands to the SC bank. As a result, the stress on batteries is reduced and enhances their cycle life. The most widely adopted load power allocation techniques are discussed in the following section.

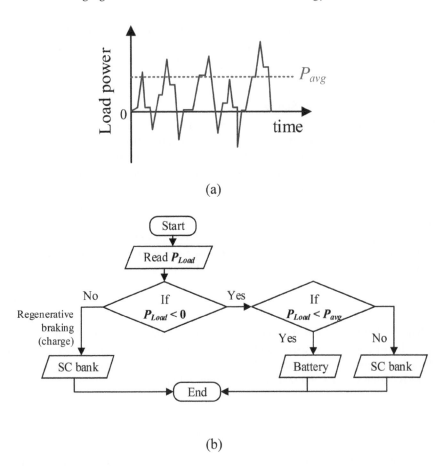

FIGURE 4.44   Rule-based power sharing: (a) load power profile and (b) control flow.

4.5.4.1   Rule-Based Power Allocation

The load power is differentiated using a rule-based control which splits the power profile based on the average power (P_{avg}). The P_{avg} for a specific power profile is shown in Figure 4.44(a) and the rule-based control flow is shown in Figure 4.44(b). The flow states that if the load power is less than zero, a bidirectional DC–DC converter associated with an SC bank operates in buck mode and absorbs all the available power. On the other hand, if the P_{load} is greater than zero and less than P_{avg}, only battery is used to deliver the load power. Whereas, if the P_{load} is greater than zero and less than P_{avg}, then both battery pack and SC bank serve the load.

4.5.4.2   Frequency-Based Power Allocation

The power required by the load is separated and allocated using a frequency-based approach [45] as shown in Figure 4.45.

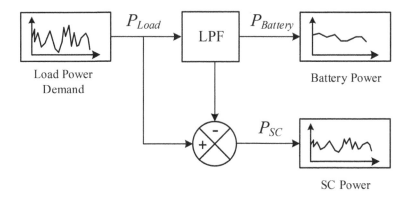

FIGURE 4.45   Frequency-based load power allocation.

It comprises a low pass filter (LPF) which eliminates the high-frequency power components and allocates it to the battery pack. On the other hand, the high-frequency power is allocated to the SC bank. However, this approach requires the tuning of the minimum frequency of an LPF.

4.6   CASE STUDIES ON AN ELECTRIC VEHICLE

A detailed case study on a small EV is performed to analyze the suitability of the aforementioned converters and respected power management schemes. A parallel-active interface is used in implementing the HESS for an EV case study. This comprises two DC–DC converters for the battery pack and the SC bank.

The sizing of batteries and SCs are varied with the selected EV parameters that include, weight, the diameter of the wheels, cross-sectional area, etc. The parameters of the selected EV for the case study are given in Table 4.7. The energy and power ratings of an electric propulsion system are selected by considering the forces acting on an EV.

TABLE 4.7
Electric Vehicle Parameters for the Case Study

Parameter	Units	Values
Curb weight	Kg	740
Cross-sectional Area	m^2	2.6
Diameter of the wheel	M	0.62
Gear ratio	–	10.2

4.6.1 Battery and SC Bank Sizing

The total energy required from the battery-SC hybrid to drive an EV is obtained as in:

$$E_{ESS} = \int_0^t (P_{ESS})\, dt \qquad (4.24)$$

The power profile of the highway fuel economy test (HWFET) pattern is analyzed for identifying the peak power and constant power of an EV. The regeneration mode is considered in this study to verify the charging capability. From the battery sizing equations, the energy required from the battery pack to cover a 100 km range is equal to 13.5 kWh. Thus, the battery pack is selected with the energy in the range of 14 kWh. The battery with a lower voltage of 48 V and 292 Ah is selected for the study, as a unidirectional boost converter is utilized to step up the battery voltages.

On the other hand, the SC bank is designed to supply beyond the average power (P_{avg}) i.e. only peak demands as shown in the power profile. Therefore, an SC bank rated 62.5 V, 400 F with the energy of 0.22 kWh is considered to meet the peak power demands. The maximum and minimum voltages of the SC bank are 62.5 and 20 V, respectively. A constant DC link voltage of 80 V is maintained using the power converter interface and drives the load.

4.6.2 HESS Control Schemes

The effective power management of HESS is the key to control the power electronics and load power allocations, which efficiently delivers the power from ESS to the EV propulsion system.

4.6.2.1 Power Converter Control

The boost converter and bidirectional buck-boost converter is modeled and designed based on the ratings of battery and SC banks. A proportional-integral (PI)-based controller is used for the closed-loop control, which regulates a constant DC link voltage at the output. Each converter is employed with an individual PI controller. The system parameters of PE converters in HESS-based EV are given in Table 4.8.

4.6.2.2 Power Allocation Control

The frequency-based load power allocation technique is implemented in this study. A second-order Butterworth filter with a frequency of 820 mHz is employed that separates the load power and allocates between batteries and SCs. The simulation results of frequency-based power allocation are shown in Figure 4.46.

4.6.3 HESS Drive Cycle Analysis

The overall HESS system is implemented using the battery pack and SC bank in MATLAB/Simulink platform with all the aforementioned sizing and controls.

TABLE 4.8
HESS System Parameters

Parameter	Units	Values
Inductor (Boost converter)	mH	1
Inductor (Buck-boost converter)	mH	0.8
DC link capacitor	µF	470
DC link voltage	V	80
Switching frequency	kHz	10

The simulations results for an EV under the HWFET and New European driving cycles (NEDC) in terms of their SoC variation are shown in Figure 4.47 (a) and (b), respectively.

The results indicate that the transient load demands are served by the SC bank and reduce the burden on the batteries. Thus, the battery pack is discharged slowly and enhances the overall driving range of an EV.

On the other hand, the SC bank recuperates the power from regenerative braking in the course of the full driving cycle.

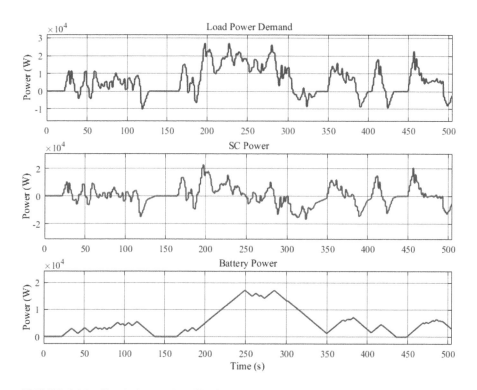

FIGURE 4.46   Simulation results of load power allocation.

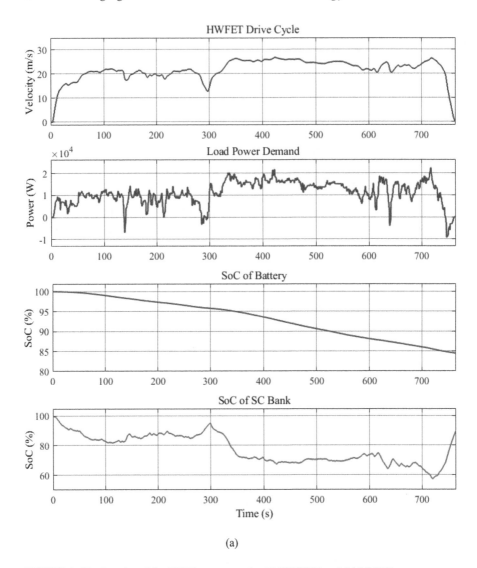

(a)

FIGURE 4.47   Results of the HESS system under (a) HWFET and (b) NEDC.

4.7   CONCLUDING REMARKS

This chapter provides the fundamentals, classification, and characteristics of the SCs. The electrical equivalent models of SCs to determine the performance, control strategy, lifetime estimation, and future predictions were also discussed. Various types of charging and discharging methods for different types of applications were described in this chapter. The majority of voltage cell equalizers to balance series SC cell voltages were also discussed and compared with many of the different advantages/disadvantages. The voltage equalizers in the past and present are focused on the cost and simplicity. This chapter analyzed the state-of-the-art bank switching architectures, including multi-shift switching, three-level, and four-branch switching along with their main merits and demerits considering a case study during the discharge operation. The analysis presented in this chapter demonstrates that the bank-switching concept is a viable approach to enhance the SC cell energy utilization and power delivery duration. Therefore, many more research initiatives are required to utilize the more useable capacity of SCs in a discharge cycle. The fundamentals, design, sizing, and power electronic interfaces of battery-SC based HESS are extensively discussed. The HESS modes of operation, the performance of various power electronic interfaces, and important concepts are presented through case studies on a small EV under standard urban and highway driving schedules. The technical details given in this chapter are also applicable to other industrial and transportation applications such as railway electric systems, electric ship propulsion systems, solid-state transformer-based systems, agricultural machinery, shovel trucks, more electric aircraft, forklifts, excavators, mining shovels, microgrids, grid regulation services, and consumer electronics, which employ SCs as a primary and secondary power source.

REFERENCES

1. S. Manzetti and F. Mariasiu, "Electric vehicle battery technologies: From present state to future systems," *Renew. Sustain. Energy Rev.*, vol. 51, pp. 1004–1012, Nov. 2015.
2. S. S. Williamson, A. Khaligh, S. C. Oh and A. Emadi, "Impact of energy storage device selection on the overall drive train efficiency and performance of heavy-duty hybrid vehicles," In *Proceedings of IEEE Vehicle Power and Propulsion Conference*, Chicago, IL, pp. 1–10, 2005.
3. J. R. Miller and A. F. Burke, "Electrochemical capacitors: Challenges and opportunities for real-world applications," *Electrochem. Soc. Interf.*, vol. 17, no. 1, pp. 53–57, Spring 2008.
4. D. Ronanki, S. A. Singh, and S. S. Williamson, "Comprehensive topological overview of rolling stock architectures and recent trends in electric railway traction systems," *IEEE Trans. Transport. Electrific.*, vol. 3, no. 3, pp. 724–738, Sept. 2017.
5. J. Rocabert, R. Cap-Misut, R. S. Muoz-Aguilar, J. I. Candela, and P. Rodriguez, "Control of energy storage system integrating electrochemical batteries and supercapacitors for grid-connected applications," *IEEE Trans. Ind. Appl.*, vol. 55, no. 2, pp. 1853–1862, Mar.–Apr. 2019.
6. A. Ostadi and M. Kazerani, "A comparative analysis of optimal sizing of battery-only, ultracapacitor-only, and battery–ultracapacitor hybrid energy storage systems for a city bus," *IEEE Trans. Vehi. Tech.*, vol. 64, no. 10, pp. 4449–4460, Oct. 2015.
7. A. Misra, "Energy storage for electrified aircraft: The need for better batteries, fuel cells, and supercapacitors," *IEEE Electrific. Mag.*, vol. 6, no. 3, pp. 54–61, Sept. 2018.
8. D. Ronanki and S. S. Williamson, "New modulation scheme and voltage balancing control of modular multilevel converters for modern electric ships," *IET Power Electron.*, vol. 12, no. 13, pp. 3403–3410, Nov. 2019.
9. D. Ronanki and S. S. Williamson, "A simplified space vector pulse width modulation implementation in modular multilevel converters for electric ship propulsion systems," *IEEE Trans. Transport. Electrific.*, vol. 5, no. 1, pp. 335–342, Mar. 2019.
10. H. Liu, C. Mao, J. Lu, D. Wang, "Electronic power transformer with supercapacitors storage energy system," *Electric Power Sys. Research*, vol. 79, no. 8, pp. 1200–1208, 2009.

11. D. Ronanki and S. S. Williamson, "Topological overview on solid-state transformer traction technology in high-speed trains," In *Proceedings of IEEE Transportation Electrification Conference and Exposium (ITEC)*, Long Beach, CA, 2018, pp. 32–37.

12. A. Emadi, S. S. Williamson, and A. Khaligh, Power electronics intensive solutions for advanced electric, hybrid electric, and fuel cell vehicular power systems, *IEEE Trans. Power Electron.*, vol. 21, no. 3, pp. 567–577, May 2006.

13. N. Khan, N. Mariun, M. Zaki and L. Dinesh, "Transient analysis of pulsed charging in supercapacitors," In *Proceedings of IEEE TENCON'00*, vol. 3, pp. 193–199, Sept. 2000.

14. R. P. Deshpande, *Ultracapacitors*, New York: McGraw-Hill Education, 2013.

15. "Datasheet, Maxell Ultracapacitors." *Maxwell Technologies*, Accessed: Feb. 2019. Available at http://www.maxwell.com/products/ultracapacitors/.

16. "Datasheet, VINATech. Supercapacitor Solution." *VINATech*. Accessed: Jan. 2019. Available at http://www.vina.co.kr/eng/product/supercap.html.

17. A. Berrueta, A. Ursúa, I. S. Martín, A. Eftekhari, and P. Sanchis, "Supercapacitors: Electrical characteristics, modeling, applications, and future trends," *IEEE Access*, vol. 7, pp. 50869–50896, 2019.

18. "Datasheet, Bootscap Ultracapacitors," Technical Report Document No. 1009364 Rev.3, *Maxwell Technologies*, 2009.

19. P. Kurzweil, B. Frenzel and R. Gallay, "Capacitance characterization methods and ageing behaviour of supercapacitors," In *Proceedings of International Seminar on Double Layer Capacitors*, Deerfield Beach, FL., USA, p. 14–25, 2005.

20. L. Shi and M. L. Crow, "Comparison of ultracapacitor electric circuit models," In *Proceedings of IEEE Power and Energy Society General Meeting*, Pittsburgh, PA, pp. 1–6, July 2008.

21. B. Ambrosio Cultura II, "Design, Modeling, Simulation and Evaluation of a distributed energy system," *Doctoral Thesis*, University of Massachusettes Lowell, 2011.

22. Y. Wang, J. E. Carletta, T. T. Hartley and R. J. Veillette, "An ultracapacitor model derived using time-dependent current profiles," In *Proceedings Midwest Symposium on Circuits and Systems*, pp. 726–729, Aug. 2008.

23. J. R. Miller, "Battery-capacitor power source for digital communication applications: simulations using advanced electrochemical capacitors," In *Proceedings of the Symposium on Electrochemical Capacitors II*, 29–95, pp. 246–254, 1995.

24. L. Zubieta, and R. Bonert, "Characterization of double-layer capacitors for power electronics applications," *IEEE Trans. Industry Appl.*, vol. 36, no. 1, pp.199–205, Jan. 2000.

25. R. Faranda, M. Gallina and D. Son, "A new simplified model of double layer capacitors," In *Proceedings of IEEE Conference ICCEP*, pp. 706–710, May 2007.

26. A. Singh, N. A. Azeez and S. S. Williamson, "Dynamic modeling and characterization of ultracapacitors for electric transportation," In *Proceedings of International Symposium on Industrial Electronics (ISIE)*, Buzios, pp. 275–280, 2015.

27. D. Linzen, S. Buller, E. Karden, and R. W. D. Doncker, "Analysis and evaluation of charge-balancing circuits on performance, reliability, and lifetime of supercapacitor systems," *IEEE Trans. Ind. Appl.*, vol. 41, no. 5, pp. 1135–1141, Sep./Oct. 2005.

28. S. R. Raman, X. D. Xue and K. W. E. Cheng, "Review of charge equalization schemes for Li-ion battery and super-capacitor energy storage systems," In *Proceedings International Conference Advances in Electronics Computers and Communications*, Bangalore, pp. 1–6, 2014.

29. N. Sidhu, L. Patnaik and S. S. Williamson, "Power electronic converters for ultracapacitor cell balancing and power management: A comprehensive review," In *Proceedings of 42nd Annual Conference of the IEEE Industrial Electronic Society*, Florence, pp. 4441–4446, 2016.

30. K. Maneesut and U. Supatti, "Reviews of supercapacitor cell voltage equalizer topologies for EVs," In *Proceedings of 14th International Conference on Electrical Engineering/ Electronics, Computer, Telecommunications and Information Technology (ECTI-CON)*, Phuket, pp. 608–611, 2017.

31. Q. Zhang and G. Li, "Experimental study on a semi-active battery-supercapacitor hybrid energy storage system for electric vehicle application," *IEEE Trans. Power Electron.*, vol. 35, no. 1, pp. 1014–1021, Jan. 2020.

32. M. Uno, "Series-parallel reconfiguration technique for supercapacitor energy storage systems," In *Proceedings of TENCON-IEEE Region 10 Conference*, Singapore, pp. 1–5, 2009.

33. S. Sugimoto, S. Ogawa, H. Katsukawa, H. Mizutani, M. Okamura, "a study of series-parallel changeover circuit of a capacitor bank for an energy storage system utilizing electric double-layer capacitors," *Electr. Eng. in Japan*, vol. 145, no. 3, pp. 34–42, 2003.

34. X. Fang, N. Kutkut, J. Shen, I. Batarseh, "Analysis of generalized parallel-series ultracapacitor shift circuits for energy storage systems," *Renewable Energy*, vol. 36, pp. 2599–2604, 2011.

35. J. Nie, X. Xiao, Z. Nie, P. Tian and R. Ding, "A novel three-level changeover circuit of super-capacitor bank for energy storage systems," In *Proceedings of 38th Annual Conference of the IEEE Industrial Electronic Society*, Montreal, QC, pp. 144–149, 2012.

36. N. Sidhu, L. Patnaik, N. A. Azeez and S. S. Williamson, "Bank switching technique in supercapacitor energy storage systems for line voltage regulation in pulsed power applications," In *Proceedings of 44th Annual Conference of the IEEE Industrial Electronic Society*, Washington, DC, pp. 5027–5031, 2018.

37. S. C. Lee, O. Kwon, S. Thomas, S. Park, and G. Choi, "Graphical and mathematical analysis of fuel cell/battery passive hybridization with K factors," *Appl. Energy*, vol. 114, 135–145, February 2014.

38. S. Vazquez, S. M. Lukic, E. Galvan, L. G. Franquelo, and J. M. Carrasco, "Energy storage systems for transport and grid applications," *IEEE Trans. Ind. Electron.*, vol. 57, no. 12, pp. 3881–3895, Dec. 2010.

39. O. Onar and A. Khaligh, "Dynamic modeling and control of a cascaded active battery/ ultra-capacitor based vehicular power system," In *Proceedings of IEEE Vehicle Power and Propulsion Conference*, Harbin, pp. 1–4, 2008.

40. A. Khaligh and Z. Li, "Battery, ultracapacitor, fuel cell, and hybrid energy storage systems for electric, hybrid electric, fuel cell, and plug-in hybrid electric vehicles: State of the art," *IEEE Trans. Vehi. Tech.*, vol. 59, no. 6, pp. 2806–2814, Jul. 2010.

41. B. G. Dobbs and P. L. Chapman, "A multiple-input DC–DC converter topology," *IEEE Power Electron. Letters*, vol. 1, no. 1, pp. 6–9, Mar. 2003.

42. M. Marchesoni and C. Vacca, "New DC–DC converter for energy storage system interfacing in fuel cell hybrid electric vehicles," *IEEE Trans. Power Electron.*, vol. 22, no. 1, pp. 301–308, Jan. 2007.

43. M. H. Kheraluwala, R. W. Gasgoigne, D. M. Divan and E. Bauman, "Performance characterization of a high power dual active bridge DC/DC converter," In *Proceedings of IEEE Industrial Applied Society Annual Meeting*, Seattle, WA, USA, pp. 1267–1273, vol. 2, 1990.

44. R. Bindu and S. Thale, "Sizing of hybrid energy storage system and propulsion unit for electric vehicle," In *Procedings of IEEE Transportation Electrification Conference (ITEC-India)*, Pune, pp. 1–6, 2017.

45. Z. Jin, L. Meng, J. C. Vasquez and J. M. Guerrero, "Frequency-division power sharing and hierarchical control design for DC shipboard microgrids with hybrid energy storage systems," *IEEE Applied Power Electronic Conference and Exposium (APEC)*, Tampa, FL, pp. 3661–3668, 2017.

5 Front-End Power Converter Topologies for Plug-In Electric Vehicles

Chandra Sekar S. and Asheesh K. Singh
Motilal Nehru National Institute of Technology, Allahabad

Sri Niwas Singh
Indian Institute of Technology Kanpur

Vassilios G. Agelidis
Technical University of Denmark

CONTENTS

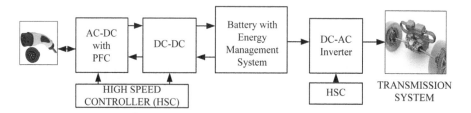

FIGURE 5.1 Architecture of an electric vehicle system.

5.1 INTRODUCTION

Numbers of electric vehicles (EVs) and plug-in hybrid EVs (PHEVs) have increased in recent years and expected to ascend soon. To reduce fuel import rate and pollution, the governments are promoting the EVs into the automobile market with subsidies and tax exclusions. However, still, the EVs do not have the substantial market place in the automobile sectors due to constraints in driving range, charging time, expensive and massive battery requirement, and deficient charging station (CS) infrastructures. Further, nation-wide infrastructure development remains a huge task, even in developed countries. The United Kingdom aims to increase the charging infrastructure through building charging ports in newly designed homes, office buildings, and parking-adjacent street lights. The EVs/PHEVs are having different charging options, such as (i) the level 1/level 2 onboard charger and (ii) off-board chargers. Structure of an EV with the onboard charger is shown in Figure 5.1. Charging several vehicles may become an extra burden to the electricity grid during peak demand schedule. In various literature, the issues related to power demand due to uncontrolled EV/PHEV charging systems are addressed [1]. Also, the vehicle-to-home (V2H) capability of EV/PHEV has been used for power backup systems, but the problems of uncontrolled charging were not addressed. Additionally, power backup systems market in India is increasing, with the main focus on replacing diesel engine generator systems (DEGs) with uninterruptible power supply (UPS) systems. To replace the DEGs, the UPS shall have the capabilities, such as high energy storage and better power quality (PQ) delivery. By using EV/PHEV battery packs for UPS systems, the following advantages can be utilized:

- Battery packs used in EV/PHEVs have a balance between power and energy densities. In EVs, the energy density and power density are considered for driving range and acceleration, respectively. Similarly, in UPS applications, the energy density and power density are considered for the long backup duration and delivering high PQ in dynamic load behavior conditions, respectively.
- UPS system, with same battery pack, can be used as household battery swapping station.
- Demand level in local distribution transformer (DT) can be reduced by switching the UPS to backup mode, in the case of critical low voltages.
- Charge power (P_{EV}) optimization/smart charging can be done through bidirectional DC–DC converter (BDC).

A typical V2H UPS system is developed in MATLAB®/Simulink® to demonstrate the functionalities mentioned above. Vehicle-to-grid (V2G) concepts have also emerged to compensate for the peak demand in the distribution system. The UPS application, as mentioned earlier with off-board chargers and the onboard chargers with V2G capability needs efficient, reliable, and controllable power electronic converters. For both the claims, a combination of front-end AC–DC converter which connects the EV charger to the electricity grid and a BDC converter with battery energy management capability is the essential power electronic converters. The front-end converter converts the input voltage into pulsating DC with power factor correction (PFC), and the DC–DC stage converts the pulsating DC to regulated DC voltage for storage application. In this chapter, in Section 5.2 and 5.3, various power converters topologies for the AC–DC converters and the BDC converters are carefully examined with their operations and advantages/disadvantages. In Section 5.4, a typical V2H UPS system is configured, and Section 5.5 gives the simulation results with discussions.

5.2   AC–DC PFC CONVERTER TOPOLOGIES

The sequences of switching pulses, used in power electronics converters, draw non-sinusoidal current from the grid and generate harmonic currents. This nonlinear load effect reduces the degree of utilization of the power from the grid. Thus, PFC is an essential part of the UPS system to obtain unity power factor, and the AC–DC converter with PFC plays a vital role to transfer the electrical power from grid to vehicle and vice versa. For power quality improvement, various circuit topologies have been derived. Front-end AC–DC converter is classified into two types: (1) Two-stage approach and (2) single-stage approach. The single-stage approach is more fit for the low power application and fits for lead-acid battery storage system due to the low ripple in the output current. Two-stage systems are fit for high-power application and suitable for lithium-ion battery. A single-stage AC–DC PFC converter is illustrated in Figure 5.2 [2]. This section reviews the existing front-end AC–DC converter topologies for PHEVs with its general operations, advantages, and disadvantages. The selection considerations for the suitable PFC converters are given in Table 5.1.

5.2.1   Full Bridge with Enabling Window Controlling Method

A full-bridge converter with enabling window control (EWC) method, shown in Figure 5.3, reduces the switching loss without increasing the size of the inductor [3]. Also, unity power factor and higher efficiency is obtained from this technique. In this circuit, nearly all the power is transferred in the middle part of the half-line cycle with a lower frequency to avoid the switching loss. The instantaneous power in this case is given as:

$$P_{in}(t) = P_o(1-\cos(2\omega t)) \tag{5.1}$$

where, P_{in} is the input power, P_o is the output power.

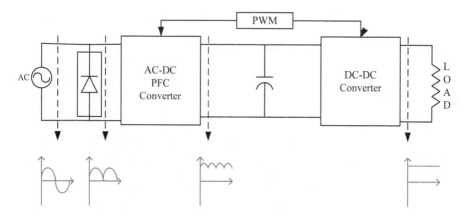

FIGURE 5.2   Block diagram of single-stage rectifier.

TABLE 5.1
Selection Consideration of Active PFC Converter

Factors	Types
Power flow	Unidirectional and bidirectional
Nature of DC output	Isolated and non-isolated
Type of output DC voltage	Constant, variable
Total number of quadrants	One, two or four
Essential of DC output	Buck, boost and buck-boost
Power quality in DC output	Voltage ripple, sag, swell, regulation
Types of load	Nonlinear, linear
Noise level	EMI, RFI, etc.

Power transferred in the higher half-line cycle P_{halfline} is:

$$P_{\text{halfline}}(t) = \frac{1}{\Pi} \int_{\frac{\Pi}{4}}^{\frac{3\Pi}{4}} P_{\text{in}}(\omega t)\mathrm{d}(\omega t) \tag{5.2}$$

The power loss of the circuit can be divided into conduction and dynamic loss. Dynamic loss includes switching loss, output capacitance loss, and driver loss. Dynamic loss is always directly proportional to the switching frequency, but the frequency should not be less than 20 kHz, and the next half-line cycle power will transfer with 100 kHz switching frequency to get a higher power factor. Here, the switching losses occur in the higher power frequency area; hence, there the switching losses are less due to the low switching frequency. However, the core loss is slightly higher than the conventional circuit. This may not affect efficiency. The core loss ($P_{\text{core\_loss}}$) in the window area calculated as follows:

$$P_{\text{core\_loss}} = 4.169(5\Delta B)^2 \left(\frac{1}{5f_{\text{s}}}\right)^{1.46} \tag{5.3}$$

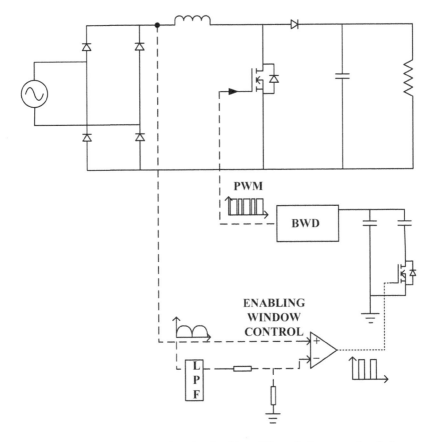

FIGURE 5.3   Power factor correction circuit with enabling window control technique.

where, f_s is the switching frequency.

The core loss of the EWC method is double the time of conventional bridge circuit shown in reference [4]. EWC method having two different switching is done by PFC control circuit with the change of oscillation capacitor or oscillation resistor and it improves the efficiency during light load conditions.

5.2.2   FULL-BRIDGE SEPIC PFC CONVERTER

A single-ended primary inductor converter (SEPIC) circuit is shown in Figure 5.4 with class E resonant converter for light-emitting diode (LED) drivers [5]. SEPIC converter consists of two inductors, two capacitors, three diodes, and one switch. The working mode of SEPIC converter consists of discontinuous conduction mode (DCM) and the class E resonant DC–DC converter made by chock inductance, and the switch is common to both circuits. The class E converter aims to reduce the DC voltage and is given to the LED driver.

As given in reference [5], in mode 1, the signal from the driver turns ON switch S, and L_1, L_2 current increases linearly due to the inductor voltage equal to the line

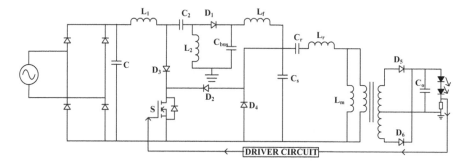

FIGURE 5.4   Full-bridge SEPIC PFC converter.

voltage and diode D_6 turns ON to energies the capacitor from the inductor. In mode 2, D_6 turns OFF and current through inductor increasing linearly until the turn OFF signal comes from the driver. In mode 3, switch S turns OFF, so the current in L_1 and L_2 decreases and conducts through a freewheeling diode to charge the capacitor C_{bus}. In mode 4, diode D_1 turns OFF and transfers the power and ends with freewheeling stage; mode 5 and mode 6 are also working on the freewheeling stage to perform the discontinuous mode operation. The input current (i_{in}) equation in reference [6]:

$$i_{in}(t) = \frac{V_{in}D^2 T_s}{2L_{eq}} = \frac{V_{in}D^2 T_s}{2L_{eq}} \sin(\omega t) \tag{5.4}$$

Equation (5.4) reflects the input voltage and input current flowing without any phase shift to achieve the power factor correction in the DCM.

Here, V_{in} is the input voltage, D is the duty cycle, T_s is the switching period, and L_{eq} is the equivalent inductance value.

In class E resonant circuit, C_s has high value and C_r has lower value to avoid the voltage stress. From the experimental results shown in reference [6], the soft switching dissipation decreased, the harmonic component is within the IEC 6100-3-2 standard, and the power factor attained is almost unity.

5.2.3   ZERO-VOLTAGE SWITCHING (ZVS) INTERLEAVED BOOST PFC CONVERTER

A simple PFC converter with the elementary topology of zero-voltage switching (ZVS) interleaved boost converter for plug-in EV charging application with a passive auxiliary circuit is given in Figure 5.5 [7]. Usually, switching loss occurs due to the hard switching of the metal–oxide–semiconductor field effect transistor (MOSFET), and diode reverses recovery time. To remove the switching losses, a variety of auxiliary circuits are designed to overcome this problem. Usually, an auxiliary circuit resides of passive and active components. Auxiliary circuit classified as resonant, non-resonant and dual auxiliary circuit [8–11]. Here, the ZVS is achieved by the auxiliary circuit. By charging and discharging the output capacitor with the reactive current, the control system of this converter utilizes the reactive current for soft switching. In this design, the boost converter decreases the input ripple current by

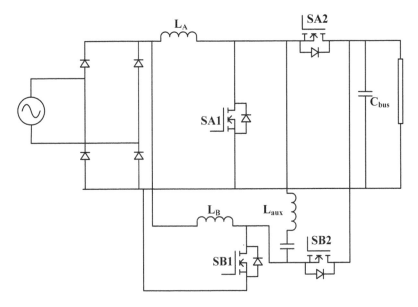

FIGURE 5.5   Zero-voltage switching interleaved boost PFC converter.

operating with a 180° phase shift. In mode 1, when the gate signal applies switch, SA1 turned ON with zero voltage. It attains the ZVS, and in mode 2, auxiliary circuit charges the capacitance of switch SB1 and discharge the capacitance of switch SB2 across the inductor. The average voltage is zero, and the current remains constant.

Auxiliary inductor voltage (V_{au}) is given by,

$$V_{au}(t) = \frac{V_o}{(t_2 - t_1)}(t - t_1) \tag{5.5}$$

where, V_o is the output voltage and t, t_1 and t_2 are the switching intervals

Current (i_{au}) across the circuit:

$$i_{au}(t) = I_{au,p} - \frac{V_o}{2(t_2 - t_1)L_{aux}}(t - t_1)^2 \tag{5.6}$$

where, $I_{au,p}$ is the current at the primary side of the auxiliary inductor (L_{aux}) in Figure 5.5.

In mode 3, after charging and discharging of the capacitor, again it gives the gate signal to turn ON SA2; when the auxiliary voltage is $-V_o$. In mode 4, the output capacitance of both switches will charge zero to V_o and discharge V_o to zero. During mode 5, the voltage across the inductor is always zero. Here, the Inductor current (i_{LA}) is given by,

$$i_{LA}(t) = I_v + \frac{V_{in}}{L_A}(t - t_o) \tag{5.7}$$

where, V_{in} is the input voltage, L_A is the inductor value, and t & t_o are the switching intervals.

In mode 6, again the output capacitor of switch SA1 charging from zero to the output voltage and the capacitance of switch SB2 discharge during this period inductor current is constant. In the mode 7, auxiliary circuit voltage is equal to the output voltage. At last in mode 8, the switch SA1 is discharging from V_o to zero and SA2 is charging from zero to V_o and across the inductor the current is excess, and it charges the output-side capacitor to boost the output voltage. Control loop includes with current loop, voltage loop, and switching frequency loop to attain the ZVS. This frequency loop control is an additional loop control, and they designed an auxiliary inductor and obtained higher efficiency compared to the conventional converter.

5.2.4 QUASIS ACTIVE FULL-BRIDGE PFC CONVERTER

A single-stage switch quasis active PFC converter with a combination of two DC–DC modules is shown in Figure 5.6 [12]. Here, the input power is not transferred directly; initially, some segment of input power fetched to the load due to DC–DC module 1. The remaining input power stored in a capacitor and transferred by the DC–DC module 2. The second segment of input power fetching is acting as a regenerative snubber circuit and both the DC modules perform on same duty cycle and load. Here, a snubber circuit which consists of two diodes, one capacitor and one transformer winding. With these modules operate in DCM and the extra winding that operates in continuous conduction mode (CCM), the quasi active PFC can achieve almost unity power factor. This topology consists of five modes of operation. The efficiency of the converter increased by the direct power transfer (DPT) concept. Efficiency can be calculated by:

$$P_{o1} = \eta_1 k.P_{in}; P_{11} = \eta_1(1-k).P_{in}; P_{o2} = \eta_2 P_{11} \qquad (5.8)$$

where, η_1, η_2 are the efficiencies of main and auxiliary stage, P_{o1} is the power output at the first stage, P_{o2} is the power output at the second stage, and P_{11} is the

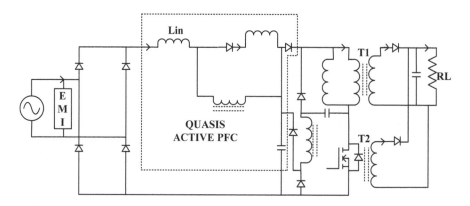

FIGURE 5.6   Quasis active PFC converter.

power at auxiliary stage. Gross efficiency can be upgraded by expanding the direct power k component:

$$k = \frac{P_{o,pk}}{P_o} = \frac{V^2_m d3^2}{4L_m P_o} T_s \tag{5.9}$$

For higher power factor, the transformer should operate on DCM, and the secondary current should be zero before the switching cycle end [13–17]. To decrease the ripple, turns of active PFC should be higher than the DC–DC module, and the capacitor must be selected to hold the output voltage while inductor stores the peak energy of the leakage inductance.

$$I_{Lm,peak} = \sqrt{2P_{in}/f_s L_{m1}} \tag{5.10}$$

where, $I_{Lm,peak}$ is the peak value of leakage current due to mutual inductance L_{m1}, P_{in} is the input power and f_s is the switching frequency.

From the experimental result shown in reference [12], the higher efficiency can be reached by the direct power transfer concept with the bulk capacitor selection.

5.2.5 A SIMPLE BIDIRECTIONAL AC–DC CONVERTER

A simple BDC converter is shown in Figure 5.7, it reduces 75% of switches compared to conventional ones. A simplified pulse width modulation (PWM) and feedforward control strategy achieved with Field Programmable Gate Array (FPGA) in reference [18].

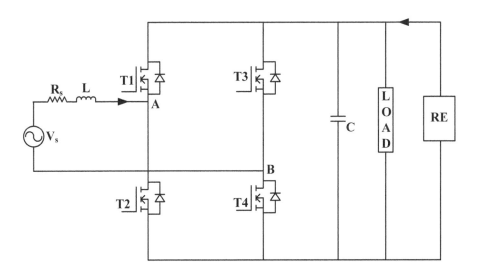

FIGURE 5.7 Simple bidirectional converter.

In the rectifier mode, when it is operating in the half-cycle, the inductor current is growing in the charging mode. When applying Kirchhoff law:

$$V_s - L\frac{d}{d_t}i_L - V_{dc} = 0 \tag{5.11}$$

where, V_s is the supply voltage, i_L is the current through the inductor, V_{dc} is the DC voltage, and L is the inductance value.

Where $V_s - V_{dc}$ inductor turns to discharged state and when the negative half cycle occurs, the inductor current reduces in the discharging mode. The other equation obtained from Kirchhoff law,

$$V_s - L\frac{d}{d_t}i_L + V_{dc} = 0 \tag{5.12}$$

where, $V_s + V_{dc}$ inductor turns to charged state, the converter operates with PWM in rectifier mode to maintain the DC regulation. In the inverter mode, when the inductor current is less than zero, inductor charge the inductor current; when it is greater than zero, inductor discharges the inductor current to maintain the bidirectional flow with simple PWM strategy. When comparing to the dual-loop control system, its efficiency is lesser than the simple feedforward [19]. Also, from the simulation and hardware result, efficiency is higher than the unipolar and bipolar PWM techniques.

5.2.6   BRIDGELESS HIGH PFC AC–DC CONVERTER

A bridgeless AC–DC rectifier with PFC is shown in Figure 5.8 [20]. PF and efficiency are improved with the low harmonic distortion for low power application. The discontinuous voltage mode is giving soft turn OFF to insulated-gate bipolar transistor to reduce the conduction loss. In this topology, they added one inductor and capacitor to acquire better thermal performance. In the analysis of operation, where the stage 1 switch turned ON and the capacitor discharge, the diode D_o becomes reverse biased; also, switch current equal to the output current. In stage 2, again switch is in ON condition till the capacitor is discharged, and where the input current is equal to the switch current. Now, D_o conducts in forward bias. In stage 3, switch becomes turned OFF and the capacitor starts charging by the input current capacitor until it reaches the peak voltage. Input capacitor's maximum voltage (V_{cm}) can be obtained as:

$$V_{cm} = \frac{i_{L_1}}{C_1}(1\text{-}D)T_s \tag{5.13}$$

where, C_1 is the capacitance value, D is the duty ration, T_s is the switching period, and i_{L1} is the inductor current.

While selecting the value of the capacitor, we should consider the factor K and expression is obtained as:

$$C_1 = C_2 = \frac{KT_s}{2R_L} \tag{5.14}$$

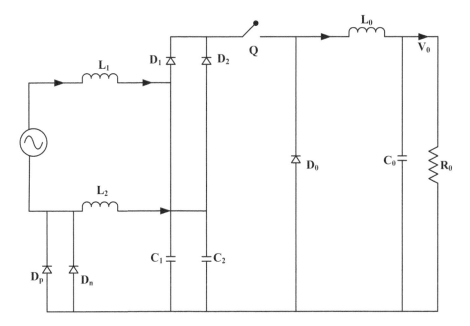

FIGURE 5.8 Bridgeless high PFC converter.

where, C_1, C_2 are the capacitance values, T_s is the switching period, and R_L is the load resistance.

Inductor selection should fill the constant requirement of current source throughout the switching cycle DCVM having the disadvantages of operating in higher voltage stress. Simulation and hardware prototype was done with buck mode and it attained a higher power factor and efficiency with lower total harmonic distortion (THD).

5.2.7   LLC Resonant Full-Bridge Converter

A resonant LLC converter for PHEV charger is shown in Figure 5.9. A constant maximum power method is implemented for charging the nonlinear loads like lithium-ion battery [21]. As shown in reference [21], for PHEVs, full-bridge topology is much suited then half-bridge topology for the high-power applications. Duty cycle operates on 0.5 and 180° phase shift to create symmetric waveform. Here, the LLC alone not responsible for the voltage conversion; it also depends on load and inductance. Another main advantage of the resonant converter is that the soft-switching performance decreases the switching losses. The switching speed is adjusted to attain the maximum value to buck the input voltage. Here, the conversion gain (F_{nw}) can be expressed as:

$$F_{nw} = \frac{\Pi}{2}\sqrt{\frac{l}{1+l}}\frac{1}{\cos^{-1}\left(\dfrac{1}{M_{min}(1+l)}\right)} \tag{5.15}$$

where, l is the inductance ratio, M_{min} is the minimum value of gain.

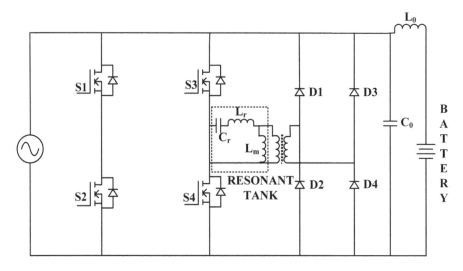

FIGURE 5.9   LLC resonance full-bridge converter.

In the boost mode ZVS operation, the worst case is also designed perfectly. The transformer ratio should be selected in minimum. Moreover, the output comes under regulation. Here, to calculate critical conversion gain and the maximum characteristics impedance $Z_{0,max}$, a critical condition expressed as follows:

$$M_{crit} = \sqrt{1 + \sqrt{\frac{l}{1 + l}}} \tag{5.16}$$

where, M_{crit} is the critical value of gain.

$$I_{out.crit} = \frac{P_{out,max}}{M_{crit} V_{in}, \dfrac{M_{in}}{\eta}} \tag{5.17}$$

where, $I_{out.crit}$ is the critical value of output current. V_{in} is the input voltage, $P_{out,max}$ is the maximum output power, and η is the efficiency.

But on conducting mathematical modeling and experimental result, in peak analysis, the efficiency of 98.2% is obtained with a lithium-ion battery and resistive loads.

5.2.8   BRIDGELESS INTERLEAVED (BLIL) PFC CONVERTER

A BLIL PFC converter for a plug-in EV with a minimum size of the charger is shown in Figure 5.10 [22]. In comparison to the normal interleaved and boost PFC converters, this topology needs only two extra switches and four slow and two fast diodes. Steady-state analysis of this topology requires two operations; positive and negative

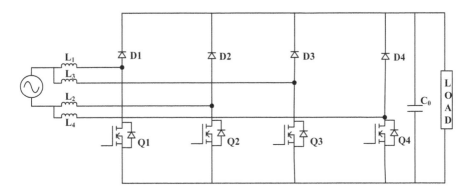

FIGURE 5.10   BLIL PFC converter.

half cycle periods. When switch 1, 2 are ON and switch 3, 4 are OFF, the inductors L_1 and L_2 store the energy. The ripple current (Δi_{L1}) is expressed as:

$$\Delta i_{L1} = \frac{1}{L_1 + L_2} V_i (1-D) T_s \qquad (5.18)$$

where, L_1, L_2 are the inductance values, V_i is the input voltage, D is the duty ratio, and T_s is the switching period.

   After that, inductor currents in 3 and 4 linearly decrease and given to the load by D3. In that negative half-cycle operation, the duty cycle should less than 0.5. When four switches are in OFF, inductance L_1, L_2 start increasing linearly and the ripple in the switch and inductor is equal.

$$\Delta i_{L1} = \frac{1}{L_1 + L_2} V_i (D) T_s \qquad (5.19)$$

   There is some gain that will be attained because of the elimination of the full-bridge converter. The mathematical and experimental result was conducted, and the THD presented is less than 5%. It achieves a higher peak efficiency of 98.9% with high power factor.

5.2.9   IMPROVED BUCK PFC CONVERTER

Improved buck PFC converter with one auxiliary switch and two diodes, as shown in Figure 5.11, enhances the power factor with constant on-time control [23]. This converter works with a higher duty cycle to get the higher voltage gain. During the positive cycle, V_o is higher than the V_{in}, and the converter operates as a buck-boost mode. When the input voltage is larger than V_o, it acts as a buck mode, and when the negative cycle occurs, it can act in a negative buck-boost and buck operation. During boost operation, Q2 is ON, and Q1 is OFF, and the inductor charged. The inductor discharged when Q2 is OFF. Likewise, during the buck operation, Q1 is ON, and

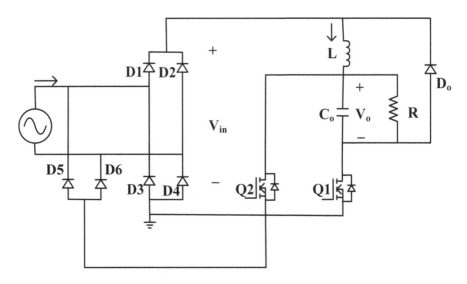

FIGURE 5.11   Buck PFC converter.

L is charged and Q1 is OFF, L is discharged through the load. To operate the criti-cal continuous conduction mode (CCCM); here, they are implementing the constant on-time control with a voltage reference V_{ph} to achieve either buck or boost mode. An error amplifier compares the reference and output voltage signals. In this work, improved the power factor and reduction in THD up to 18% is achieved.

5.3   DC–DC BIDIRECTIONAL CONVERTER TOPOLOGIES

BDC is commissioned to action the power transferring processes from battery/ultra-capacitor to attain V2G or G2V purposes with the inverter [24, 25]. The choice of this converter is essential to achieve the overall system efficiency. The two categories of BDCs are: (i) non-isolated, and (ii) isolated bidirectional converters. This section presents the various topologies used in the BDCs.

The main difference between the two converters is that the isolated converter is having a high-frequency transformer between the input and output side. It brings higher power density and causes the leakage inductance stored in the transformer. Here, the stored energy creates voltage spikes on the switches, which causes voltage stress. A resistor-capacitor-diode (RCD) snubber circuit is recommended to control the switch voltage stress [26].

On the other hand, the non-isolated converter topology has a straightforward structure, low cost, high reliability, high efficiency. The non-isolated half-bridge converter is better but has low efficiency with a wide voltage conversion range. To avoid this issue, various PWM techniques are available. Three-level bidirec-tional DC–DC converters are suggested to the EV applications, and using this topology, the inductor size reduced by one-third of the half-bridge bidirectional converter.

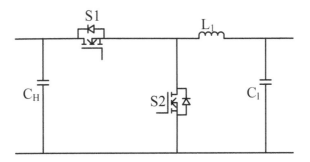

FIGURE 5.12   Conventional buck-boost converter.

5.3.1   Non-Isolated Bidirectional Converter

The non-isolated bidirectional converters have very simple in construction, less cost, and work under different voltage range [27]. It can be classified into multi-level type, switched capacitor type, cuk type and sepic/zeta type, buck-boost type, coupled inductor type, three-level, and conventional buck/boost types. Various non-isolated bidirectional converters are shown in Figure 5.12–5.17. In that, multilevel and switched capacitor types having more switches and capacitor to attain the high voltage gain, also the control circuits of these types are very complicate [28]. Using cuk and sepic/zeta type converter gives low conversion efficiency due to the two power stages also cannot provide the wide voltage conversion range. The coupled inductor type can achieve large voltage gain by adjusting the number of turn ratios in the coupled inductor. The Circuit configuration of this type is complicated [29]. Three-level converter type having the stress on the switches is half compared to the conventional buck/boost type. In this type converter, the conversion range is narrow. The conventional bidirectional buck-boost converter has a simple circuit here the

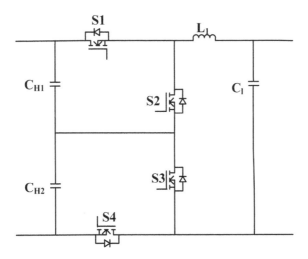

FIGURE 5.13   Modified buck-boost bidirectional converter.

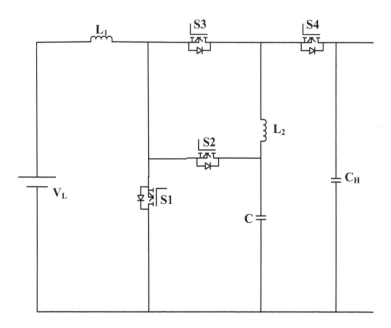

FIGURE 5.14   Cascaded bidirectional boost converter.

limitation is that the effect of the switches and the equivalent series resistance of the inductor and capacitor will limit the step-up voltage gain. Figure 5.12 represents the modified conventional buck-boost converter. This topology is implemented with various control techniques in references [29–33]. The voltage conversion ratio in this converter is higher than the previous conventional bidirectional buck/boost converter, which has 92.3–94.8% step-down voltage efficiency and step-up efficiency is 91.2–94.1%. Reference [34] presents a modified buck/boost converter in Figure 5.13 and Figure 5.14 presents the circuit, which acts as a cascaded boost converter in step-up mode and a cascaded buck converter in step-down mode. In step-up mode, it operates as CCM and in step-down mode, it operates as DCM [34]. In this topology, step-up voltage gain is high compared to conventional buck/boost topology.

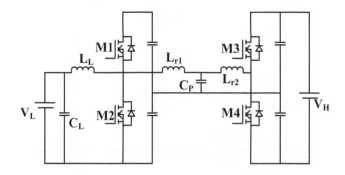

FIGURE 5.15   Bidirectional LCL resonant converter.

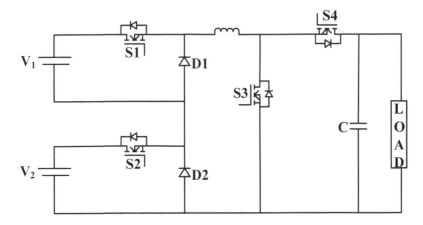

FIGURE 5.16   Multi-input bidirectional converter.

A non-isolated soft-switching bidirectional LCL resonant DC–DC converter is shown in Figure 5.15 [35]. During bidirectional conversion processes, the diode and switches are getting more stress. It results in high reverse recovery loss and electromagnetic interference problems. Also, the conversion ratio is very less. However, the advantages of the non-isolated converters are the compactness and higher efficiency. To improve the power density, the high-frequency operation is required in DC–DC converters, but the switching losses are very high. To overcome this, soft switching is required. Coupled inductor converter design makes the system more complex with high voltage gain. Usually, non-isolated hard switching limits the switching

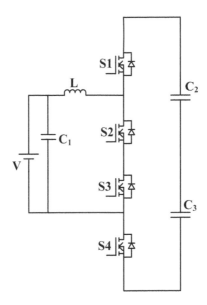

FIGURE 5.17   Three-level bidirectional converter.

frequency, and the switched capacitor is given high efficiency compared to other topologies. Still, the number of components increases the converter cost and limits the switching frequency due to the hard switching.

Figure 5.15 represents the non-isolated soft-switching bidirectional LCL resonant DC–DC converter, which is having a half-bridge in the front end and LCL resonant with voltage doublers combined with another end [36]. The main advantages of this topology are the ZVS turn ON in both direction, and ZCS turn ON and turn OFF in both directions, low voltage stress without any additional snubber circuit, high step-up, and step-down ratio and reduced the volume of the magnetic. It attains 95.5% efficiency in boost operation and 95% efficiency in the buck operation.

Reference [37] proposed a multiple single-input DC–DC converter having advantages of simple and more compact design and reduce the overall cost of the system. Also, the regulated voltage output improves system reliability.

Always the connection of an isolated converter is magnetically coupled circuit, but non-isolated converters are coupled based on the electrical circuit connected. Using isolated converter energy transformed from source to load using the technique of time-domain multiphase is a commonly used one. This additional requirement of circuits makes the system bulky, costly. In non-isolated, the electrically coupled system having a modular structure, low cost, and the absence of transformer make it simpler and more attractive [38]. This electrically coupled circuit can combine various input power source either in parallel or series. The main limitation of parallel-connected source topologies is at a time only one input can be connected to avoid the power coupling effect. For multiple power transfer, the series-connected combination gives the solution. In a series connection using a bypass diode, it can be modified to parallel connection. Still, it increases the count of the overall components which increases the cost of the system.

The series connection, shown in Figure 5.16, operates in bidirectional with the buck, boost, or buck/boost operations. The simple construction makes very less fault occurring capability to improve the reliability of the system. In this system, all the input sources are connected to the load through a single inductor [39]. Here, the operation of the converter is based on a single switching cycle. The passive elements stored the energy for some particular time then discharges to the load in the remaining time. The power flow is controlled by the inductor only. In this topology, the inductor charged by multiple input sources instead of a single input source.

A three-level bidirectional DC–DC converter is suitable for the photovoltaic energy conversion system proposed in [40]. Here, the advantages of these topologies are low voltage stress, and the ripple frequency of the inductor is twice the converter switching frequency. A zero-voltage transition (ZVT) three-level DC–DC converter to induce to operate in very high switching frequency to obtain the high-power density; also, it improves the efficiency. For the low cost and high-efficiency operation, the bidirectional two-quadrant switches are preferred. However, it gives low efficiency when the light load occurs. Also, due to the power losses associated with the parasitic capacitor, the passive components need large volume and increased input current ripple frequency. Interleaving topology reduces the size of those passive elements. Three-level converters, shown in Figure 5.17, are suitable for higher-level application and implementing soft switching techniques without sacrificing the efficiency to attain the power density with higher switching frequency. The interleaving method is a primary way to achieve the soft-switching method. In this topology, two identical soft switching are implemented for each pair of switches, and it achieved the ZVT for every switch during turn-on operation mode. This circuit contains two resonant inductors, one resonant capacitor, and an auxiliary switch to avoid the turn ON switches by creating a resonant between the inductor and capacitor.

For EVs, the commonly used topology is the hard switching cascaded buck-boost converter in CCM mode with conventional PWM technique [41]. This conventional PWM technique causes the switching loss because of the reverse recovery of diodes, which gives low efficiency. The PWM technique is changed from the conventional to constant frequency soft switching modulation, which performs the ZVS improve the efficiency of this converter. Also, the use of silicon carbide diodes reduces the reverse recovery losses by 67% [41].

5.3.2   ISOLATED BIDIRECTIONAL CONVERTERS

In the BDC converter, usually, isolation is given by a high-frequency transformer, as shown in Figure 5.18–5.22. This method gives additional cost and losses, but it is essential when (i) the high voltage and low voltage side negatives not grounded together, and (ii) the voltage ratio is much more enough to tackle very high current and high voltage simultaneously [42–46]. There are a number of topologies available with voltage fed on both the high voltage and low voltage side. Also, it is essential to implement the current source between the circuit to get the smooth power shift. Fixing the inductor on the low voltage side needs a high current carrying magnetic component, and fixing the inductor on high voltage side needs high voltage switches because of the voltage stress, the bidirectional DC–DC converter acting as an inverter and rectifier. In the rectification mode, the current conducts by diodes and in the inverter mode switches conducts the current. In case the switches are MOSFETs, to obtain less voltage drop under synchronous rectification, it conducts in the reverse direction. For high power application, the magnetic components and interconnect parasitic are quite different and result in significant variation in losses. For low power application, interconnection is not an issue, but the number of

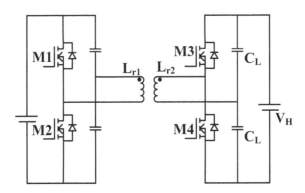

FIGURE 5.18   Bidirectional DC–DC two voltage-fed half-bridge converter.

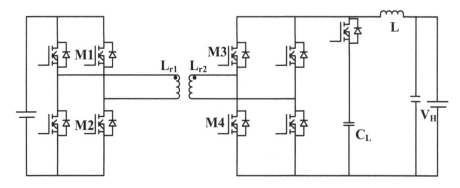

FIGURE 5.19   Bidirectional DC–DC voltage-fed and current-fed full-bridge converter with active clamp circuit.

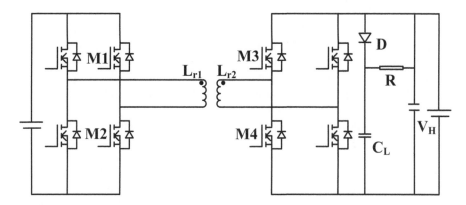

FIGURE 5.20   Bidirectional DC–DC voltage-fed and current-fed full bridge converter with RCD snubber circuit.

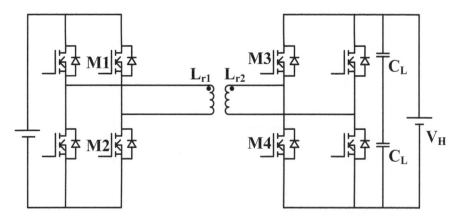

FIGURE 5.21   Bidirectional DC–DC two voltage-fed full-bridge converter.

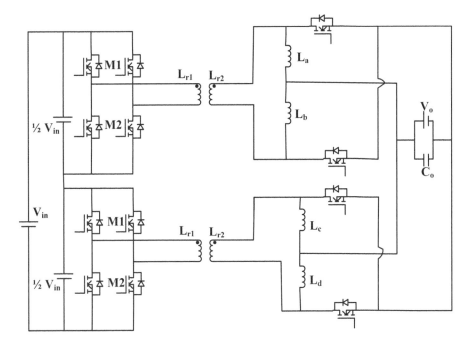

FIGURE 5.22 Cascaded isolated bidirectional converter.

switching devices must be reduced. Half-bridge converter has a problem of floating gate driver, which can be rectified by replacing the push-pull converter. This converter requires a center tapped transformer, which makes the circuit complex in the design of high-power applications [47]. Various half-bridge current source converters were proposed to eliminate the floating gate drive and to minimize the size of the inductor and switches.

Always the current source converters considered as an isolated boost converter. The essential operation is to connect all switches in a short circuit to store energy in the inductor. While opening half of the switches, the energy will be transmitted through the transformer to the output side. From the figures, the voltage fed converter always suffering from various limitations of very high input pulsating current and limited soft-switching range, circulation current through the switches, and the magnetic devices. Also, the efficiency of high voltage amplification is very low for high input current applications [48]. Current-fed converters proved meritorious more than the voltage fed converters. It has the advantages of the less input current ripple, negligible duty cycle loss, easy to control, and RCD snubber circuit or some other snubber circuit to absorb the voltage spikes when the devices turn OFF. The active clamp snubber circuit gives more efficiency and performs zero voltage switching for the switches.

Recently, the usage of the high-frequency transformer replaced the traditional line frequency transformer for the present and future generations [43, 49]. The merits of a high-frequency transformer are light in weight, and the volume is less and cheaper. And this type of transformer avoids the distortion of voltage and current

TABLE 5.2

Comparison of Various Topologies for High Power Design

Characteristics	Phase-Shifted Bridge	Hard-Switched PWM	Active Resonant	Series/Parallel Resonant	Dual Active Bridge
Switching frequency (kHz)	20	5	20	20	20
Control complexity	Simple	Simple	Complex	Moderate	Moderate
Constant frequency	Yes	Yes	Yes	No	Yes
Circulation current (A)	Yes	No	No	Yes	Yes
Peak IGBT current (A)	335	211	211	317	978
IGBT stress	Moderate	High	Low	Moderate	Moderate
No. of active devices	4	4	8	4	8
Resonant inductor	In transformer	No	Yes	Biggest	In transformer
Resonant capacitor	No	No	No	Largest	No
Output rectifier stresses	Moderate	High	Moderate	Lowest	Low
Ripple current	Low	Low	Low	Low	High

Abbreviations: IGBT, insulated-gate bipolar transistor; PWM, pulse width modulation.

waveforms, where the switching frequency is higher than the total noise of the power conversion system, such as 20 kHz. A comparison of various design considerations of isolated converter topologies for the high-power applications is given in Table 5.2.

5.4 CONFIGURATION OF AN EV-BASED ONLINE UPS

The online UPS is available with multiple configurations for low, medium, and high-power applications [50]. The proposed UPS configuration has a front-end PFC rectifier unit, a sinusoidal PWM (SPWM) voltage source inverter (VSI), and a BDC, connected to DC-link capacitor. The mode of operation of the UPS depends on the operation of BDC. The BDC operates in two modes, (i) grid-mode, and (ii) backup-mode, as shown in Figure 5.23. In grid-mode, the BDC performs in buck operation, and charges the battery, whereas, in backup-mode, the BDC performs in boost operation, to supply the VSI connected with domestic non-EV loads. Figure 5.23 shows a typical 1-ϕ configuration of the V2H Online UPS system.

Among various topologies, continuous conduction mode (CCM) PFC boost rectifier is found simpler, reliable, and more suitable for medium and high-power applications [51, 52]. The custom design of the PFC boost rectifier is concentrated in this brief.

A well-known SPWM voltage source inverter with LC (low pass) filter is used, with design parameters from [53–55].

A non-isolated BDC is used to store and retrieve the energy, as shown in Figure 5.24. When the S_2 is in ON condition, the battery discharges the power, and it performs boosting action. In charging mode, S_3 is in ON condition, and BDC operates under a single switch buck converter. The power circuit design is adopted from [56, 57].

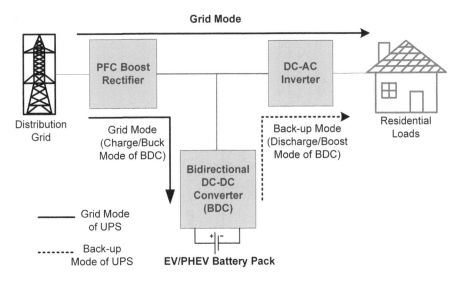

FIGURE 5.23 Modes of operation of uninterruptible power supply system.

In Figure 5.24, EV/PHEV with an additional Li-ion battery is shown. Either one of them can be connected to the BDC at a time.

5.4.1 Power Circuit Design

The value of inductor L_S is determined on the basis of suitable maximum inductor current ripple (%ripple). It is advised that for CCM mode of operation maximum inductor current ripple should be 20–40% of the average input current [51, 52]. The L_s has been chosen from the following equations adopted from reference [51]:

$$L_s = \frac{1}{\%\text{ripple}} \cdot \frac{V_{(S.\text{min})}^2}{P_0} \left(1 - \frac{\sqrt{2}.V_{S.\text{min}}}{V_{DC}}\right) \cdot \frac{1}{f_{S_1}} \tag{5.20}$$

$$I_{L.\text{pk}} = \frac{\sqrt{2}.P_0}{V_{S.\text{min}}} \cdot \left(1 + \frac{\%\text{ripple}}{2}\right) \tag{5.21}$$

Where,

P_0 is maximum power rating of the proposed UPS system,

f_{S_1} is switching frequency of S_1, and

$I_{L.\text{pk}}$ is inductor peak current.

The value of C_L determines (i) voltage ripple of DC link voltage (ΔV_{DC}) and (ii) hold-up time (t_{hold}) between grid mode and battery mode operation of the UPS. To satisfy the above two requirements the C_L is selected using the following equation:

$$C_L \, {}^3\text{max}\left(\frac{P_o}{2\pi f_{\text{line}} \cdot \Delta V_{DC} \cdot V_{DC}}, \frac{2P_o t_{\text{hold}}}{V_{DC}^2 - V_{DC.\text{min}}^2}\right) \tag{5.22}$$

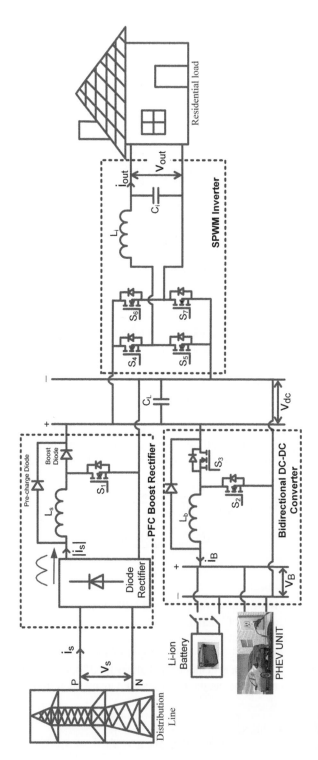

FIGURE 5.24   Single-phase V2H online uninterruptible power supply configuration for plug-in hybrid electric vehicle charging/discharging.

Similarly, selection of L_b is done based on the following equation adopted from reference [56]:

$$L_b = \frac{1}{\%RippleCurrent} \cdot \frac{V_{B.min}^2}{P_o} \left(1 - \frac{V_{B.min}}{V_{DC}}\right) \cdot \frac{1}{f_{sw}} \qquad (5.23)$$

The pre-charge diodes in boost converters are used to reduce the initial surge current stress on boost diode.

Selection of filter inductor (L_i) is based on the following equation adopted from [54]:

$$L_i = \frac{V_{DC}}{4f_s \left(\dfrac{P_o}{V_{out}}\right)(\sqrt{2})(\%ripple)} \qquad (5.24)$$

Where,

f_s is the switching frequency of inverter power switches, and

%ripple is tolerable percentage of ripple current by the inductor core.

The value of C_i is selected based on required cutoff frequency (f_c) of the LC filter. For the UPS applications, the cutoff frequency must be less than 5–10% of the switching frequency [53]. To obtain lower percentage of THD in output voltage, f_c and C_i are set as based on the following relation [54]:

$$\left| \frac{1}{f_c^2 \cdot \dfrac{X_{L_i}}{X_{C_i}} - 1} \right| \le \left(\%filtering\,required\right) \qquad (5.25)$$

A precharge diode is preferred to reduce the initial surge current stress on boost diode.

5.5   SIMULATION RESULTS AND DISCUSSIONS

To examine the proposed V2H online UPS configuration, a simulation model is developed in MATLAB/Simulink platform with the parameters shown in Table 5.3.

The Figure 5.25 illustrates the response of the proposed system under load variations and UPS backup mode operation. The supply voltage is kept OFF between 0.2 and 0.3 second. Figure 5.25(a) shows that the input voltage and current are in-phase to maintain the power factor close to unity. During mode transitions, it displays the fast response and excellent current tracking performances, within a short interval. Figure 5.25(b) shows that variation in DC link voltage is maintained within ± 5% of 400 V, by BDC with Li-ion battery. Figure 5.25(c) shows that i_B in charging/discharging mode maintains the current ripple within ~2% (~ 1 A) of maximum charge current (of 1 C = 50 A) without affecting the performance of the Li-ion battery. In Figure 5.25(c), at 0.2 second, even though a sudden decrease in i_B is experienced due to change in BDC operation mode, configuration of the power converters is found capable to recover and maintain the DC link voltage within ± 5%. Thus, due to less variation in DC link voltage, the inverter voltage (V_{out}) is maintained under considered load variations, which is clearly depicted in Figure 5.25(d).

TABLE 5.3

Parameters and Specifications

Input and Output		Power Circuit		Battery Specifications	
V_s rms	230 V	L_s	6 mH	Nominal voltage	210 V
i_s rms Max.	17.7 A	C_L	4700 μF	Rated capacity	50 Ah
P_{EV} Max.	4000 W	L_b	6 mH	Internal resistance	0.174 Ω
$P_{non\text{-}EV}$ Max.					
$f_{S1\text{-}S7}$	15 kHz	L_i	35 mH	$V_{cut\text{-}off}$	157.5 V
V_{out} rms	230 V	C_i	5 μF	$V_{full\text{-}charge}$	244 V

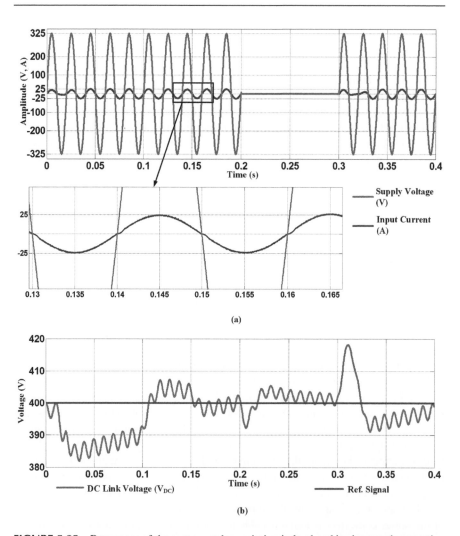

(a)

(b)

FIGURE 5.25 Responses of the system under variation in load and backup mode operating conditions. (a) Supply voltage (V_s) and input current (i_s) with PFC; (b) DC link voltage (V_{dc}). (Continued)

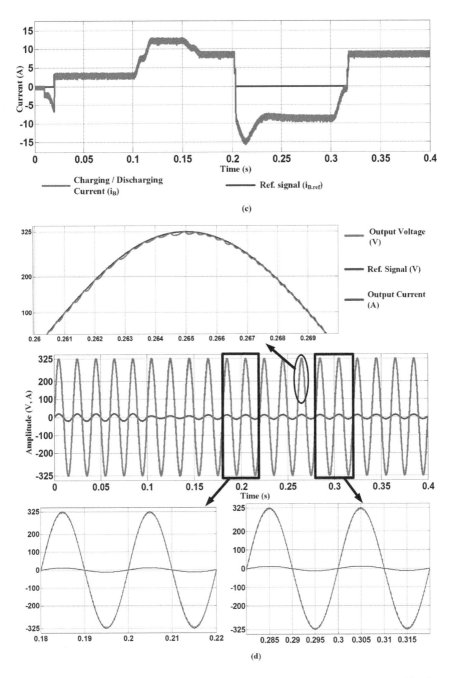

FIGURE 5.25 (Continued) Responses of the system under variation in load and backup mode operating conditions. (c) battery charging/discharging current (i_B); (d) inverter output voltage (V_{out}) and current (I_{out}). (Continued)

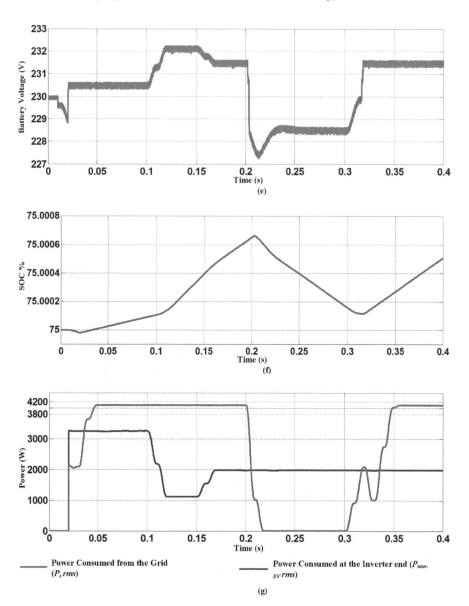

FIGURE 5.25 (Continued)   Responses of the system under variation in load and backup mode operating conditions. (e) battery voltage (V_B); (f) rate of change in SOC (%); (g) power consumption of overall system (P_s rms) and power consumption by residential loads at inverter output end (P_{non-EV}).

It shows that there is no deviation in inverter output during switching between grid and backup mode operations. Figure 5.25(e) and (f) illustrate the battery voltage and rate of change of SOC with respect to i_B variations. Finally, the cumulative power consumption during the given P_{non-EV} and $i_{B.ref}$ variations is shown in Figure 5.25(g).

Therefore, the combination of front-end PFC converter and a BDC achieves a higher response time with proper design considerations and it also makes a reliable online UPS system using the EV batteries.

5.6 CONCLUSION

This chapter provides various front-end AC–DC power factor correction topologies with their operation, design considerations, and validations to explore the broad scope of these topologies. The capabilities of improved power quality and efficiency depend on application-specific selection the topology. Also, the DC–DC converters were explored for cost effectiveness with higher efficiency. The fast responding capabilities of these front-end power converter topologies are demonstrated with a household EV-based online UPS system configuration developed on the MATLAB/ Simulink platform. The online UPS configuration for utilizing EV/PHEV battery packs has been analyzed extensively. The system shows excellent performances such as fast response time, higher efficiency, reliability on load and input voltage variations and capability of switching between grid and backup modes without any deviation in inverter output. In future, this technique could be more suitable for developing countries like India where making CS over large scale for EV/PHEV chargers is a difficult task and at the same time need of EVs is evident in light of increased pollution and dependency on fossil fuels.

ACKNOWLEDGMENT

This work was carried out under a project sponsored by Scheme for Promotion of Academic and Research Collaboration (SPARC), Ministry of Human Resource Development (MHRD), Government of India. (Project Code: P1542)

REFERENCES

1. K. Clement-Nyns, E. Haesen, and J. Driesen, "The impact of charging plug-in hybrid electric vehicles on a residential distribution grid," *IEEE Transactions on Power Systems*, vol. 25, no. 1, pp. 371–380, February 2010.
2. Sheldon S. Williamson, Akshay K. Rathore, and Fariborz Musavi, "Industrial electronics for electric transportation: current state of the art and challenges", *IEEE Transactions on Industrial Electronics*, vol. 62, no. 5, pp. 3021–3032, March 2015.
3. Xudan Liu, Dehong Xu, Changsheng Hu, Heng Yue, Yashun Li, Ping Lin, and Hangwen Pan, "A high efficiency single phase AC-DC converter with enabling window control and active input bridge", *IEEE Transactions on Power Electronics*, vol. 27, no. 6, pp. 2912–2924, December 2011.
4. Alireza Abasian, Hosein Farzaneh-fard, and Seyed Madani, "Single stage soft switching AC-DC converter without any extra switches", *IET Power Electronics*, vol. 7, no. 3, pp. 745–752, March 2014.
5. Yijie Wang, Jiaoping Huang, Guangyao Shi, Wei Wang, and Dianguo Xu, "A single stage single switch led driver based on integrated SEPIC circuit and class E converter", *IEEE Transactions on Power Electronics*, vol. 31, no.8, pp. 5814–5824, October 2015.

6. Sin-woo Lee and Hyunk lark Do, "Soft switch two switching resonant AC-DC converter with high power factor", *IEEE Transcation on Industrial Electronics*, vol. 63, no. 4, pp. 2083–2091, December 2015.

7. Majid Pahlevaninezhad, Pritam Das, Josef Drobnik, Praveen K. Jain, and Alireza Bakhshai, "A ZVS interleaved boost AC-DC converter used in plug-in electric vehicles", *IEEE Transactions on Power Electronics*, vol. 7, no. 8, pp. 3513–3528, August 2012.

8. J. Deng, K. Shi, A. Zhao and D. Xu, "universal zero-voltage-switching technique for multi-phase AC/DC converter," *2019 IEEE Applied Power Electronics Conference and Exposition (APEC)*, Anaheim, CA, USA, 2019.

9. K. Shi, A. Zhao, J. Deng, and D. Xu, "Zero-voltage-switching SiC-MOSFET three-phase four-wire back-to-back converter," *IEEE Journal of Emerging and Selected Topics in Power Electronics*, vol. 7, no. 2, pp. 722–735, June 2019.

10. W. Han and L. Corradini, "Wide-Range ZVS control technique for bidirectional dual-bridge series-resonant DC–DC converters," *IEEE Transactions on Power Electronics*, vol. 34, no. 10, pp. 10256–10269, October 2019.

11. M. Abbasi and J. Lam, "A multimode bridge-less SiC-Based AC/DC step-up converter with a dual active auxiliary circuit for wind energy conversion systems with MVDC grid," *2019 IEEE Energy Conversion Congress and Exposition (ECCE)*, Baltimore, MD, USA, 2019.

12. Hussain S. Athab, Dylan Dah-Chuan Lu, Amirnaser Yazdani, and Bin Wu, "An efficient single switch efficient quasis active PFC converter with continuous input current and low DC bus voltage stress," *IEEE Transaction on Industrial Electronics*, vol. 61, no. 4, pp. 1735–1749, April 2014.

13. Yijie Wang, Yueshi Guan, Xinyu Liang, Wei Wang, and Dianguo Xu, "Two stage led stret light system based on a novel single stage AC-DC converter", *IET Power Electronics*, vol. 7, no. 6, pp. 1374–1384, June 2014.

14. Majid Pahlevaninezhad, Pritam Das, Josef Drobnik, Praveen K. Jain, and Alireza Bakhshai, "A control approach based on the differential flatness theory for an AC-DC converter used in electric vehicle", *IEEE Transactions on Power Electronics*, vol. 27, no. 4, pp. 2085–2103, April 2012.

15. J. Duarte, L. Ricardo Lima, L. Oliveira, L. Michels, C. Rech, and M. Mezaroba, "Single stage high power factor step-up/step down isolated AC-DC converter", *IET Power Electronics*, vol. 5, no. 8, pp. 1351–1358, September 2012.

16. Hugo Santos Ribeiro and Beatriz Vieira Borges, "High performance voltage fed AC-DC full bridge single stage PFC with reduced DC bus capacitor", *IEEE Transactions on Power Electronics*, vol. 29, no. 6, pp. 2680–2692, June 2014.

17. N. D. Dao and D. Lee, "Modulation of bidirectional AC/DC converters based on half-bridge direct-matrix structure," *IEEE Transactions on Power Electronics*, vol. 35, no. 12, pp. 12657–12662, Dec. 2020.

18. Yi-hung Liao, "A novel reduced switching loss bidirectional AC-DC converter PWM strategy with feedforward control for grid-tied microgrid systems", *IEEE Transactions on Power Electronics*, vol. 29, no. 3, pp. 1500–1513, March 2014.

19. R. I. Bojoi, L. R. Limongi, D. Roiu, and A. Tenconi, "Enhanced power quality control strategy for single-phase inverters in distributed generation systems", *IEEE Transactions on Power Electronics*, vol. 26, no. 3, pp. 798–806, March 2011.

20. Abbas A. Fardoun, Esam H. Ismail, Nasrullah M.Khraim, Ahmad J. Sabzali, and Mustafa A. Al-Saffar, "Bridgeless high powerfactor buck converter operating in discontinues capacitor voltage mode", *IEEE Transaction on Industry Applications*, vol. 50, no. 5, pp. 3457–3467, October 2014.

21. Junjun Deng, Siqi Li, Sideng Hu, Chunting Chris Mi, and Ruiqing Ma, "Design methodology of LLC resonant converters for electric vehicle battery chargers", *IEEE Transactions on Vehicular Technology*, vol. 63, no. 4, pp. 1581–1592, May 2014.

22. Fariborz Musavi, Wilson Eberle, and William G. Dunford, "A high performance single phase bridgeless interleaved PFC converter for plug-in hybrid electric vehicle chargers", *IEEE Transactions on Industry Applications*, vol. 47, no. 4, pp. 1833–1843, August 2011.

23. X. Xie, C. Zhao, L. Zheng, and S. Liu, "An improved Buck PFC converter with high power factor", *IEEE Transactions on Power Electronics*, vol. 28, no. 5, pp. 2277–2284, May 2013, doi: 10.1109/TPEL.2012.2214060.

24. Kan Zhou and Lin Cai, "Randomized PHEV charging under distribution grid constraints", *IEEE Transactions on Smart Grid*, vol. 5, no. 2, pp. 879–887, March 2014.

25. Mansour Tabari and Amirnaser Yazdani, "Stability of a DC distribution system for power system integration of plug in electric vehicles", *IEEE Transactions on Smart Grid*, vol. 5, no. 5, pp. 2564–2573, September 2014.

26. C. C. Lin, L. S. Yang, and G. W. Wu, "Study of a non-isolated bidirectional DC-DC converter", *IET Power Electronics*, vol. 6, no. 1, pp. 30–37, January 2013.

27. H. Ardi, R. Reza Ahrabi, and S. Najafi Ravadanegh, "Non-isolated bidirectional DC–DC converter analysis and implementation", *IET Power Electronics*, vol. 7, no. 12, pp. 3033–3044, 12 2014.

28. A. K. Rathore, D. R. Patil, and D. Srinivasan, "Non-isolated bidirectional soft-switching current-Fed LCL resonant DC-DC converter to interface energy storage in DC microgrid", *IEEE Transactions on Industry Applications*, vol. 52, no. 2, pp. 1711–1722, March-April 2016.

29. L. Kumar and S. Jain, "Multiple-input DC-DC converter topology for hybrid energy system", *IET Power Electronics*, vol. 6, no. 8, pp. 1483–1501, September 2013.

30. S. Dusmez, A. Khaligh, and A. Hasanzadeh, "A zero-voltage-transition bidirectional DC-DC converter", *IEEE Transactions on Industrial Electronics*, vol. 62, no. 5, pp. 3152–3162, May 2015.

31. R. M. Schupbach and J. C. Balda, "Comparing DC-DC converters for power management in hybrid electric vehicles," *Electric Machines and Drives Conference, 2003. IEMDC'03. IEEE International*, 2003, pp. 1369–1374, vol.3.

32. S. Waffler, M. Preindl and J. W. Kolar, "Multi-objective optimization and comparative evaluation of Si soft-switched and SiC hard-switched automotive DC-DC converters," *Industrial Electronics, 2009. IECON '09. 35th Annual Conference of IEEE*, Porto, 2009, pp. 3814–3821.

33. K. Zhiguo, Z. Chunbo, Y. Shiyan and C. Shukang, "Study of bidirectional DC-DC converter for power management in electric bus with supercapacitors," *2006 IEEE Vehicle Power and Propulsion Conference*, Windsor, 2006, pp. 1–5.

34. L. Wang, D. Zhang, Y. Wang, B. Wu, and H. S. Athab, "Power and voltage balance control of a novel three-phase solid-state transformer using multilevel cascaded H-bridge inverters for microgrid applications", *IEEE Transactions on Power Electronics*, vol. 31, no. 4, pp. 3289–3301, April 2016.

35. R. T. Naayagi, A. J. Forsyth, and R. Shuttleworth, "Bidirectional control of a dual active bridge DC-DC converter for aerospace applications", *IET Power Electronics*, vol. 5, no. 7, pp. 1104–1118, August 2012.

36. B. Zhao, Q. Song, W. Liu, and Y. Sun, "Overview of dual-active-bridge isolated bidirectional DC–DC converter for high-frequency-link power-conversion system", *IEEE Transactions on Power Electronics*, vol. 29, no. 8, pp. 4091–4106, August 2014.

37. K. Filsoof and P. W. Lehn, "A bidirectional multiple-input multiple-output modular multilevel DC–DC converter and its control design", *IEEE Transactions on Power Electronics*, vol. 31, no. 4, pp. 2767–2779, April 2016.

38. B. Zhao, Q. Song, W. Liu, and Y. Sun, "A synthetic discrete design methodology of high-frequency isolated bidirectional DC-DC converter for grid-connected battery energy storage system using advanced components", *IEEE Transactions on Industrial Electronics*, vol. 61, no. 10, pp. 5402–5410, October 2014.

39. Z. U. Zahid, Z. M. Dalala, R. Chen, B. Chen, and J. S. Lai, "Design of bidirectional DC–DC resonant converter for vehicle-to-grid (V2G) applications", *IEEE Transactions on Transportation Electrification*, vol. 1, no. 3, pp. 232–244, October 2015.

40. P. Xuewei and A. K. Rathore, "Novel bidirectional snubberless naturally commutated soft-switching current-fed full-bridge isolated DC-DC converter for fuel cell vehicles", *IEEE Transactions on Industrial Electronics*, vol. 61, no. 5, pp. 2307–2315, May 2014.

41. C. J. Shin and J. Y. Lee, "An electrolytic capacitor-less bi-directional EV on-board charger using harmonic modulation technique", *IEEE Transactions on Power Electronics*, vol. 29, no. 10, pp. 5195–5203, October 2014.

42. C. Li *et al.*, "Design and Implementation of a bidirectional isolated Ćuk converter for Low-voltage and high-current automotive DC source applications", *IEEE Transactions on Vehicular Technology*, vol. 63, no. 6, pp. 2567–2577, July 2014.

43. T. J. Liang and J. H. Lee, "Novel high-conversion-ratio high-efficiency isolated bidirectional DC–DC converter,, *IEEE Transactions on Industrial Electronics*, vol. 62, no. 7, pp. 4492–4503, July 2015.

44. L. Wang, Z. Wang, and H. Li, "Asymmetrical duty cycle control and decoupled power flow design of a three-port bidirectional DC-DC converter for fuel cell vehicle application", *IEEE Transactions on Power Electronics*, vol. 27, no. 2, pp. 891–904, February 2012.

45. F. Z. Peng, Hui Li, Gui-Jia Su and J. S. Lawler, "A new ZVS bidirectional DC-DC converter for fuel cell and battery application", *IEEE Transactions on Power Electronics*, vol. 19, no. 1, pp. 54–65, January 2004.

46. S. Inoue and H. Akagi, "A bidirectional isolated DC–DC converter as a core circuit of the next-generation medium-voltage power conversion system", *IEEE Transactions on Power Electronics*, vol. 22, no. 2, pp. 535–542, March 2007.

47. L. Zhu, "A novel soft-commutating isolated boost full-bridge ZVS-PWM DC-DC converter for bidirectional high power applications," *Power Electronics Specialists Conference, 2004. PESC 04. 2004 IEEE 35th Annual*, 2004, pp. 2141–2146, Vol. 3.

48. H. Tao, A. Kotsopoulos, J. L. Duarte, and M. A. M. Hendrix, "Transformer-coupled multiport ZVS bidirectional DC–DC converter with wide input range", *IEEE Transactions on Power Electronics*, vol. 23, no. 2, pp. 771–781, March 2008.

49. F. Krismer and J. W. Kolar, "Efficiency-optimized high-current dual active bridge converter for automotive applications" *IEEE Transactions on Industrial Electronics*, vol. 59, no. 7, pp. 2745–2760, July 2012.

50. M. Aamir, K. A. Kalwar, and S. Mekhilef, "Review: uninterruptible power supply (UPS) system", *Renewable and Sustainable Energy Reviews*, vol. 58, pp. 1395–1410, May 2016.

51. S. Abdel-Rahman. (2013, Jan.). CCM PFC boost converter design. Design Note. DN 2013–01. Infineon Technologies. [Online]. Available: http://www.mouser.ec/pdfdocs/2-7.pdf.

52. Noon, P. James, "Designing high-power factor off-line power supplies." *Texas Instruments Power Supply Design Seminar*, 2003, Vol. 1500.

53. Texas Instruments Incorporated (2017, Nov.). Voltage source inverter design guide, Dallas, Texas. [Online]. Available: http://www.ti.com/lit/ug/tiduay6c/tiduay6c.pdf.

54. S. B. Dewan and P. D. Ziogas, "Optimum filter design for a single-phase solid-state UPS system", *IEEE Transactions on Industry Applications*, vol. IA-15, no. 6, pp. 664–669, Nov. 1979.

55. J. Kim, J. Choi and H. Hong, "Output LC filter design of voltage source inverter considering the performance of controller," PowerCon 2000. *2000 International Conference on Power System Technology Proceedings (Cat. No.00EX409)*, Perth, WA, 2000, pp. 1659–1664, Vol. 3.

56. Texas Instruments Incorporated (2015, Aug.). Basic Calculation of a Buck Converter's Power Stage, Dallas, Texas. [Online]. Available: http://www.ti.com/lit/an/slva477b/slva477b.pdf.

57. J. Ejury. (2013, Jan.). Buck converter design. Infineon Design Note DN. 2013-01. Infineon Technologies. [Online]. Available: http://www4.hcmut.edu.vn/~ndtuyen/Dowload/BuckConverterDesignNote.pdf.

6 Advanced and Comprehensive Control Methods in Wind Energy Systems

Md. Moinul Islam
System Development – Transmission Planning, Electric
Reliability Council of Texas

Dhiman Chowdhury
Department of Electrical Engineering,
University of South Carolina

CONTENTS

6.1 INTRODUCTION

Renewable energy solutions like solar power access from photovoltaic modules, energy harness from wind turbines etc. conform to sustainable means of reliable and clean electrical energy across the world. The integration of wind energy conversion systems (WECSs) into the electrical grid is increasing rapidly as they are economically feasible, environmentally clean, and are safe renewable power sources compared to conventional coal and nuclear power plants. However, WECSs are considered as fluctuating power sources due to uncertain nature of the wind, wind shear, and tower shadow effect. Due to intermittent nature of the WECSs, they introduce mechanical oscillations and large torque ripples that have negative impacts on the grid. Therefore, it is necessary to control the WECSs to minimize the power fluctuations and capture the maximum power out of various wind speeds. Regarding this, a comprehensive discussion of various control techniques such as pitch control method, inertial control, direct current vector control, sliding mode control, model predictive control system, and coordinated control of WECSs is presented in this chapter.

State-of-the-art WECSs can be realized as fixed-speed and variable-speed types. In a fixed-speed WECS, wind turbine always spins at the same rotor/generator speed during the entire operation disregarding the wind speed (or any change in wind speed). Therefore, the tip-speed ratio changes with wind speed and the rotor aerodynamic behavior is optimal at a referenced speed. In case of variable-speed WECS, wind turbine spins in proportion with the changes in wind speed, thus the rotor/generator speed is allowed to vary accordingly. As a result, a constant tip-speed ratio is maintained and optimal aerodynamic behavior is obtained for the complete range of wind speed variation – between cut-in and rated (nominal) speed during operation. However, above the nominal speed the rotor/generator speed is kept constant. Variable-speed WECS needs to be controlled actively for effective power and torque generation and maintenance.

As an active control method designed for variable-speed WECS, pitch control implies regulation of the blade angle in regard to execute controlled aerodynamic power extraction from wind. Inertial control method takes into account the physical properties and locomotion characteristics of the wind turbine blade to utilize its kinetic energy and inertia for power regulation. Direct current vector control method is a more developed and effective technique than the indirect current control to have improved regulation of maximum wind power extraction, system reactive power and grid voltage support. Nevertheless, more advanced and complex control algorithms are designed and implemented in WECS frameworks which are capable of dealing with complex performance optimization constraints, highly nonlinear perturbations, dynamic uncertainties during wind turbine operation, and overall system disturbances. There are a number of novel control strategies proposed and applied by the researchers and practitioners such as sliding mode control, model reference adaptive controller, fuzzy-logic based mechanism, modal analysis of dynamics, feedback linearization control, genetic algorithms, particle swarm optimizer, model predictive control, multiple model predictive controller, combined active and reactive power control, coordinated control etc. Among these, in this chapter, fundamental concepts of sliding mode, model predictive and coordinated control schemes are articulated.

6.2   WIND ENERGY CONVERSION SYSTEMS

The mechanical power output from the wind turbine can be expressed as follows:

$$P_w = \frac{1}{2}\rho V_w^3 \pi R^2 C_P(\lambda,\beta) \qquad (6.1)$$

where, P_w is the mechanical power extraction from the wind turbine, ρ is the air density (kg/m^3), V_w is the wind speed (m/s), R is the radius of the wind turbine blade (m), $C_P(\lambda, \beta)$ is the power coefficient which is a function of the tip speed ratio λ and blade pitch angle β ($^\circ$) and is defined as:

$$\lambda = \frac{\omega_m R}{V_w} \qquad (6.2)$$

$$C_P(\lambda,\beta) = 0.645\left\{0.000912\lambda + \frac{-5 - 0.4(2.5 + \beta) + 116\lambda_i}{e^{21\lambda_i}}\right\} \qquad (6.3)$$

$$\lambda_i = \frac{1}{\lambda + 0.08(2.5 + \beta)} - \frac{0.035}{1 + (2.5 + \beta)^3} \qquad (6.4)$$

where, ω_m is the rotational speed of the wind turbine blade (rad/s).

The plot of wind power vs. rotational speed is incorporated in Figure 6.1 for different wind speed. Figure 6.1 is obtained using power coefficient curve for pitch angle $\beta = 8^\circ$ in Figure 6.2. Figure 6.1 demonstrates that the maximum power point changes owing to the change in rotational speed at various wind speed. Therefore, it is necessary to track the maximum power point at variable wind speed. This is achieved by the wind turbine controller that decreases the pitch angle to extract more aerodynamic power if wind speed is below the rated speed and increases the pitch angle above rated speed to reduce aerodynamic power as illustrated in Figure 6.2. Thus, regulating the pitch angle wind turbine controller minimizes the power fluctuations caused by the variable wind speed. The characteristics of wind power with respect to wind speed are demonstrated in Figure 6.3. As shown in the figure, the output power of WECS is approximately zero below cut-in wind speed (5 m/s), and below rated wind speed, wind turbine controller tracks the maximum power point shown in Figure 6.1. The wind turbine operates at rated power above the rated wind speed, however, once wind speed reaches cut-out wind speed (25 m/s), WECS is disconnected form the system to prevent mechanical damage.

6.2.1   FIXED-SPEED WIND ENERGY CONVERSION SYSTEMS

In fixed-speed WECSs, wind turbine generator is connected to the grid through a soft-starter and a transformer as shown in Figure 6.4. A squirrel cage induction generator (SCIG) is solely used in a fixed-speed WECS where rotational speed of the

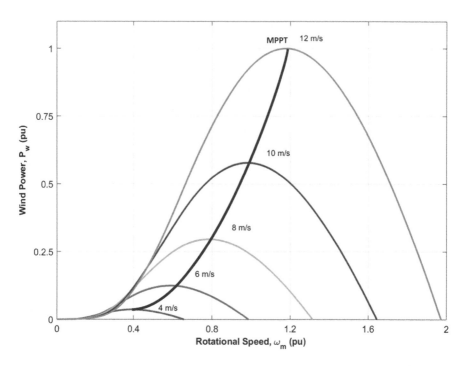

FIGURE 6.1   Wind power vs. rotational speed characteristics at various wind speeds.

generator is determined by the gear ratio, grid frequency and number of the stator poles. For example, a four pole SCIG connected to 60 Hz system rotates at:

$$Ns = \frac{120 * f}{P} = \frac{120 * 60}{4} = 1800 \ rpm \tag{6.5}$$

As the wind speed changes, rotational speed of the fixed-speed WECS system varies within 1% of the rated speed. Since the rotational speed varies within a small range, this type of WECS is referred to fixed-speed WECS. A gearbox is utilized to adjust the speed difference between wind turbine and the generator so that generator can output rated power at the rated speed of the generator. During the startup of the SCIG, a high inrush current is produced which in turn causes the grid voltage to drop. Therefore, a soft-starter is employed to minimize the inrush current during the start-up. Once generator reaches the rated speed, soft-starter is bypassed by a switch, and a three-phase capacitor bank is connected at the SCIG terminal to minimize the reactive power drawn from the grid by the generator.

A fixed-speed WECS is less expensive. However, as the wind speed changes, power delivered by the WECS fluctuates and disturbances are introduced into the grid. Also, in SCIG, generator is locked to the power system frequency and cannot speed up when there is a change in wind speed. Therefore, a large amount of force is developed that causes mechanical damage to the system when power grid pushes back. Therefore, this type of WECS requires a strong and solid

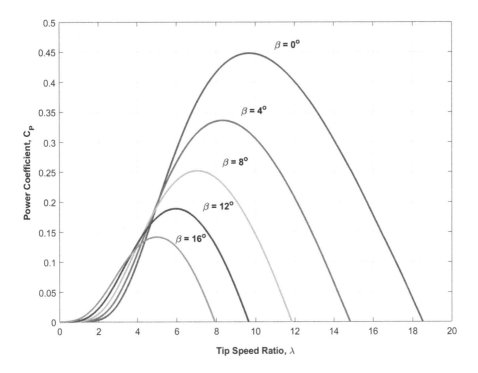

FIGURE 6.2   Power coefficient vs. tip speed ratio for different values of pitch angle.

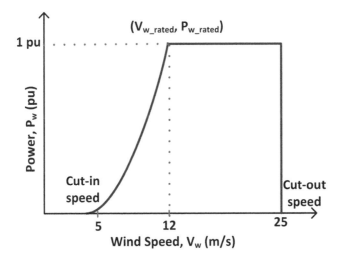

FIGURE 6.3   Typical characteristics of wind power as wind speed changes.

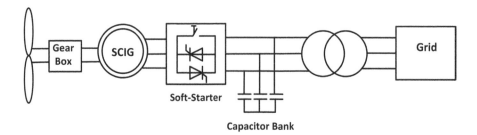

FIGURE 6.4 Integration of fixed-speed squirrel cage induction generator wind energy conversion system into the grid.

built mechanical design to absorb the stresses due to fluctuating wind speed. Furthermore, it has low energy conversion efficiency as rated power is delivered to the grid at a fixed speed of the wind turbine, leading to low energy conversion at other wind speed.

6.2.2   Variable-Speed Wind Energy Conversion Systems

In variable-speed WECSs, rotational speed of the generator changes with the wind speed to track the maximum power point. A frequency converter is utilized to adjust the frequency to the desired grid frequency. Compared to the fixed-speed WECSs, variable-speed WECSs have a number of advantages. For example, it increases the energy efficiency as it captures more power at variable wind speed. Mechanical stress caused by the wind gusts is reduced due to adaptable speed capability of the generator in accordance with the wind speed. Also, variable-speed WECS allows the construction of larger wind turbines. The major drawback of the variable-speed WECS is it requires a power converter to control the rotational speed of the generator which makes it more expensive and complex than the fixed-speed WECS. Nonetheless, it is possible to control the grid-side active and reactive power since power converter decouples the grid from the generator.

6.2.2.1   Wound Rotor Inductor Generator

The basic difference between SCIG and wound rotor induction generator (WRIG) is that in SCIG rotor circuit is not accessible externally. Therefore, induced current in the rotor circuit, which is accountable for torque generation, is stringently a function of the slip speed. Whereas, in WRIG as shown in Figure 6.5, variable resistance is connected to the rotor circuit which is electrically accessible via slip rings on the mechanical shaft. Therefore, changing the rotor resistance by a power converter, it is possible to adjust the rotor current, hence, the electromagnetic torque of the generator. Thus, the change in the rotor resistance changes the torque/speed characteristics and enables variable speed operation of the WRIG. The range of variable speed operation is usually limited to 10% approximately above the synchronous speed. This type of WECSs can capture more power due to variable speed operation; however, power losses in the rotor resistance of WRIG make them less attractive compared to other types of WECSs. Like SCIG, variable-speed WRIG WECSs also

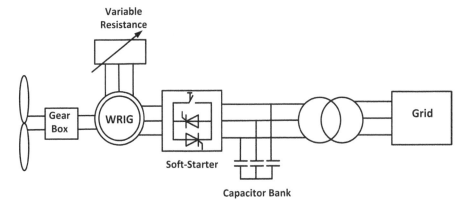

FIGURE 6.5 Integration of variable-speed wound rotor inductor generator wind energy conversion system into the grid.

require a soft-starter and a capacitor bank for reactive power compensation as incorporated in Figure 6.5.

6.2.2.2 Doubly-Fed Induction Generator

The most commonly used variable-speed WECS is the doubly fed induction generator (DFIG) which has a number of advantages over traditional induction generator. A typical configuration of the DFIG WECS is shown in Figure 6.6. The configuration of DFIG WECS is similar to the WRIG except that in DFIG rotor circuit is replaced by a voltage source converter (VSC) and it doesn't require the soft-starter and capacitor bank for reactive power compensation. In this configuration, stator is directly connected to the grid, whereas, rotor is connected to the grid through a back-to-back voltage source converter (VSC). The VSC controls the rotor circuit and the induction generator is able to deliver or absorb reactive power from the grid. Therefore,

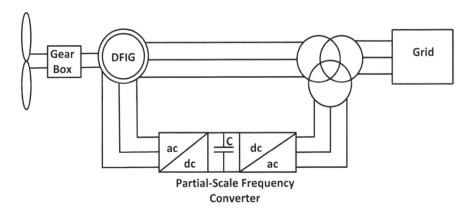

FIGURE 6.6 Integration of variable-speed doubly fed induction generator wind energy conversion system into the grid through a partial-scale frequency converter.

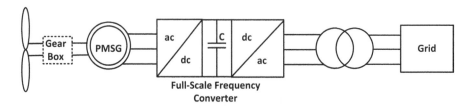

FIGURE 6.7 Integration of variable-speed permanent magnet synchronous generator wind energy conversion system into the grid through a full-scale frequency converter.

DFIG can provide grid voltage support during an extreme disturbance event, for example enhances low-voltage ride-through (LVRT) capability during a power system fault. In addition, as wind speed changes, DFIG can remain synchronized to the grid controlling the rotor voltage and current by the VSC. DFIG is also more cost effective than other variable-speed WECSs, since only a fraction of the mechanical power, usually 25–30% is delivered to the grid though the VSC, and remaining power is delivered to the grid through stator. Therefore, smaller VSC is necessary that reduces the cost of DFIG WECS.

6.2.2.3   Permanent Magnet Synchronous Generator (PMSG)

In this configuration, a wind turbine generator is connected to the grid through a full-scale frequency converter as shown in Figure 6.7. Either squirrel cage induction generator (SCIG), wound rotor synchronous generator (WRSG) or permanent magnet synchronous generator (PMSG) can be utilized in this kind of WECSs with a power rating up to several megawatts. The generator is decoupled from the grid via power converter and can operate in full speed range. The converter power rating is usually same as the generator power rating. Power converter controls both active and reactive power for smooth grid connection. The major drawback of this type of WECS is complexity and higher cost compared to other types of WECSs owing to large power converter. In this configuration, the need for the gear box can be eliminated with a large number of poles to improve the efficiency and reduce the initial cost and maintenance. However, as the number of the pole increases, generator is required to be designed with larger diameter that increases the cost of the generator and installation.

6.3   PITCH CONTROL METHOD

The integration of a variable-speed WECS to the electrical grid is shown in Figure 6.8. The wind turbine has a pitchable blade (indicated by dashed line) to control the aerodynamic power extracted from the wind, and is connected to the generator through a low-speed shaft, a high-speed shaft and a mechanical component such as gear box in between them. The low-speed shaft is propelled by the aerodynamic power which in turn rotates high-speed shaft of the generator.

The block diagram of the WECS with pitch controller is illustrated in Figure 6.9 where P_g^* is the reference power and P_g is the power delivered to the grid from the

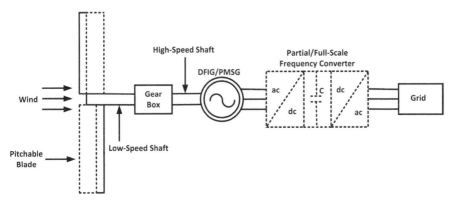

FIGURE 6.8 Integration of variable-speed wound rotor inductor generator wind energy conversion system into the grid through a full-scale frequency converter.

generator. The difference in power is measured by the pitch control system that generates pitch angle command β_{cmd} for pitch actuator. Pitch actuator is basically a hydraulic servo system or electromechanical device that rotates the wind turbine blade. The pitch control system operates in four different regions in accordance with Figure 6.2 as follows:

- Below cut-in wind speed (5 m/s), pitch angle $\beta = 90°$, therefore, output power of the generator $P_g = 0$ (ignoring losses).
- In between cut-in and rated wind speed, pitch angle varies from 0° to 10° to track the maximum power according to maximum power point tracking (MPPT) curve provided in Figure 6.2.
- In between rated and cut-out wind speed pitch angle varies between 10° to 90° to maintain rated output power of the generator. If P_g decreases, pitch angle decreases to increase the mechanical power output from the turbine, and vice-versa.
- Above cut-out wind speed (25 m/s), pitch angle $\beta = 90°$, therefore, output power of the generator $P_g = 0$ (ignoring losses).

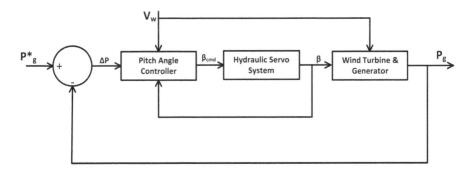

FIGURE 6.9 Variable-speed WECS with pitch angle controller.

6.4 INERTIAL CONTROL METHOD

Like conventional generators, wind turbine has large amount of kinetic energy stored in the rotating mass of its blade. The kinetic energy stored in a rotating mass, therefore, a wind turbine is given by:

$$E = \frac{1}{2} j\omega^2 \tag{6.6}$$

where, j and ω are the inertia and rotational speed of the wind turbine, respectively. From definition of energy in electrical power engineering, E can be expressed as:

$$E = St = SH \tag{6.7}$$

where, S is the nominal apparent power and H is the inertia constant of wind turbine. Inertia constant H has time dimension and provides an indication of how long a wind turbine can deliver power to the grid utilizing its stored kinetic energy. The inertia constant for large power plants typically lies between 2 and 9 second and for smaller wind turbines it is approximately 2–6 second. In case of variable-speed WECS, this energy does not provide grid frequency support since rotational speed of the wind turbine is separated from the grid through power electronic converter. Therefore, an additional control system is needed to utilize this kinetic energy for grid frequency support.

The block diagram of wind turbine inertial control with primary frequency support is provided in Figure 6.10. The controller has three parts: Upper part of the controller controls the mechanical speed of the rotor to operate at the optimal tip speed ratio for maximum power tracking as illustrated in Figure 6.2. Based on the difference between measured rotational speed ω_m and reference rotational speed ω_m^*, it generates power reference point for the rotor side converter (RSC). Converter in turn controls the generator currents based on the power set point.

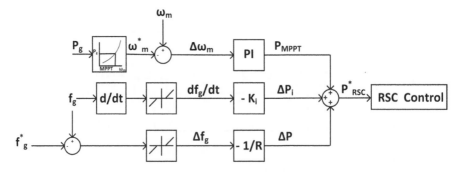

FIGURE 6.10 Block diagram of wind turbine inertial control with primary frequency support.

Second part of the controller represents two additional control loops known as rate of change of frequency (ROCOF) loop and droop loop that adjust the power set point based on the rate of change of grid frequency and grid frequency deviation that are proportional to the controller constants $\frac{1}{R}$ and K_f, respectively. Limiters are used to keep the ROCOF and frequency deviation within acceptable limits.

As observed controller block diagram in Figure 6.10, reference power P_{RSC}^* for RSC controller consists of three terms: P_{MMPT}, ΔP_i and ΔP, where P_{MMPT} is utilized for maximum power point tracking (MPPT), ΔP_i and ΔP are output of the ROCOF loop and droop loop, respectively that regulate the P_{RSC}^*. The droop controller is activated if grid frequency f_g exceeds the certain limit with respect to the reference frequency f_g^*.

The output ΔP_i and ΔP of the ROCOF controller and droop controller are expressed as:

$$\Delta P_i = -\frac{1}{R}\frac{df_g}{dt} \tag{6.8}$$

$$\Delta P = -K_f \Delta f = -K_f (f_g - f_g^*) \tag{6.9}$$

Prior to the disturbance, ROCOF and droop loops are deactivated, therefore,

$$\Delta P_i = \Delta P = 0 \tag{6.10}$$

And, only P_{MMPT} is utilized as reference power for the RSC controller, thus:

$$P_{RSC}^* = P_{MPPT} \tag{6.11}$$

Figure 6.11 shows the simulation results with and without frequency support controller following a loss of a synchronous generator in the simulated system that consists of 16.5 MW of aggregated DFIG WECS, two synchronous generators of 20 MW and 80 MW, and a 120 MW load. Note that in the beginning of a disturbance, ROCOF controller dominates, whether at frequency nadir droop controller dominates to provide frequency support for the grid. As illustrated in the figure, due to the loss of 120 MW synchronous generator, grid inertia decreases and frequency drops. As the frequency drops below 49.9 Hz, droop controller is activated resulting in higher frequency (solid line) than the frequency response (dashed line) without the controller. During the frequency drop, kinetic energy is released from the wind turbine and more power is delivered for grid frequency support. As wind turbine releases kinetic energy, rotational speed of the turbine decreases. At $t = 25$ s, system reaches in steady state and wind turbine operates under normal condition. However, power drops below initial value since wind turbine is not at its optimal speed anymore due to transfer of aerodynamic power into the grid, and additional power is required to bring it up to the optimal speed. Therefore, frequency is lower than the frequency response (solid line) without the controller (dashed line) as normal operation resumes at $t = 25$ s.

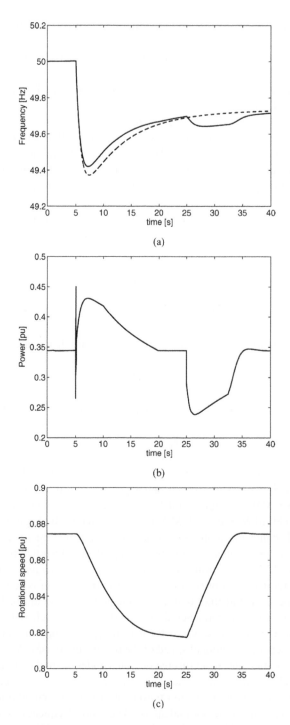

FIGURE 6.11 Response of the grid frequency, power output, and rotational speed of the WECS.

6.5   DIRECT CURRENT VECTOR CONTROL METHOD OF DFIG

DFIG WECS has mainly three control parts: rotor side converter (RSC) controller, grid side converter (GSC) controller and wind turbine controller as incorporated in Figure 6.12. The primary functions of these controllers in DFIG WECS are:

- The RSC controller either extracts maximum power from the wind turbine or complies with the wind turbine generator control demand.
- The GSC controller keeps the DC link voltage constant through exchanging power with the grid and regulates reactive power absorbed from the grid to maintain constant grid voltage.
- The wind turbine controller consists of speed controller and power controller. If wind speed is below the rated speed, speed controller provides power reference for the RSC to track the maximum power point. Whereas, at high wind speed, power controller increases or decreases the pitch angle to maintain the rated power of the wind turbine. The reference power for each wind turbine is sent from the SCADA system, and wind turbine controller makes sure that the reference power level is achieved.

6.5.1   GSC Controller

GSC has a DC link capacitor and connected to the grid through six-pulse voltage source converter (VSC) and grid filter resistance R_g and inductance L_g as illustrated

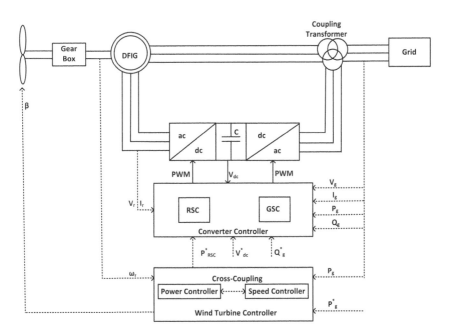

FIGURE 6.12 DFIG WECS with wind turbine controller, rotor side converter (RSC) controller, and grid side converter (GSC) controller.

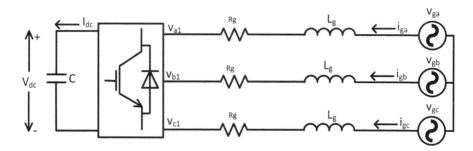

FIGURE 6.13 One line diagram of DC link capacitor and rotor side converter (RSC) connected to the grid.

in Figure 6.13. The equivalent single-phase circuit is provided in Figure 6.14 where v_{ga}, i_{ga}, and v_{a1} are the grid voltage, current and GSC output voltage, respectively. Applying KVL, the grid voltage v_g is expressed as

$$v_g = R_g i_g + L_g \frac{di_g}{dt} + v_1 \tag{6.12}$$

The voltage and current signals in Equation (6.12) are time varying. To obtain time-invariant differential equations, Park's transformation is employed to convert time-varying signals from abc to dq0 reference frame. For example, dq0 components of the grid voltage are obtained as follows:

$$\begin{bmatrix} v_{gd} \\ v_{gq} \\ v_{g0} \end{bmatrix} = \frac{2}{3} \begin{bmatrix} cos\theta & \cos\left(\theta - \frac{2\pi}{3}\right) & \cos\left(\theta + \frac{2\pi}{3}\right) \\ -sin\theta & -\sin\left(\theta - \frac{2\pi}{3}\right) & -\sin\left(\theta + \frac{2\pi}{3}\right) \\ \frac{1}{2} & \frac{1}{2} & \frac{1}{2} \end{bmatrix} \begin{bmatrix} v_{ga} \\ v_{gb} \\ v_{gc} \end{bmatrix} \tag{6.13}$$

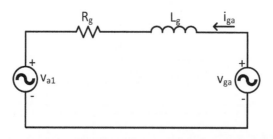

FIGURE 6.14 Equivalent single-phase circuit of grid side converter connected to the grid.

$$= \frac{2}{3} T_\theta \begin{bmatrix} V_{ga} \\ V_{gb} \\ V_{gc} \end{bmatrix} \tag{6.14}$$

The inverse transformation of the grid voltage from dq0 to abc can be derived as

$$\begin{bmatrix} v_{ga} \\ v_{gb} \\ v_{gc} \end{bmatrix} = \begin{bmatrix} \cos\theta & -\sin\theta & 1 \\ \cos\left(\theta - \dfrac{2\pi}{3}\right) & -\sin\left(\theta - \dfrac{2\pi}{3}\right) & 1 \\ \cos\left(\theta + \dfrac{2\pi}{3}\right) & -\sin\left(\theta + \dfrac{2\pi}{3}\right) & 1 \end{bmatrix} \begin{bmatrix} v_{gd} \\ v_{gq} \\ V_{g0} \end{bmatrix} \tag{6.15}$$

$$= T_\theta^{-1} \begin{bmatrix} v_{gd} \\ v_{gq} \\ V_{g0} \end{bmatrix} \tag{6.16}$$

where, V_{gd}, V_{gq}, and V_{g0} are d-axis, q-axis, and zero sequence components of grid voltage, respectively; T_θ is the transformation matrix, and θ is the transformation angle which is basically phase angle difference between fixed abc axis and rotating dq axis as shown in Figure 6.15; and ω is the synchronous speed of the dq reference frame. The transformation angle θ is obtained via a phase locked loop (PLL).

In case of voltage source converter, the synchronous reference frame is aligned with the grid voltage. Therefore, q-axis component of the grid voltage becomes zero and d-axis component refers to the magnitude of grid voltage. Since, zero sequence component of the grid voltage $v_{g0} = \frac{1}{3}\left(v_{ga} + v_{gb} + v_{gc}\right) = 0$ for a balanced three-phase system, Equation (6.13) can be rewritten in per unit (pu) as follows:

$$\begin{bmatrix} v_{gd} \\ v_{gq} \end{bmatrix} = \begin{bmatrix} \cos\theta & \cos\left(\theta - \dfrac{2\pi}{3}\right) & \cos\left(\theta + \dfrac{2\pi}{3}\right) \\ -\sin\theta & -\sin\left(\theta - \dfrac{2\pi}{3}\right) & -\sin\left(\theta + \dfrac{2\pi}{3}\right) \end{bmatrix} \begin{bmatrix} v_{ga} \\ v_{gb} \\ v_{gc} \end{bmatrix} \tag{6.17}$$

In the same manner, dq transformation of the grid current can be derived, and Equation (6.12) can be expressed in dq reference frame as follows:

$$\begin{bmatrix} v_{gd} \\ v_{gq} \end{bmatrix} = R_g \begin{bmatrix} i_{gd} \\ i_{gq} \end{bmatrix} + L_g \frac{d}{dt}\begin{bmatrix} i_{gd} \\ i_{gq} \end{bmatrix} + j\omega L_g \begin{bmatrix} -i_{gq} \\ i_{gd} \end{bmatrix} + \begin{bmatrix} v_{1d} \\ v_{1q} \end{bmatrix} \tag{6.18}$$

$$v_{gdq} = R_g i_{gdq} + L_g \frac{d}{dt} i_{gdq} + j\omega L_g i_{gdq} + v_{1dq} \tag{6.19}$$

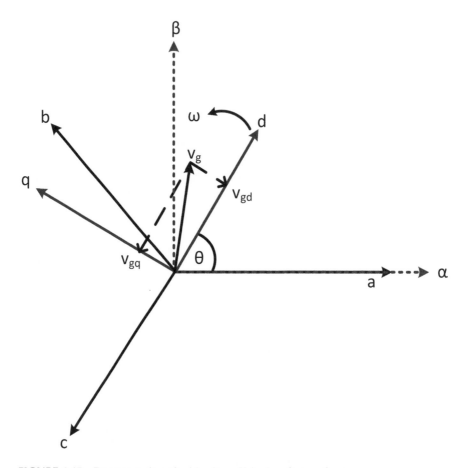

FIGURE 6.15   Representation of grid voltage V_g in dq reference frame.

Ignoring the time-varying component, it can be rewritten for steady-state condition as:

$$v_{gdq} = R_g i_{gdq} + j\omega L_g i_{gdq} + v_{1dq} \tag{6.20}$$

$$v_{gd} + jv_{gq} = R_g i_{gdq} + j\omega L_g i_{gdq} + v_{1d} + jv_{1q} \tag{6.21}$$

$$v_{gd} = R_g i_{gdq} + j\omega L_g i_{gdq} + v_{1d} + jv_{1q}[v_{gq} = 0] \tag{6.22}$$

$$i_{gdq} = \frac{v_{gd} - v_{1d}}{j\omega L_g} - \frac{v_{1q}}{\omega L_g} \tag{6.23}$$

$$i_{gd} - ji_{gq} = \frac{v_{gd} - v_{1d}}{j\omega L_g} - \frac{v_{1q}}{\omega L_g} \tag{6.24}$$

$$i_{gd} - ji_{gq} = -j\frac{v_{gd} - v_{1d}}{\omega L_g} - \frac{v_{1q}}{\omega L_g} \tag{6.25}$$

Equating real and imaginary parts of Equation (6.25):

$$i_{gd} = -\frac{v_{1q}}{\omega L_g} \tag{6.26}$$

$$i_{gq} = -\frac{v_{gd} - v_{1d}}{\omega L_g} \tag{6.27}$$

In dq reference frame, the instantaneous active and reactive power absorbed by the GSC from the grid are expressed as:

$$p_g = \left(v_{gd} i_{gd} + v_{gq} i_{gq} \right) \tag{6.28}$$

$$q_g = \left(v_{gq} i_{gd} - v_{gd} i_{gq} \right) \tag{6.29}$$

In Equations (6.28)–(6.29), it is observed that both active and reactive power depend on the d-axis and q-axis quantities, therefore, it is not possible to control them independently. For independent control, the d-axis and q-axis quantities require to be decoupled. The decoupling of d-axis and q-axis quantities is achieved by aligning the grid voltage V_g with d-axis in Figure 6.15 as discussed earlier. Therefore, the active and reactive power in Equations (6.28) and (6.29) are expressed in voltage-oriented reference frame as:

$$p = v_{gd} i_{gd} [v_{gq} = 0] \tag{6.30}$$

$$= -v_{gd} \frac{v_{1q}}{\omega L_g} \tag{6.31}$$

$$q = -v_{gd} I_{gq} \, [v_{gq} = 0] \tag{6.32}$$

$$= v_{gd} \frac{v_{gd} - v_{1d}}{\omega L_g} \tag{6.33}$$

Considering power flow is positive in the direction of grid current i_g, active power P will flow into the grid from GSC, and reactive power flow will from the grid into GSC if grid voltage is greater than GSC terminal voltage i.e. $v_{gd} > v_{d1}$. Also, active and reactive power are proportional to the d-axis and q-axis currents, respectively based on the assumption that grid voltage is constant. The block diagram of GSC controller is shown in Figure 6.16 that consists of a multi-loop control with a slower outer loop and a faster inner current loop. The q-axis loop is used for reactive power or grid voltage control, and d-axis loop is used for control of active power or DC link voltage.

The actual control action obtained from Equation (6.18) is as follows:

$$v_{1d} = -\left(R_g i_{gd} + L_g \frac{d}{dt} i_{gd} \right) + \omega L_g i_{gq} + v_{gd} \tag{6.34}$$

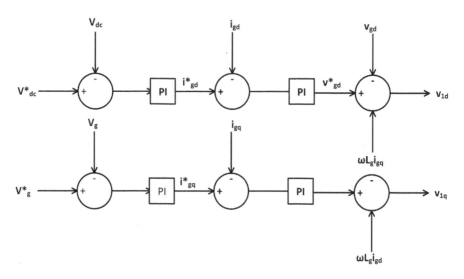

FIGURE 6.16 Block diagram of GSC controller.

$$= -v_{gd}^* + \omega L_g i_{gq} + v_{gd} \tag{6.35}$$

$$v_{1q} = \left(R_g i_{gq} + L_g \frac{d}{dt} i_{gq} \right) - \omega L_g i_{gd} \tag{6.36}$$

$$= v_{gq}^* - \omega L_g i_{gd} \tag{6.37}$$

where, $v_{gd}^* = R_g i_{gd} + L_g \frac{d}{dt} i_{gd}$ and $v_{gq}^* = R_g i_{gq} + L_g \frac{d}{dt} i_{gq}$ in above equations are transfer functions between input voltage and output current, and other terms are considered as compensating terms.

6.5.2 ROTOR SIDE CONVERTER (RSC) CONTROL

The equivalent circuit of a DFIG in synchronous reference frame is illustrated in Figure 6.17. Applying KVL in the equivalent circuit, the stator and rotor voltage can be written as:

$$v_s = i_s R_s + \frac{d\lambda_s}{dt} + j\omega_s L_{ls} \tag{6.38}$$

$$v_r = i_r R_r + \frac{d\lambda_r}{dt} + j\omega_{sl} L_{lr} \tag{6.39}$$

where, R_s and L_{ls} are stator resistance and inductance, respectively, R_r and L_{lr} are rotor resistance and inductance respectively, ω_s and ω_{rl} are synchronous speed and angular slip frequency of the generator, respectively. Generator angular slip frequency is the difference in frequency of synchronous speed and rotors speed,

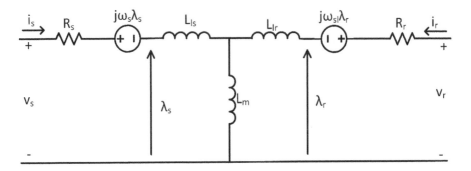

FIGURE 6.17   Equivalent circuit of DFIG in synchronous reference frame.

$\omega_{sl} = \omega_s - \omega_r$, where ω_r refers to the rotor speed. λ_s and λ_r refer to stator and rotor flux linkage respectively and defined as:

$$\lambda_s = L_s i_s + L_m i_r \tag{6.40}$$

$$\lambda_r = L_r i_r + L_m i_s \tag{6.41}$$

where, L_s, and L_r denote stator and rotor self-inductance and L_m corresponds to mutual inductance. The stator and rotor self-inductance are summation of leakage inductance and mutual inductance, and expressed as:

$$L_s = L_{sl} + L_m \tag{6.42}$$

$$L_r = L_{rl} + L_m \tag{6.43}$$

where, L_{ls} and L_{lr} are stator and rotor inductance, respectively.

The equivalent circuits of stator and rotor in dq reference frame are shown in Figure 6.18, and the respective stator and rotor voltage in dq reference can be expressed as:

$$v_{sd} = i_{sd} R_s + \frac{d\lambda_{sd}}{dt} - j\omega_s \lambda_{sq} \tag{6.44}$$

$$v_{sq} = i_{sq} R_s + \frac{d\lambda_{sq}}{dt} + j\omega_s \lambda_{sd} \tag{6.45}$$

$$v_{rd} = i_{rd} R_r + \frac{d\lambda_{rd}}{dt} + j\omega_{sl} \lambda_{rq} \tag{6.46}$$

$$v_{rq} = i_{rq} R_r + \frac{d\lambda_{rq}}{dt} + j\omega_{sl} \lambda_{rd} \tag{6.47}$$

And, flux linkage equations in dq reference frame are:

$$\lambda_{sd} = L_{sl} i_{sd} + L_m i_{rd} \tag{6.48}$$

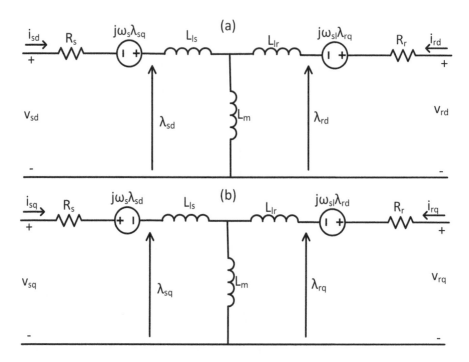

FIGURE 6.18 Rotor and stator equivalent circuits in dq reference frame.

$$\lambda_{sq} = L_{sl}i_{sq} + L_m i_{rq} \tag{6.49}$$

$$\lambda_{rd} = L_{rl}i_{rd} + L_m i_{sd} \tag{6.50}$$

$$\lambda_{rq} = L_{rl}i_{rq} + L_m i_{sq} \tag{6.51}$$

The dq components of stator flux linkage are represented in Figure 6.19. Similar to GSC active and reactive control, for independent control of active and reactive power of RSC stator flux is aligned with d-axis, therefore, $\lambda_{sd} = \lambda_s$, and $\lambda_{sq} = 0$. Other assumptions that are also considered for RSC controller are

- Stator resistance R_s is neglected. Since it is quite low, voltage drop across stator resistance is zero i.e. $i_{sd}R_s = i_{sq}R_s = 0$.
- Grid voltage V_g and frequency ω are constant considering DFIG is connected to stiff grid.
- Magnetizing current of the stator is determined by the grid.

Based on the above assumptions, stator voltage in Equations (6.44)–(6.45) and flux linkage in Equations (6.48)–(6.49) in the dq reference frame can be rewritten as:

$$v_{sd} = 0 \tag{6.52}$$

$$v_{sq} = \omega_s \lambda_s = v_s = constant \tag{6.53}$$

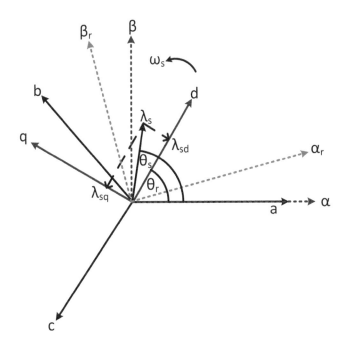

FIGURE 6.19   Representation of stator flux linkage in dq reference frame.

$$i_{sd} = \frac{\lambda_{sd} - L_m i_{rd}}{L_{sl}} \qquad (6.54)$$

$$i_{sq} = -\frac{L_m}{L_{sl}} i_{rq} \qquad (6.55)$$

The per unit instantaneous active and reactive power in stator winding of DFIG in dq reference frame are given by:

$$p_s = v_{sd} i_{sd} + v_{sq} i_{qs} = v_{sq} i_{qs} = -v_s \frac{L_m}{L_{sl}} i_{rq} \qquad (6.56)$$

$$q_s = v_{sq} i_{sd} - v_{sd} i_{sq} = v_{sq} i_{sd} = \frac{v_s^2 L_m}{\omega_s L_{sl}} - \frac{v_s L_m}{L_{sl}} i_{rd} \qquad (6.57)$$

As observed in Equations (6.55)–(6.56), stator active and reactive power can be independently controlled by rotor q-axis and d-axis currents, respectively. To obtain actual control action of RSC, stator currents are substituted into rotor flux Equations in (6.50)–(6.51), and can be rewritten as:

$$\lambda_{rd} = L_{rl} i_{rd} + L_m i_{sd} \qquad (6.58)$$

$$= L_{rl} i_{rd} + L_m \left(\frac{\lambda_{sd} - L_m i_{rd}}{L_{sl}} \right) \qquad (6.59)$$

$$\sigma L_{rl} i_{rd} + \frac{L_m V_s}{L_{sl} \omega_s} \tag{6.60}$$

And,

$$\lambda_{rq} = L_{rl} i_{rq} + L_m i_{sq} \tag{6.61}$$

$$= L_{rl} i_{rq} + L_m \left(-\frac{L_m}{L_s} i_{rq} \right) \tag{6.62}$$

$$= \left(1 - \frac{L_m^2}{L_{rl} L_{sl}} \right) L_{rl} i_{rq} \tag{6.63}$$

$$= \sigma L_{rl} i_{rq} \tag{6.64}$$

The actual control action is then obtained by substituting dq components of rotor flux in Equations (6.46)–(6.47). The rewritten dq components of rotor voltage are:

$$v_{rd} = i_{rd} R_r + \sigma L_{rl} \frac{di_{rd}}{dt} - \omega_{sl} \sigma L_{rl} i_{rq} \tag{6.65}$$

$$= v_{rd}^* - \omega_{sl} \sigma L_{rl} i_{rq} \tag{6.66}$$

$$v_{rq} = i_{rq} R_r + \sigma L_{rl} \frac{di_{rq}}{dt} + \omega_{sl} \left(\sigma L_{rl} i_{rd} + \frac{L_m}{L_{sl}} \lambda_{sd} \right) \tag{6.67}$$

$$= v_{rq}^* + \omega_{sl} \left(\sigma L_{rl} i_{rd} + \frac{L_m}{L_{sl}} \lambda_{sd} \right) \tag{6.68}$$

$$= v_{rq}^* + \omega_{sl} (\sigma L_{rl} i_{rd} + L_m i_{sm}) \tag{6.69}$$

where, $i_{sm} = \frac{\lambda_{sd}}{L_{sl}}$ is the current flowing through magnetizing reactance due to stator current, and $v_{rd}^* = i_{rd} R_r + \sigma L_{rl} \frac{di_{rd}}{dt}$ and $v_{rq}^* = i_{rq} R_r + \sigma L_{rl} \frac{di_{rq}}{dt}$ are transfer function between input voltage and output current, and other terms as observed in Equations (6.66) and (6.69) are referred to compensating terms.

The block diagram of RSC controller is provided in Figure 6.20 based on the control action in (6.66) and (6.69), and simulation results performed on a 9 MW DFIG WECS employing the direct current vector control method are incorporated in Figure 6.21–6.23.

Figure 6.21 shows the decrease in grid voltage v_{ga} from 1 pu to 0.9 pu following a disturbance at $t = 0.1$ s. Owing to the disturbance power P_g injected into grid drops below the rated power, therefore, DFIG wind turbine power controller decreases the pitch angle to generate more power. As more power is extracted from the wind, rotor speed of the generator increases. It is also observed that once disturbance is cleared at $t = 0.13$ s, the pitch angle is increased to limit the power as it goes beyond the rated power.

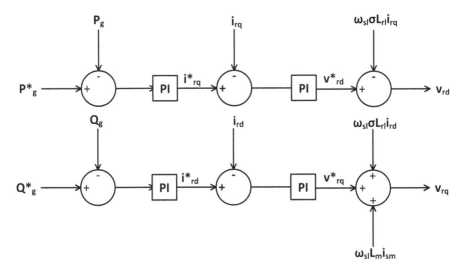

FIGURE 6.20   Block diagram of RSC controller.

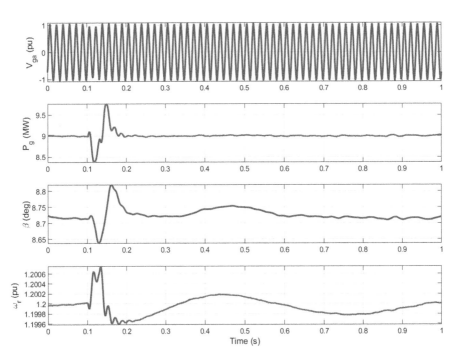

FIGURE 6.21   Grid voltage, active power flow into grid, pitch angle, and rotor speed following the disturbance at $t = 0.1$ s.

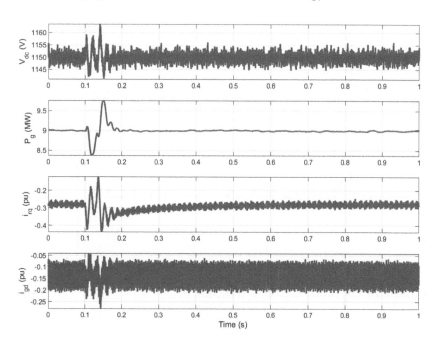

FIGURE 6.22 DC link voltage, active power flow into the grid, rotor q-axis and grid d-axis currents that controls the active power following the disturbance at $t = 0.1$ s.

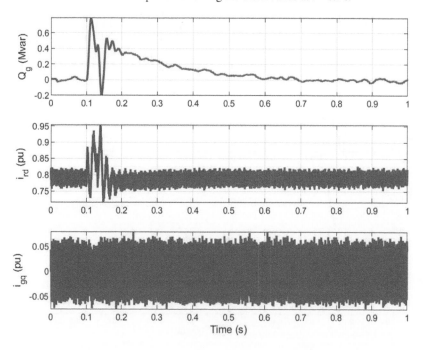

FIGURE 6.23 Reactive power flow into grid, rotor d-axis current and grid q-axis currents that control the reactive power following the disturbance at t = 0.1 s.

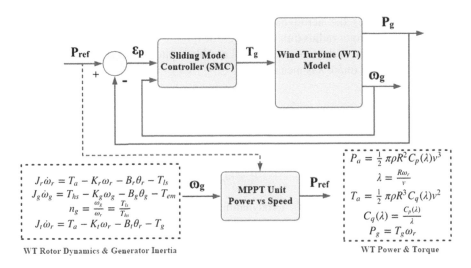

FIGURE 6.24   Sliding mode control scheme for a fixed pitch VS-WECS.

Figure 6.22 illustrates the change in DC link voltage following the disturbance and response of the RSC q-axis current i_{rq} and GSC d-axis current i_{gd} to minimize the power fluctuation and keep the DC link voltage constant. As the DC link voltage V_{dc} and power delivered to grid P_g decrease, RSC and GSC increase rotor q-axis current I_{rq} and stator d-axis current I_{gd}, respectively to deliver more power into the grid. Since, active power is proportional to RSC I_{rq} and GSC I_{gd} demonstrated in Equations (6.30) and (6.56).

In Figure 6.23, more reactive power Q_g is delivered into the grid to increase the grid voltage. To deliver more reactive power for grid voltage support, RSC increases rotor d-axis current i_{rd} and GSC increases stator q-axis current i_{gq} since reactive power is proportional to i_{rd} and i_{gq} discussed earlier in Equations (6.32) and (6.57).

6.6   SLIDING MODE CONTROL METHOD

Variable-speed wind energy systems (VS-WECSs) employ aerodynamic regulations along with power electronic technologies to control power, speed and torque. It is imperative to have stable dynamic characteristics in a VS-WECS in combination with improved system efficiency. In this regard, nowadays a state-of-the-art control methodology, referred to as sliding mode control (SMC), is being applied in VS-WECS frameworks which present robustness to the parametric uncertainties of the wind turbine and the generator. SMC also ameliorates the grid disturbances and overall network sustainability issues.

Layout of the SMC framework for a fixed pitch variable-speed wind turbine model is shown in Figure 6.24. The aerodynamic (rotor) power fed into the turbine model is expressed as

$$P_a = \frac{1}{2}\pi\rho R^2 C_p\left(\lambda\right)v^3 \tag{6.70}$$

where, P_a implies the aerodynamic power (watt), ρ represents the air density (kg/m³), R is the rotor radius (m), C_p denotes the turbine power conversion efficiency which is a function of the tip-speed ratio λ and blade pitch angle β and v is the measured wind speed (m/s). λ is measured as

$$\lambda = \frac{R\omega_r}{v} \tag{6.71}$$

Here ω_r is the rotor angular speed (rad/s). By definition, the rotor power is:

$$P_a = \omega_r T_a \tag{6.72}$$

where, T_a is the wind turbine's aerodynamic torque (N.m). Additionally, the torque coefficient is given by:

$$C_q(\lambda) = \frac{C_p(\lambda)}{\lambda} \tag{6.73}$$

As a follow-up, the aerodynamic torque can be expressed as:

$$T_a = \frac{1}{2}\pi \rho R^3 C_q(\lambda) v^2 \tag{6.74}$$

The rotor dynamics along with the generator inertia contents can be mathematically defined as:

$$J_r \dot{\omega}_r = T_a - K_r \omega_r - B_r \theta_r - T_{ls} \tag{6.75}$$

and

$$J_g \dot{\omega}_g = T_{hs} - K_g \omega_g - B_g \omega_g - T_{em} \tag{6.76}$$

Here in the differential equations, J_r and J_g are rotor and generator inertia (kg.m²) contents, respectively, K_r and K_g are rotor and generator external damping factors (N.m/rad.s), respectively, B_r and B_g represent rotor and generator external stiffness (N.m/rad.s) parameters, respectively, θ_r and θ_g denote the rotor and generator angular orientations (rad), respectively, ω_g is the generator angular speed (rad/s), and T_{ls}, T_{hs} and T_{em} represent low speed, high speed and generator electromagnetic torque (N.m) quantities, respectively. The gearbox ratio of the model is given by:

$$n_g = \frac{\omega_g}{\omega_r} = \frac{T_{ls}}{T_{hs}} \tag{6.77}$$

Typically, external stiffness is found very low and thus insignificant to include in the model equation. However, the generated power can be obtained as:

$$P_g = \omega_r T_g \tag{6.78}$$

where, T_g is the generator torque at the rotor side (N.m).

Conventionally, wind turbines are cost-effective means of electrical energy generation. In order to optimize the power accessibility with respect to the rated speed, a standardized control law formulated to keep the turbine to operate at the maximum of the C_p value is as follows:

$$T_g = k\omega^2 \text{ with } k = \frac{1}{2}\pi\rho R^3 \frac{C_{pmax}}{\lambda_{opt}^3} \tag{6.79}$$

where k is a constant to determine and λ_{opt} is the optimal tip-speed ratio.

This standard control law has a number of limitations, since there is no accurate and universally expected way to measure k due to frequent changes in blade aerodynamics. Moreover, if k is assumed to be determined correctly, the turbine may not operate at C_{pmax} because of the speed fluctuations. However, SMC for wind power regulation reduces the negative impacts due to the uncertainty lies in k determination and untoward operating characteristics of the model. SMC is a dynamic and robust nonlinear control strategy that can provide a compromising trade-off between conversion efficiency and turbine operating oscillations. This is basically an adaptive algorithm measuring a gain that increases as long as any tracking error is recorded. The tracking error in this adaptive control algorithm is defined as:

$$\varepsilon_p = P_{ref} - P_g \tag{6.80}$$

where, P_{ref} is the reference power generated by the maximum power point tracking (MPPT) unit. Applying derivatives and previous definition, it becomes

$$\dot{\varepsilon}_p = \dot{P_{ref}} - \dot{T_g}\omega_r - \dot{\omega_r}T_g \tag{6.81}$$

For the dynamic SMC scheme,

$$\dot{T_g} = \frac{(B+\lambda)sgn(\varepsilon_p)}{\omega_r} \tag{6.82}$$

Here

$$\dot{B} = |\varepsilon_p| \text{ and } \lambda > 0 \tag{6.83}$$

Thus, it can be obtained as:

$$\dot{\varepsilon}_p = \dot{P_{ref}} - \dot{\omega_r}T_g - (B(t)+\lambda)sgn(\varepsilon_p) \tag{6.84}$$

Let us consider,

$$d = \dot{P_{ref}} - T_g\dot{\omega_r} \text{ and } |d| < B_1 \tag{6.85}$$

where, B_l is a positive unknown constant. The SMC scheme implies:

$$\dot{\varepsilon}_p = -\left(B(t) + \lambda\right) sgn\left(\varepsilon_p\right) + d \tag{6.86}$$

Lyapunov function of the following type can be employed to prove the stability of the SMC baseline algorithm as:

$$V = \frac{1}{2}\varepsilon^2 + \frac{1}{2}\left(B - B_1\right)^2 \tag{6.87}$$

In this follow-up, mathematical computations yield that the time derivative of the function satisfies

$$\dot{V} \leq -\lambda |\varepsilon| \tag{6.88}$$

which conveys that the tracking error converges asymptotically to zero. Furthermore, in general practice, it is recommended to operate the wind turbine at a less efficiency than the maximum to facilitate an energy buffer for any sudden grid change. Thereby, the reference power, generator optimal torque and speed can be adopted as follows:

$$P_{ref} = 0.9 T_{opt} \omega_{opt} \tag{6.89}$$

$$T_{opt} = \frac{1}{2}\pi\rho R^3 \frac{C_{pmax}}{\lambda_{opt}} v^2 \tag{6.90}$$

$$\omega_{opt} = \frac{\lambda_{opt} v}{R} \tag{6.91}$$

6.7   MODEL PREDICTIVE CONTROL METHOD

In Figure 6.25, a first-order WECS model is considered for designing the control framework using model predictive control (MPC) scheme. The primary drawbacks of the classical control strategies based on PI regulators are: (a) in the partial load region, PI regulator fails in terms of fine tuning considering energy maximization and transient load reduction; (b) considerable amount of power and turbine operating torsional torque overshoots occur at the rated wind speed due to fast switching between partial and full load structures; and (c) significant pitch activity of no use and severe power variations can happen because of controlling only the pitch during the full load condition. However, MPC can successfully handle multiple-input-multiple-output (MIMO) plant scenarios like WECSs, where multivariate control scheme is required for design and time constraints. For an MPC implication, basically a dynamic plant model is needed of which the physical constraints and variables need to be known. There is a fixed sampling time for emulating the overall system and at each sampling window, a prediction model works to estimate future outputs within a horizon. Then, an optimization problem is solved for measuring the future set of

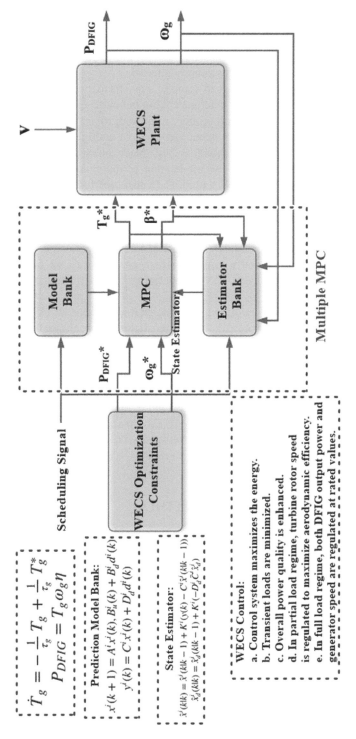

FIGURE 6.25 Control framework for WECS applying model predictive controller.

outputs. In this approach, a predefined figure of merit is determined to assess the prediction performance of the algorithm. When the prediction is executed, the first input is then sent to system optimal sequence of operation and in this manner, the entire calculation is carried out in subsequent steps. It is to be noted that the prediction horizon gets shifted with each sampling interval.

However, MPC algorithm needs to compile a quadratic programming problem online to accurately predict the outcomes. This behavior limits its use in relatively slow dynamic systems. Another more improved, robust, and fast executable scheme is multiple MPC. The rudimentary concept of multiple MPC is to develop a multi-variable controller to manipulate both blade pitch angle and generator torque in a wind turbine plant. The implication yields inclusion of design constraints in the manifested controller unit. The entire operating region is divided into subregions with associated linearized models. These subregions essentially represent all the dynamics of the whole system cumulatively. Then, a linear MPC scheme is designed for each model. The fundamental constituents of the modified MPC network are: prediction model bank, solver for optimization purpose, system variables' state estimator, and a criterion which enables the switching between control premises in accordance with the operating condition changes i.e. speed, partial or full load scales.

In the prediction model bank, the dynamic characterization of WECS is consisted of linearized model formulation based on state-space configuration. Here, all the state variables, signals for optimal controller (mainly to dampen high frequency oscillations), disturbance parameters (white Gaussian noise), and control outputs are configured. The constraints optimization unit assumes to know estimates of the state variables and disturbance parameters. The estimator evaluates states using an observer approach with adaptive filtering technique. Here mainly the adaptive gain is measured and updated. A scheduling signal i.e. generator speed or pitch angle or average wind speed controls the switching between MPCs. This signal happens to be measured online during operation.

6.8   COORDINATED CONTROL METHOD

As an immaculate control scheme for WECS, a simple coordinated controller of DC link voltage and pitch angle with a PMSG can be applied for power smoothing purpose. The WECS can be realized as an AC-DC-AC power electronic converter system. The DC link is composed to generate a DC voltage command to lower the output power fluctuations of PMSG. The power fluctuations due to the frequency outliers are minimized by the pitch angle control. With this coordinated control scheme, the turbine physical blade stress is mitigated, DC link capacitor size is reduced and output power gets smoothened.

Figure 6.26 presents the generic system diagram of a WECS (power converter network) controlled via a coordinated regulation scheme. In the DC link control, the charge/discharge power of the capacitor is given by:

$$\Delta P_{dc} = P_c^* - P_g \tag{6.92}$$

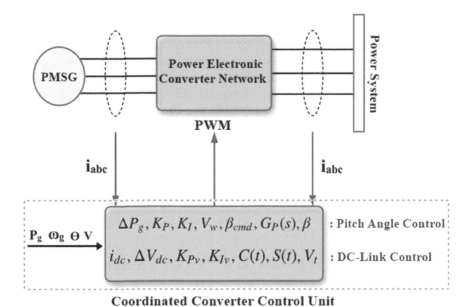

FIGURE 6.26 Coordinated control scheme for a WECS framework (K_P and K_I are PI constants for pitch angle control loop, V_w is the wind speed, $G_P(s)$ is the servo system plant transfer function, i_{dc} is the DC link current, ΔV_{dc} is the smoothing index, K_{Pv} and K_{Iv} are PI constants for DC link controller, $C(t)$ is the carrier signal, $S(t)$ is switching signal and V_t is the grid voltage).

where, P_c^* is the smoothing command. The smoothing index of the control loop is given by:

$$\Delta V_{dc} = \frac{\Delta P_{dc}}{i_{dc}} \tag{6.93}$$

Besides the combined control of pitch angle and DC link voltage, coordinated control method can be used for mitigating sub-synchronous interaction (SSI) between DFIG-based wind farms and series capacitor compensated transmission systems for a number of wind farm test scenarios. SSI is reduced by applying a supplemental control signal in the reactive power controller at the grid-side converter of DFIG and full-scale frequency converter wind turbines along with the reactive power control scheme of custom HVDC modular multilevel converter wind farms. Wind turbines are connected to the grid line through HVDC links to record the SSI phenomena.

6.9 DISCUSSIONS

Demand of renewable energy keeps growing rapidly due to ever-increasing population and prevalent environmental degradation caused by the traditional power generation means worldwide. WECS is ratified as one of the most significant and mature resources of renewable energy which has received and is receiving considerable attention by the research and engineering fraternity for enormous and consistent industrial applications. The noticeable advancements of regulatory techniques to determine and extract most optimized performance constraints, dynamic models, stable operational premises, and maximum power generation quantities with controllable features in WECS frameworks comply with the design and integration of these control strategies in the conventional wind farms. Therefore, this chapter reports a number of robust and effective control algorithms for variable-speed WECS plants, namely, pitch angle, blade inertial, direct current vector, sliding mode, model predictive, and coordinated control schemes. As a design approach for any of these controllers, it can be stated that the primary goal is to extract wind power considering random wind speed fluctuations ensuring the maximum power point tracking with stability.

REFERENCES

1. D. Chowdhury, A. S. M. K. Hasan, and M. Z. R. Khan, "Islanded DC microgrid architecture with dual active bridge converter-based power management units and time slot-based control interface", *IEEJ Transactions on Electrical and Electronic Engineering*, vol. 15, no. 6, pp. 863–871, March 2020, doi.org/10.1002/tee.23128.
2. J. Lee, E. Muljadi, P. Srensen, and Y. C. Kang, "Releasable kinetic energy-based inertial control of a DFIG wind power plant", *IEEE Transactions on Sustainable Energy*, vol. 7, no. 1, pp. 279–288, January 2016, doi: 10.1109/TSTE.2015.2493165.
3. J. Morren, S. W. H. de Haan, W. L. Kling, and J. A. Ferreira, "Wind turbines emulating inertia and supporting primary frequency control", *IEEE Transactions on Power Systems*, vol. 21, no. 1, pp. 433–434, February 2006, doi: 10.1109/TPWRS.2005.861956.
4. R. Sakamoto, T. Senjyu, N. Urasaki, T. Funabashi, H. Fujita and H. Sekine, "Output power leveling of wind turbine generators using pitch angle control for all operating regions in wind farm," Proceedings of the 13th International Conference on, Intelligent Systems Application to Power Systems, Arlington, VA, 2005, pp. 1–6, doi: 10.1109/ISAP.2005.1599291.
5. S. Li, T. A. Haskew, K. A. Williams, and R. P. Swatloski, "Control of DFIG wind turbine with direct-current vector control configuration", *IEEE Transactions on Sustainable Energy*, vol. 3, no. 1, pp. 1–11, January 2012, doi: 10.1109/TSTE.2011.2167001.
6. H. A. Mohammadpour, M. M. Islam, E. Santi, and Y. Shin, "SSR damping in fixed-speed wind farms using series FACTS controllers", *IEEE Transactions on Power Delivery*, vol. 31, no. 1, pp. 76–86, February 2016, doi: 10.1109/TPWRD.2015.2464323.
7. S. Li and T. A. Haskew, "Analysis of decoupled d-q vector control in DFIG back-to-back PWM converter," 2007 IEEE Power Engineering Society General Meeting, Tampa, FL, 2007, pp. 1–7, doi: 10.1109/PES.2007.385461.
8. M. M. Islam, E. Hossain, S. Padmanaban, and C. W. Brice, "A new perspective of wind power grid codes under unbalanced and distorted grid conditions", *IEEE Access*, vol. 8, pp. 15931–15944, 2020, doi: 10.1109/ACCESS.2020.2966907.
9. X. Zhao, Z. Yan, Y. Xue, and X. Zhang, "Wind power smoothing by controlling the inertial energy of turbines with optimized energy yield", *IEEE Access*, vol. 5, pp. 23374–23382, 2017, doi: 10.1109/ACCESS.2017.2757929.
10. M. Hwang, E. Muljadi, J. Park, P. Sørensen, and Y. C. Kang, "Dynamic droop–based inertial control of a doubly-fed induction generator", *IEEE Transactions on Sustainable Energy*, vol. 7, no. 3, pp. 924–933, July 2016, doi: 10.1109/TSTE.2015.2508792.

11. J. Mohammadi, S. Vaez-Zadeh, S. Afsharnia, and E. Daryabeigi, "A combined vector and direct power control for DFIG-based wind turbines", *IEEE Transactions on Sustainable Energy*, vol. 5, no. 3, pp. 767–775, July 2014, doi: 10.1109/TSTE.2014.2301675.

12. T. L. Van, T. H. Nguyen, and D. Lee, "Advanced pitch angle control based on fuzzy logic for variable-speed wind turbine systems", *IEEE Transactions on Energy Conversion*, vol. 30, no. 2, pp. 578–587, June 2015, doi: 10.1109/TEC.2014.2379293.

13. M. R. Islam, Y. Guo, and J. Zhu, "A review of offshore wind turbine nacelle: technical challenges, and research and development trends", *Renewable and Sustainable Energy Reviews*, vol. 33, pp. 161–176, May 2014, doi: 10.1016/j.rser.2014.01.085.

14. B. Beltran, T. Ahmed-Ali, and M. E. H. Benbouzid, "Sliding mode power control of variable-speed wind energy conversion systems", *IEEE Transactions on Energy Conversion*, vol. 23, no. 2, pp. 551–558, June 2008, doi: 10.1109/TEC.2007.914163.

15. B. Yang, T. Yu, H. Shu, J. Dong, and L. Jiang, "Robust sliding-mode control of wind energy conversion systems for optimal power extraction via nonlinear perturbation observers", *Applied Energy*, vol. 210, pp. 711–723, January 2018, doi: 10.1016/j.apenergy.2017.08.027.

16. H. De Battista, P. F. Puleston, R. J. Mantz, and C. F. Christiansen, "Sliding mode control of wind energy systems with DOIG-power efficiency and torsional dynamics optimization", *IEEE Transactions on Power Systems*, vol. 15, no. 2, pp. 728–734, May 2000, doi: 10.1109/59.867166.

17. A. Merabet, K. T. Ahmed, H. Ibrahim, and R. Beguenane, "Implementation of sliding mode control system for generator and grid sides control of wind energy conversion system", *IEEE Transactions on Sustainable Energy*, vol. 7, no. 3, pp. 1327–1335, July 2016, doi: 10.1109/TSTE.2016.2537646.

18. I. Munteanu, S. Bacha, A. I. Bratcu, J. Guiraud, and D. Roye, "Energy-reliability optimization of wind energy conversion systems by sliding mode control", *IEEE Transactions on Energy Conversion*, vol. 23, no. 3, pp. 975–985, September 2008, doi: 10.1109/TEC.2008.917102.

19. M. Soliman, O. P. Malik and D. T. Westwick, "Multiple model predictive control for wind turbines with doubly fed induction generators", *IEEE Transactions on Sustainable Energy*, vol. 2, no. 3, pp. 215–225, July 2011, doi: 10.1109/TSTE.2011.2153217.

20. Y. Shi, X. Xiang, Y. Zhang and D. Sun, "Design of stochastic model predictive control for wind energy conversion system," 2017 International Workshop on Complex Systems and Networks (IWCSN), Doha, 2017, pp. 108–114, doi: 10.1109/IWCSN.2017.8276513.

21. H. Zhao, Q. Wu, J. Wang, Z. Liu, M. Shahidehpour, and Y. Xue, "Combined active and reactive power control of wind farms based on model predictive control", *IEEE Transactions on Energy Conversion*, vol. 32, no. 3, pp. 1177–1187, September 2017, doi: 10.1109/TEC.2017.2654271.

22. J. Hu, Y. Li, and J. Zhu, "Multi-objective model predictive control of doubly-fed induction generators for wind energy conversion", *IET Generation, Transmission & Distribution*, vol. 13, no. 1, pp. 21–29, 8 1 2019, doi: 10.1049/iet-gtd.2018.5172.

23. H. Meng, P. Li, Z. Lin and Y. Hu, "Multi-degree of freedom optimization control for large inertia wind energy conversion system using model predictive approach," 2017 29th Chinese Control And Decision Conference (CCDC), Chongqing, 2017, pp. 2328–2332, doi: 10.1109/CCDC.2017.7978903.

24. J. Haiping, J. Xiaohong and R. Lina, "MPC-based power tracking control for a wind energy conversion system with PM synchronous generator," 2015 34th Chinese Control Conference (CCC), Hangzhou, 2015, pp. 4079–4083, doi: 10.1109/ChiCC.2015.7260267.

25. J. Zhou, S. Li, J. Li and J. Zhang, "A combined control strategy of wind energy conversion system with direct-driven PMSG," 2016 31st Youth Academic Annual Conference of Chinese Association of Automation (YAC), Wuhan, 2016, pp. 369–374, doi: 10.1109/YAC.2016.7804921.

26. A. Uehara, A. Pratap, T. Goya, T. Senjyu, A. Yona, N. Urasaki, and T. Funabashi, "A coordinated control method to smooth wind power fluctuations of a PMSG-based WECS," in *IEEE Transactions on Energy Conversion*, vol. 26, no. 2, pp. 550–558, June 2011, doi: 10.1109/TEC.2011.2107912.

27. P. Kou, D. Liang, F. Gao, and L. Gao, "Coordinated predictive control of DFIG-based wind-battery hybrid systems: using non-Gaussian wind power predictive distributions", in *IEEE Transactions on Energy Conversion*, vol. 30, no. 2, pp. 681–695, June 2015, doi: 10.1109/TEC.2015.2390912.

28. U. Karaagac, S. O. Faried, J. Mahseredjian, and A. Edris, "Coordinated control of wind energy conversion systems for mitigating subsynchronous interaction in DFIG-based wind farms", *IEEE Transactions on Smart Grid*, vol. 5, no. 5, pp. 2440–2449, September 2014, doi: 10.1109/TSG.2014.2330453.

7 Converter-Based Advanced Diagnostic and Monitoring Technologies for Offshore Wind Turbines

Yi Liu and Xi Chen

State Key Laboratory of Advanced Electromagnetic
Engineering and Technology, School of Electrical
and Electronic Engineering, Huazhong University
of Science and Technology

Md. Rabiul Islam

School of Electrical, Computer and Telecommunications
Engineering, Faculty of Engineering and Information
Sciences, University of Wollongong

CONTENTS

7.1   INTRODUCTION

Wind power, as a renewable energy, has been regarded as one of the promising alternative energy sources in the world. However, wind turbines usually have high failure rates due to their harsh operation environments. The generator and converter are main components of an electric drive train in the wind turbine to accomplish energy conversion from mechanical to electrical, whose working condition should be paid more attentions. Since the converter contains voltage and current signals of both converter and generator, converter-based advanced diagnose and monitoring technologies for electrical drive train of wind turbines are very promising. This chapter can be divided into three main sections as follows:

a. *Converter-based diagnostic and monitoring technologies for generators*
   The failure modes of generators that can be reflected by converters are usually classified as electrical faults and mechanical faults. This section will focus on diagnose and monitoring technologies for common electrical faults (e.g., winding insulation damage, winding asymmetry, winding open- and short-circuit faults) and main mechanical faults (i.e., bearing failure and air gap eccentricity).

b. *Converter-based diagnostic and monitoring technologies for power converters*
   The reliability of power electronic converters is more critical with the increasing of the power rating for wind turbines. This section will investigate

mainstream diagnose and monitoring technologies for main faults in power converters (e.g., open-circuit faults, switch device faults, DC voltage and sensor faults), whose faults can be reflected by current, voltage, and temperature in the converter.

c. *Converter-based diagnostic and monitoring technologies in system level*

In addition to the aforementioned two kinds of component-level diagnostic and monitoring technologies, in recent years, system-level diagnostic and monitoring technologies are also important and have been developed. This section will introduce some diagnose and monitoring technologies in system level based on some typical quantities and advanced principles (e.g., power curve, extreme learning machine, and intelligent algorithm).

7.2 DIAGNOSTIC AND MONITORING TECHNOLOGIES FOR WIND POWER GENERATORS

The electric machine is the key component in a wind turbine to realize the energy conversion from wind energy to electric energy. According to state of art of wind power technologies, commercialized generators can be classified into four types, which are listed in Table 7.1. Among these generators, the permanent magnet synchronous generator (PMSG) and the electrically excited synchronous generator (EESG) are usually applied in the medium and large size wind turbines for the elimination of gearbox in the drive train. Meanwhile, the double fed induction generator (DFIG) and the squirrel cage induction generator (SCIG) are suitable for medium and small size wind turbines. As shown in Figure 7.1, the appearances of synchronous generators are flat, whereas the induction generators are relatively long in axial direction.

Wind turbines are usually installed in remote areas to capture more wind energy and to save land rental cost, if compared with other turbines used in traditional power plants [1]. The windy, rainy, and sunny environment in these remote areas introduces new challenges for the operation of wind turbines, especially for the generators [2]. Such an extreme working environment may lead to higher failure

TABLE 7.1

Classifications of Wind Power Generators

Generator	Drive Train	Grid Connection	Speed
PMSG	Direct drive	Full load power converter	Variable speed
EESG	Direct drive	Full load power converter	Variable speed
DFIG	Gearbox	Partial-load power converters	Variable speed
SCIG	Gearbox	Full load power converter	Variable speed

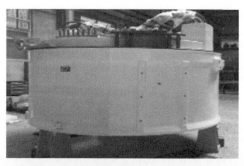

FIGURE 7.1   Four typical wind power generators.

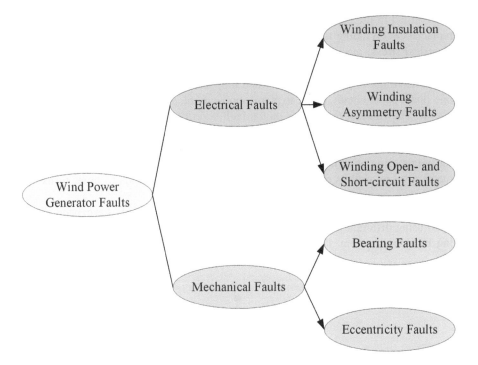

FIGURE 7.2 Classification of the faults in wind power generators.

rates, which causes revenue loss due to repair costs and unexpected downtime. Potential failures of wind power generators can be diagnosed or even prognosed from the changes in their parameters [3]. Condition monitoring is necessary to prevent the deterioration of incipient faults by early warning. As shown in Figure 7.2, the failure modes of generators in wind turbines can be classified as electrical faults and mechanical faults.

Since all generators are connected with the power converter, the fault information may be reflected by electrical signals in the power converter. Figure 7.2 shows main electrical and mechanical faults in wind power generators. Because the slip ring and brush only exist in the DFIG, this paper mainly overviews common faults of the wind generator, i.e. the winding faults, the bearing faults, and the eccentricity faults.

7.2.1 Winding Insulation Faults

The unscheduled downtime caused by failures of the insulation system can cause enormous costs and fire risks in the cabin of the wind turbines. Thus, it is desirable that a weakness in the insulation system, which can result in a severe failure, is identified in the early stages in order to perform a scheduled machine service or replacement. Winding insulation failures are usually caused by the thermal stress, electrical stress, mechanical stress, and the environment stress. Figure 7.3 briefly illustrates the different types of fault occurred due to insulation degradation in the stator windings of generators [4], which may lead to the turn to turn short-circuit, one-phase short-circuit, phase to phase short-circuit, and the phase to ground short-circuit. The value of R_f depends on the fault severity. Under fault condition, a large current will circulate through the faulty part of the stator winding. Then, significant amount of heat will be generated and the temperature will increase, which will affect the neighboring winding insulation material. Acceleration in insulation degradation mechanism as well as expansion of the fault in a few seconds to other turns and windings is inevitable.

Aiming at root causes, several online monitoring methods have been developed using different physical quantities to detect the health condition of the winding insulation system. However, online-monitoring technologies often require the

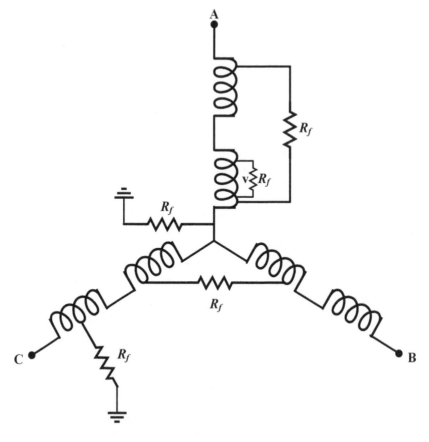

FIGURE 7.3   Different types of winding faults [4].

TABLE 7.2

Converter-Based Diagnostics and Monitoring of Winding Insulation Faults

Method	Insulation Fault Position	Features
Leakage currents	Phase to ground; phase to phase	Determine the cause of deterioration
Negative sequence current	Turn to turn	Suitable for none-ideal condition
Phase current signature	Turn to turn	Hard to distinguish from other faults
Zero sequence voltage	Turn to turn	neutral point of windings should be accessible

installation of additional equipment, which has to be installed in the upper cabinet. The additional sensors will increase the total costs and decrease the complexity of the wind turbines. Since the power converter is a necessary component in the modern wind turbines, electrical signatures, e.g., current and voltage of the connected generator can be acquired to develop nonintrusive monitoring methods. Therefore, the online monitoring techniques based on current and voltage signals in convertor are shown in Table 7.2.

7.2.1.1 Current-Based Diagnostic Technologies for Winding Insulation Faults

Based on the circumstance that the magnitude of the stator current harmonics changes with the winding insulation faults developed, current signatures can be used to detect the faults. For induction generators, some other harmonic frequency components will appear in stator currents if the turn insulation failure occurred, as illustrated by characteristic frequencies as [5]:

$$f = \left(k_1 z_r \frac{1-s}{p} \pm 2k_2 \right) f_1 \qquad (7.1)$$

where, Z_r is the number of rotor slot, s is the slip ratio, p is the number of pole-pairs, and f_1 is the output frequency of the generator.

By adding a secondary neutral point in stator windings, the turn fault can be detected at an early stage through monitoring the fault-induced circulating short-circuit current in the synchronous machine [6]. In order to adapt different load conditions, the proposed online condition monitoring approach combined with finite element model (FEM) and the discrete wavelet transform (DWT) are validated by experimental results in reference [7]. Two methods with the aid of support vector machine (SVM) were proposed in order to diagnose turn-to-turn insulation failure (short-circuit) of low voltage motor winding through magnitude and phase of load currents [8].

Several researchers also monitored the negative-sequence current component for insulation fault detection. The asymmetry introduced by a turn-to-turn fault will change the amplitude of negative-sequence current, which can be regarded as an indicator [9]. However, some other problems, such as, the supply voltage imbalances, motor and load inherent asymmetries, and measurement errors, may also cause the negative-sequence component of the current. Another way to consider the non-idealities is the use of artificial neural networks (ANNs) to determine the negative-sequence current due to a turn fault [10]. Since the neural network is trained offline over the entire range of operating conditions, the ANNs estimate the negative-sequence current of the healthy machine considering all sources of asymmetry except for the asymmetry due to a turn fault. By compared the ANN estimated value with the measured negative-sequence current, the deviation of the measured value from the estimated value is an indicator of the severity of the turn-to-turn fault.

Another kind of current signals are also available in winding fault diagnose. The concept is to measure the differential leakage currents of each phase winding from the terminal box in a noninvasive manner to assess the insulation condition during motor operation.

Indicators for insulation condition such as the capacitance and dissipation factor are calculated based on the measurements to provide a low-cost solution for online insulation condition assessment [11]. As illustrated in Figure 7.4, an analytic expression for the phase A leakage current in the differential leakage current measurement technique can be derived from the observations as

$$\vec{I}_{al} = \frac{\vec{U}_{ag}}{2Z_{pg}} + \frac{(\vec{U}_{ab} + \vec{U}_{ac})}{2Z_{pp}} \tag{7.2}$$

where U_{ag}, U_{ab}, U_{ac} are amplitudes of the voltage vectors between phase A and ground, phase B, and phase C, respectively. Z_{pg} and Z_{pp} are winding impedances of phase to ground and phase to phase, respectively.

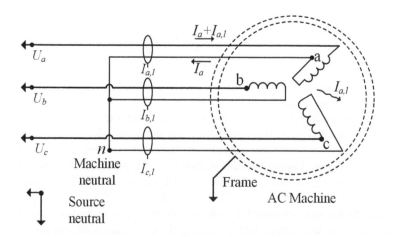

FIGURE 7.4 Differential leakage current measurement technique for online assessment of stator winding insulation condition in [11].

7.2.1.2   Voltage-Based Diagnostic Technologies for Winding Insulation Faults

Voltage sensors are necessary in the converter to calculate the power of the generator. The method utilizing the zero-sequence voltage is also proposed in [12] to detect the turn-to-turn faults in windings. Selecting the algebraic sum of the line-neutral voltages as an indicator, the sum will be zero in ideal condition meanwhile the sum will be nonzero at the faulty conditions. The sensitivity is improved by filtering the higher order harmonics of voltage sum. The premise of the zero-sequence voltage detection is the accessibility for the neutral point of the AC winding.

7.2.2   WINDING ASYMMETRY FAULTS

Rotor electrical imbalance will cause shaft vibration. Thus, shaft displacement can be an indicator of the fault. Similarly, stator electrical imbalance will cause changes in the current and power output of the electric generator. Stator electrical imbalance can be detected from the variations of the harmonic contents of electrical signals. Usually, the fault-related information is contained in rotor and stator line currents.

Ref. [13] introduced a new approach to perform the diagnosis of rotor asymmetry faults in wound rotor induction motors (WRIMs), a type of generator that is drawing increasing attention due to its use in large power applications and as a generator in wind turbine generator units. The paper explores the analysis of the external magnetic field under the starting to detect rotor winding asymmetry defects in WRIMs by using advanced signal processing techniques. Moreover, a new fault indicator based on this quantity is introduced, comparing different levels of fault and demonstrating the potential of this technique to quantify and monitor rotor winding asymmetries in WRIMs. The proposed approach relies on the detection of patterns that appear in the time–frequency maps under motor starting when the fault is present. These patterns were efficiently detected using different signal processing techniques applied to the electromotive force (EMF) signals captured during transient: On the one hand, a continuous tool, the short-time Fourier transform (STFT), was used to track the evolution of fault harmonics. On the other hand, a discrete transform, the discrete wavelet transforms (DWT), was used as an auxiliary tool in the calculation of a fault-severity indicator.

A speed-sensorless method is applied for detecting rotor asymmetries in wound rotor induction generators working under nonstationary conditions [14]. The method is based on the time–frequency analysis of rotor currents and on a subsequent transformation, which leads to the following goals: Unlike conventional spectrograms, it enables to show the diagnostic results as a simple graph, similar to a Fourier spectrum, but where the fault components are placed always at the same positions, regardless the working conditions of the generator. Moreover, it enables to assess the generator condition through a very small set of parameters. These characteristics facilitate the understanding and processing of the diagnostic results, and thus, help to design improved monitoring and predictive maintenance systems. Also these features make the proposed method, the block diagram of which is shown in Figure 7.5, very suitable for condition monitoring of wind power generators, because it fits well with the usual nonstationary working conditions of wind turbines, and

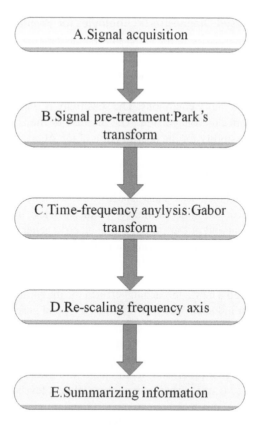

FIGURE 7.5   Block diagram of the harmonic order tracking analysis under time-varying conditions [14].

makes feasible the transmission of significant diagnostic information to the remote monitoring center using standard data transmission systems.

Ref. [15] proposes a methodology to improve the reliability of diagnosis of stator winding asymmetry and rotor winding asymmetry in wound-rotor induction generators which work under variable load conditions, as in wind turbine applications. The method is based on the extraction of the instantaneous frequency (IF) of the fault-related components of stator and rotor currents during speed changes caused by non-stationary functioning. It is shown that, under these conditions, the IF versus slip plots of the fault components are straight lines with a specific slope and y-intercept for each kind of fault. In addition, neither of these patterns are dependent on the generator features or the way that the load changes. The schematic diagrams of the rotor and stator asymmetry test rigs in [15] are shown in Figure 7.6 and Figure 7.7, respectively.

Reference [16] presents a new technique for detecting and identifying rotor electrical asymmetry faults of DFIGs from the rotor-side inverter control loop, using the error signal, to provide a future method of generator condition monitoring with enhanced detection sensitivity. Stator current and power were also investigated for rotor electrical asymmetry detection and comparison made with rotor-side inverter

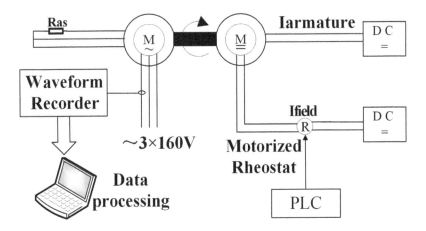

FIGURE 7.6   Rotor asymmetry test rig in [15].

control signals. An investigation was then performed to define the sensitivity of the proposed monitoring signals to fault severity changes and a comparison made with previous current, power and vibration signal methods. The results confirm that a simple spectrum analysis of the proposed control loop signals gives effective and sensitive DFIG rotor electrical asymmetry detection. The comparison of fault detection sensitivities between open- and closed-loop test rigs is shown in Table 7.3.

Reference [17] investigates the condition monitoring of wind turbine wound rotor and doubly fed induction generators with rotor electrical asymmetries by analysis of stator current and total power spectra. The research is verified using experimental data measured on two different test rigs as shown in Figure 7.8 and Figure 7.9

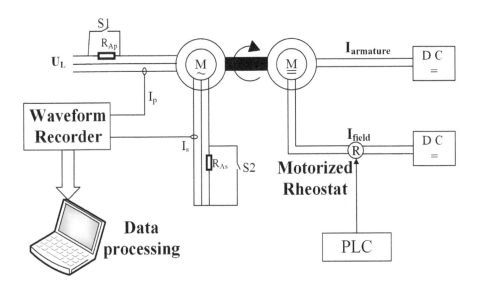

FIGURE 7.7   Stator asymmetry test rig in [15].

TABLE 7.3

Comparison of Fault Detection Sensitivities between Open- and Closed-Loop Test Rigs [16].

Test Rig System	Closed-Loop System		Open-Loop System				
Signal type	Current error signals inside RSI εi_{dr} εi_{qr}		Stator current		Stator total power		Vibration
Frequency analysis fault type	FFT 20% rotor electrical asymmetry		Frequency tracking algorithm 23% rotor electrical asymmetry				SBPF missing tooth of high speed shaft pinion
Harmonics of interests	$2sf$		$(1-2s)f$	$(3-2s)f$	$2sf$	$(2-2s)f$	$2f_{mesh,\,HS}$ and its first five sideband peaks on each side
Stator voltage	78 V				230V		
Sensitivity calculated at 1550 rev/min	14.3 dB	15.2 dB	3.0 dB	4.7 dB	6.7 dB	4.9 dB	5.1 dB
Sensitivity calculated at 1600 rev/min	14.5 dB	15.3 dB	3.7 dB	6.9 dB	7.3 dB	6.0 dB	4.6 dB

and numerical predictions obtained from a time-stepping electromagnetic model. A steady-state study of current and power spectra for healthy and faulty conditions is performed to identify fault-specific signal changes and consistent slip-dependent fault indicators on both test rigs. To enable real-time fault frequency tracking, a set of concise analytic expressions, describing fault frequency variation with operating speed, were defined and validated by measurement. A variable speed study, representative of real wind turbine operations, of current and power frequency components for healthy and faulty conditions was then carried out on one test rig, which could simulate wind conditions. The current and power fault frequency tracking previously identified achieved reliable fault detection for two realistic wind turbine generator fault scenarios of differing severity. Conclusions are drawn on the relative merits of current and power signal analysis when used for wind turbine wound rotor induction generator fault detection and diagnosis.

Determining the magnitude of particular fault signature components (FSCs) generated by wind turbine (WT) faults from current signals has been used as an effective way to detect early faults [18]. However, the WT current signals are time varying due to the constantly varying generator speed. The WT frequently operates with the generator close to the synchronous speed, resulting in FSCs manifesting themselves in the vicinity of the supply frequency and its harmonics, making their detection more challenging. To address this challenge, the detection of rotor electrical asymmetry in WT doubly fed induction generators, indicative of common winding, brush gear, or high resistance connection faults, has been investigated using a test rig under three different driving conditions, and then an effective extended Kalman filter (EKF) based method is proposed to iteratively estimate the

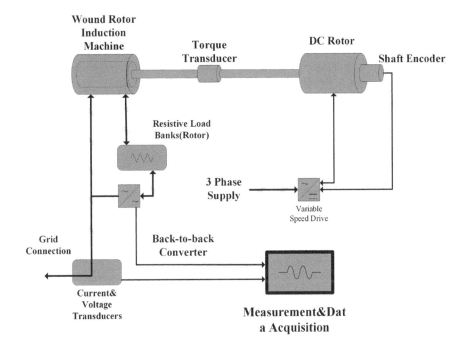

FIGURE 7.8   Test Rig 1 in reference [17].

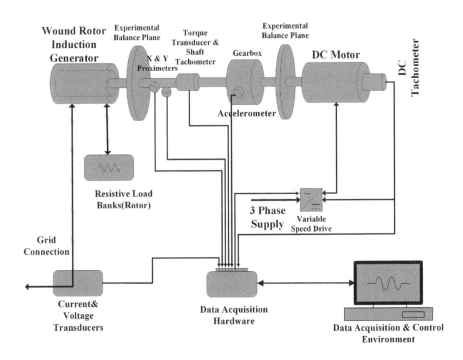

FIGURE 7.9   Test Rig 2 in reference [17].

FSCs and track their magnitudes. The proposed approach has been compared with a continuous wavelet transform (CWT) and an iterative localized discrete Fourier-transform (IDFT). The experimental results demonstrate that the CWT and IDFT algorithms fail to track the FSCs at low load operation near-synchronous speed. In contrast, the EKF was more successful in tracking the FSCs magnitude in all operating conditions, unambiguously determining the severity of the faults over time and providing significant gains in both computational efficiency and accuracy of fault diagnosis.

7.2.3   Winding Open- and Short-Circuit Faults

Short circuits of coils and inter-turn faults are one of the most common failure modes in the induction generators used in WTs. Asymmetry is usually present in the magnetic field during a winding fault. In this case, the faults can be diagnosed by monitoring their characteristic frequencies f_w described by Equation (7.3) in the electrical signals acquired from the electric generator terminals using appropriate frequency and time-frequency analysis techniques.

$$f_W = \{f \mid f = [k \pm n(1-s)/p]/f_s; \ k = 1,3; \ n = 1,2,\dots,(2p-1)\} \qquad (7.3)$$

where p is the number of pole pairs, f_s is the fundamental frequency, and s is the slip.

Stator open-circuit faults will change the spectra of stator line currents and instantaneous power. Experimental results showed that the spectrum of the instantaneous power carried more fault-related information than the stator line currents.

Rotor winding inter-turn short-circuit (RWISC) occurring in turbine generators has been plaguing power producers with respect to production safety and economy [19]. RWISC is characterized by impaired fundamental component of the exciting magnetomotive force and resultantly reduced no-load electromotive force. In reference [19], an inverse reasoning method is proposed to analyze RWISC. Firstly, it is assumed that the generator's rotor winding is running in good condition and the generator's electromagnetic power (virtual power, VP) is inversely calculated from the obtained operational data on generators. Secondly, the difference between the virtual power and the actual electromagnetic power (EMP) is calculated and then used to determine whether an RWISC occurs on the turbine generator. Three real cases of severe rotor winding shorts have demonstrated that this method proposed in reference [19] not only enables to diagnose RWISC, but also allows determining the RWISC level and its trending accurately. The method is appropriate for RWISC diagnosis, whether offline or online, and can be applied to make the generator sets in the power plant run safer, more economical, and more stable.

Figure 7.10 presents the process to diagnose RWISC, where a% indicates the difference between VP and the actual EMP. Analysis over three real RWISCs using the aforementioned method indicated that in the case of no RWISC occurring in the turbine generator, a% is always very low. For turbine generators with general 160 turns of rotor windings, even RWISC in just one turn may result in a% higher than 0.5%, so the threshold value can be set as 0.5% reasonably.

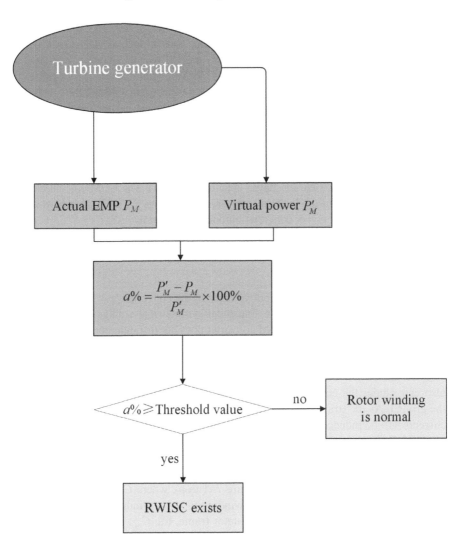

FIGURE 7.10   Process to diagnose RWISC [19].

7.2.4   BEARING FAULTS

A bearing is commonly used to support the rotating shaft of the generators, which is considered as one of the most critical components in a wind turbine [20]. With continuous rotation of the generator shaft, a bearing may wear and then cause fatigue, cracking, and even breakage of certain parts. The use of pulse width modulation (PWM) in power converters may lead to electrical corrosion of the bearing of a wind power generator, which significantly accelerates bearing wear [21]. Severe bearing failures may cause catastrophic faults in other components in a wind turbine subsystem, such as shaft cracking, gearbox failure and even damage to the generator. Therefore, early detection of bearing failures is necessary to prevent severe damage and reduce revenue loss.

As a common fault, the bearing fault distribution varies between 40% and 90% from large to small machines [22]. Some possible causes, such as electric corrosion, lubricant impurities, misalignment in assembly and the overall vibration, may lead to cracks, pits, and spalls on the rolling surface [23]. Among all the above root causes of bearing faults, the shaft voltages and the associated bearing currents are the most common causes of accelerated bearing degradation [21]. Bearing currents may be caused by electrostatic charging and magnetic flux asymmetries in electrical machines [24]. With rapid development of power electronics, the power converter has become another essential source of bearing currents [25]: Because of the common mode signal associated with PWM switching, bearing currents can be induced by the high rate of voltage change (dV/dt) and electrical discharge machining (EDM). The EDM current caused by overvoltage will lead to electrical corrosion of bearings [26]. The common mode voltage V_{com}, which the power converter outputs, is defined by:

$$V_{com} = \frac{V_a + V_b + V_c}{3} \tag{7.4}$$

where V_a, V_b, and V_c are phase output voltages of the converter.

Unlike the sum of the three-phase sinusoidal symmetrical voltages being equal to zero, the V_{com} produced by the PWM converter does not instantaneously sum to zero. The value of V_{com} is determined by both the DC bus voltage and the modulation mode [27]. For the DFIG, the rotor side converter can supply the common mode voltage V_{comr} between the rotor winding and the ground, meanwhile, the grid side converter also can supply the common mode voltage V_{coms} between the stator winding and the ground. As the stray capacitances construct the coupling circuit, common mode currents will be induced in the DFIGURE Classified according to conductivity, four kinds of conductors (the stator windings, the rotor windings, the stator core, and the rotor core) are present in the DFIGURE Between any two of the four conductors, there will be a parasitic coupling capacitance. Consequently, the complete parasitic circuit of the DFIG is as shown in Figure 7.11 (top), where C_{rwsr}, C_{rwr}, and C_{rwf} are the parasitic capacitances for the rotor winding to stator winding, the rotor winding to rotor core, and the rotor winding to stator frame, C_{rf}, C_{swr}, and C_{swf} are those for the rotor core to stator frame, the rotor core to stator winding, and the stator winding to stator frame, $C_{bearing}$ and C_{cap} are those for the bearing and its insulation cap, R_{brush} and L_{brush} are the equivalent resistance and inductance of the grounding brush, respectively. In this circuit, C_{rwf} and C_{swf} are in parallel with the common mode voltage source respectively, which should be short circuited. Moreover, the values of C_{rwsr}, C_{rwf} and C_{swr} are much smaller than those of C_{swf} and C_{rwr} in the presence of the air-gap. Thus, the voltage between the shaft and ground mainly depends on V_{comr}. However, for the generator with full power converter in the stator (e.g. the SCIG), the parasitic circuit is shown in Figure 7.11 (bottom). By simplifying the circuits, the voltage value between inner and outer raceways by neglecting grounding and brush can be expressed as

$$V_{bearing} \approx \frac{C_{rwr}C_{cap}}{(C_{rwr} + C_{rf} + C_{swr})(C_{cap} + C_{bearing}) + C_{cap}C_{bearing}} V_{com} \tag{7.5}$$

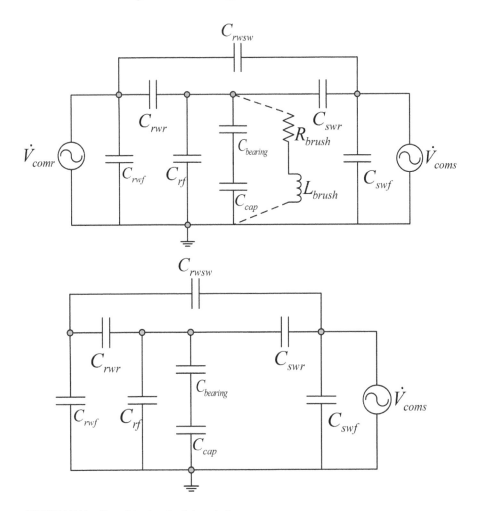

FIGURE 7.11 Parasitic circuit of the wind generator.

The threshold voltage of the bearing oil has been reported to be approximately 5 to 30 V [28]. If $V_{bearing}$ (Figure 7.12) exceeds this threshold, the film will break down. Then, the emergent energy produced by EDM current may cause denaturalization of grease and result in small pits on the surface of the raceway and balls.

Since the capacitive voltage of the DFIG bearing is greater than that of the traditional stator-side inverter-fed machines, the bearing of the DFIG will be more likely to suffer the effects of common mode current. One common way in which to ensure a minimal shaft voltage is to use the shaft grounding by means of one or more grounding brushes, which may avoid the damaging current circulating through the bearing to its pedestal [29].

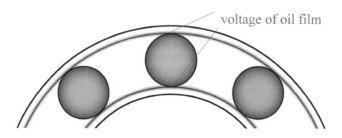

voltage of oil film

FIGURE 7.12   Illustration of bearing corrosion.

7.2.4.1   Features of Stator Current Signature Due to Bearing Faults

The grounding brush cannot completely avoid the shaft current of the generator for the nonideal impedance, and consequently long-term electrical corrosion may lead to fatigue or wear of the ball bearings. Then, the small abrasions on the bearings will develop into cracks or breakages on the surface of the raceway. When the roller ball passes through a crack, a slight collision will cause an impact wave at the natural frequency of the bearing system. The number of collisions per period can be called the fault characteristic frequency. For bearing faults in different locations (Figure 7.13), the characteristic frequencies can be expressed as [30]:

$$f_i = \frac{1}{2} N_B f_r \left(1 + \frac{D_b \cos\theta}{D_c} \right) \qquad (7.6)$$

$$f_o = \frac{1}{2} N_B f_r \left(1 - \frac{D_b \cos\theta}{D_c} \right) \qquad (7.7)$$

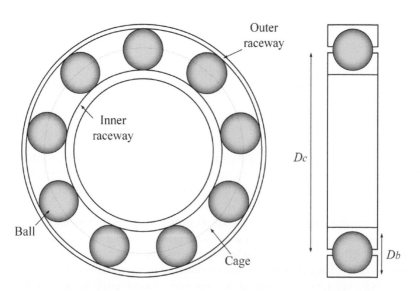

FIGURE 7.13   Configuration of the ball bearing [31].

$$f_b = \frac{1}{2} f_r \left(\frac{D_c}{D_b} \right) \left[1 - \left(\frac{D_b \cos\theta}{D_c} \right)^2 \right] \tag{7.8}$$

$$f_c = \frac{1}{2} f_r \left(1 - \frac{D_b \cos\theta}{D_c} \right) \tag{7.9}$$

where f_i, f_o, f_b, and f_c are the characteristic frequencies of the inner raceway, outer raceway, ball, and cage, respectively; f_r is the rotating frequency of the bearing, N_B the number of balls in the bearing, D_b the ball diameter, D_c the pitch or cage diameter, and θ the contact angle, respectively.

Collision in such bearings will couple to the coaxial connected generator and further causes torque variation. As a result, the torque of the generator can be modulated by different bearing failures to:

$$T(t) = T_0(t) + \sum T_v \cos(2\pi f_{fault} t + \varphi_v) \tag{7.10}$$

where T_0 is the torque produced by wind power, T_v and φ_v are the amplitude and phase of torque variation due to bearing faults, and f_{fault} is the fault characteristic frequency of the bearing, respectively [31].

Since the electromagnetic torque is the product of flux and current, the torque oscillation resulting from rotor eccentricity will lead to amplitude modulation (AM) and quadratic phase coupling (QPC) of current. The modulated phase current in stator windings of generators can be described as

$$I_{sa} = I_0 \sin(\omega_s t + \varphi_0) + \sum_{i=1}^{n} I_{i+} \cos[(\omega_s + i\omega_{fault})t + (\varphi_0 + \varphi_i)] +$$

$$\sum_{i=1}^{n} I_{i-} \cos[(\omega_s - i\omega_{fault})t + (\varphi_0 - \varphi_i)] \tag{7.11}$$

where $\omega_s = 2\pi f_s$, $\omega_{fault} = 2\pi f_{fault}$, f_s is the fundamental frequency of the stator current, I_0 and φ_0 are the amplitude and phase of the fundamental current, and $I_{i\pm}$ and $(\varphi_0 \pm \varphi_i)$ are the amplitude and phase of the modulation harmonics, respectively [32].

From (7.11), the frequency and phase of harmonic components in the stator current caused by bearing corrosion can be expressed as

$$f_{bearing} = f_s \pm i f_{fault} \tag{7.12}$$

$$\varphi_{bearing} = \varphi_0 \pm \varphi_i \tag{7.13}$$

The phases of the two sideband components in (7.13) consist of the sum and difference of φ_0 and φ_i, so this type of interaction is generally referred to as QPC. The distinctive feature of harmonic components, thereby, provides a possible solution to judgment of the bearing failure based on stator current signals.

7.2.4.2 Different Methods of Converter-Based Bearing Fault Diagnostics

A number of traditional diagnosis methods have been used in wind turbine systems to detect bearing faults, e.g. methods based on the analysis of the vibration signal, the acoustic signal, the temperature, and the lubrication oil parameter [33]. However, all of these diagnostic methods need signal acquisition devices with a high sampling rate. In addition, signals may be attenuated or interfere in the transmission process between defective components and sensors. Hence, electrical signal-based analysis methods have received more attention in recent years due to their being noninvasive and cost-effective [34]. Current signals in an electrical machine are continuously detected by current sensors in the power converter to control active and reactive power. Therefore, current-based condition monitoring and fault diagnosis may eliminate the requirements of additional hardware, which confers economic benefit and offers the potential for adoption in wind power generation systems.

Since the bearing is mechanically coupled with the electric machine, certain frequency components of the current can be monitored as an index to determine the failure due to torque variation [35]. Different data processing methods have been widely investigated to extract useful information from current signals with a low signal-to-noise ratio (SNR) [33]. Some conditioning methods are usually applied to raw signals in the time-domain to facilitate the extraction of fault features in the signals. For example, envelope analysis can outline the extremes of a signal, which is also called the amplitude modulating component of the signal [35]. The Hilbert transform is a classical method to compute the instantaneous amplitude and phase of a signal, which has been primarily used for signal demodulation combined with envelope analysis [36]. The envelop spectrum obtained from Hilbert transformation has been proved to be effective in bearing fault diagnosis based on current signals if the mechanical characteristic frequency rate is low [37]. Empirical mode decomposition and the Hilbert-Huang transform (HHT) have also been used for current-based fault diagnosis to highlight the time-frequency representation of intrinsic mode functions [38]. To classify the fault location in a bearing, the current signal (time-domain) should be transformed to the frequency-domain to extract various features. In industrial applications, Fourier analysis is probably the most popular analysis technique. Thus, the fast Fourier transform (FFT) is usually applied to obtain the frequency spectrum [39]. The Extended Park's Vector Approach (EPVA) uses the frequency spectrum of the Park's vector module to improve the SNR of the current signals by converting the fundamental component to the DC component, which successfully detects different types of bearing faults [40]. Considering the energy of such signals, the power spectrum can be introduced to evaluate the power within the unit frequency band [41]. Several improved methods have been proposed to increase the accuracy and frequency resolution of power spectra, e.g., the spectral kurtosis algorithm [35], the Welch power spectrum analysis [42], the root-MUSIC method [43] and noise cancellation [44]. However, the Fourier transform cannot indicate how the frequency content of a nonstationary signal changes. To overcome this drawback, the resampling technique and short-term Fourier transform (STFT) have been proposed for analyzing the time-varying frequency response [31, 45]. The wavelet transform is a multiscale analysis method that can improve the constant time-frequency resolution of the STFT, which is capable of grasping both the time and frequency information in a signal [46]. The optimized stationary wavelet packet transform has been proposed in [47] to decrease the computational intensity and simplify the interpretation of results. Moreover, data processing according to the model-based and probability-based method has been applied to signals with low resolution [48, 49]. Recently, AI techniques have also been introduced to bearing fault diagnosis based on electrical signals, such as artificial neural networks (ANNs), support vector machines (SVMs), expert systems, and fuzzy logic systems [50, 51]. Although these AI algorithms can identify fault modes and locations from large amounts of data, the requirement of data storage and computation is also increased. However, previous reports focus on the amplitude and frequency of current signals and ignored the important information conveyed in the phases of signal harmonics. Since bearing faults can generate nonlinear harmonics in a vibration signal, nonlinear phase coupling will be an essential feature [52]. Higher-order spectra have been developed to deal with the non-Gaussian or nonlinear processes by revealing information about amplitude and phase in a stochastic process, whereas phase relationships between frequency components are suppressed in the power spectrum as defined in terms of second-order statistics of a signal [53]. A particular case of higher-order spectra is the third-order spectrum, i.e. the bispectrum, which is the Fourier transform of the third-order statistics and incurs the lowest computation cost among such higher-order spectra. To avoid misleading results, a modulation detector is applied to revise the estimation of the bispectrum [54]. Therefore, the bispectrum estimation and its modulated signal have been utilized in diagnosis of incipient bearing faults based on the vibration signal [55]. Recently, the modulation signal bispectrum (MSB) detector has successfully identified failures caused by electrical imbalance based on electrical signals in laboratory experiments [56]. The MSB method is modified to detect bearing corrosion failure of wind turbines with limited sampling time and data in [57], which is verified by field tests.

7.2.4.3 Comparison of Four Typical Diagnostic Results

A comparison of four typical diagnose results by different data processing methods are shown in this section. Some restrictions in the comparison are as follows: (1) different algorithms are applied to the same data, (2) the current data should be converted to per unit values, (3) calculation results are expressed in decibels (dB), (4) the Hanning window is applied to all records, (5) two bearings with and without outer race fault are applied in this section, and (6) the fault frequency at a specific revolution speed is 64.9 Hz at a specific revolution speed according to Equation (7.9). Therefore, the harmonic components in the stator current caused by bearing outer race fault are |50 ± 64.9| Hz.

7.2.4.3.1 Diagnostic Results of Power Spectra

The estimation of the power spectral density has been a useful tool in digital signal processing for more than 50 years, and has been applied in bearing fault diagnosis [53]. The power spectral density is expressed as:

$$P(\omega) = \lim_{T \to \infty} \frac{|F_T(\omega)|^2}{2\pi T} \tag{7.14}$$

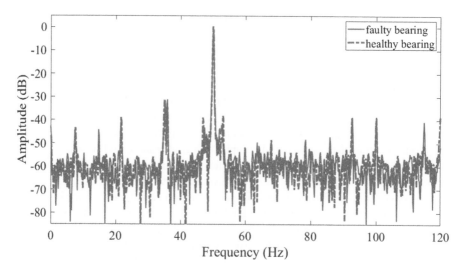

FIGURE 7.14 Comparison of power spectra for current signals [57].

where $F_T(\omega) = F(f_T(t))$, F () denotes the Fourier transform of the signal.

The amplitude comparison of the power spectra for the two segments of current data is presented in Figure 7.14. Since harmonic frequencies of the stator current caused by bearing outer race failure are 114.9 Hz and 14.9 Hz according to Equation (7.12), consideration should be given to both frequencies. In addition, the visible difference persists at the single side band of 85.9 Hz in the power spectrum, which is not caused by the mechanical modulated fault.

7.2.4.3.2 Diagnostic Results of the EPVA

The Park's vector can convert balanced frequency components in three phases to a DC signal, which improves the signal to noise ratio and merges two harmonic components [58]. The d-q components of Park's vector can be obtained from three phase current data as follows [59]:

$$i_d = \frac{\sqrt{2}}{\sqrt{3}} i_a - \frac{1}{\sqrt{6}} i_b - \frac{1}{\sqrt{6}} i_c \tag{7.15}$$

$$i_q = \frac{1}{\sqrt{2}} i_b - \frac{1}{\sqrt{2}} i_c \tag{7.16}$$

where i_a, i_b, and i_c are three phase currents in the stator windings. The value of i_c can be calculated, from the other two phase current data, as

$$i_c = -(i_a + i_b) \tag{7.17}$$

Then, the square of Park's vector in Equation (7.15), which is to be analyzed in the frequency domain, will only contain unbalanced harmonic components. This method is called the EPVA.

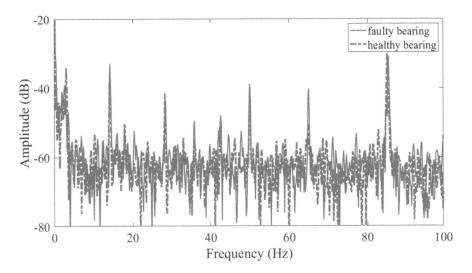

FIGURE 7.15 Comparison of current spectra by the EPVA [57].

$$I_s = \left| i_d + j i_q \right|^2 \tag{7.18}$$

Moreover, the EPVA also transforms the harmonic frequency of $f_s \pm i f_{fault}$ in the current signal to $i f_{fault}$. Therefore, the characteristic frequency of the Park's vector can be extracted using spectral analysis, which has successfully diagnosed the bearing fault by way of an analysis of the current signals; however, one more phase of current date should be monitored to calculate the Park's vector in contrast with the MSB method, which needs more data storage space. The spectral comparison of the faulty and the healthy bearings using EPVA is shown in Figure 7.15. Even though the difference of BPFO at 64.9 Hz is obvious, visible differences still remain at other frequencies such as 20.4 Hz, 35.9 Hz, and 77 Hz. These three frequencies are not characteristic frequencies of the bearing according to Equations (7.6)–(7.9). For the reason that two records are acquired before, and after, replacement of one bearing in the generator in the same condition, the differences can only be caused by the bearing. Therefore, additional peaks in the spectrum may impede fault diagnosis.

7.2.4.3.3 Diagnostic Results of Hilbert Demodulation-Based Envelop Analysis
The Hilbert demodulation-based envelope analysis is another regular data processing method used in bearing fault diagnosis. The procedure used in this method is usually described by the following four steps [37]:

Step-1, computation of the Hilbert transform of the sampled data by

$$H[f(t)] = \int\limits_{-\infty}^{+\infty} \frac{f(\tau)}{f(t-\tau)} d\tau \tag{7.19}$$

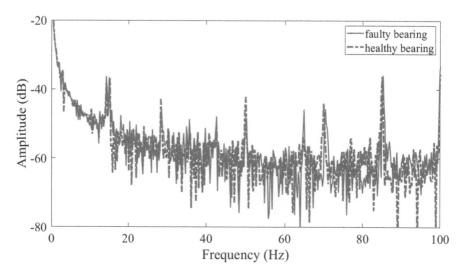

FIGURE 7.16   Comparison of current spectra by envelope analysis [57].

Step-2, computation of the upper envelope of the signal;
Step-3, filtering of the DC component of the envelope signal;
Step-4, computation of the Fourier transform of the filtered envelope signal.

Hence, it is possible to retrieve the component at the characteristic frequency of the bearing. The variations in envelope spectrum amplitudes for healthy, and faulty, cases are compared in Figure 7.16. Similar to the result arising from use of EPVA, more visible differences occur at 35.9 Hz, 50 Hz and 87 Hz.

7.2.4.3.4   Diagnostic Results of Modulation Signal Bispectrum Analysis

As illustrated above, several methods have been proposed with which to analyse the power spectra and the envelope of the electrical signal, such as use of the FFT, wavelet analysis, HHT, etc. However, all of these algorithms cannot take advantage of the special feature of symmetrical sideband components in the modulated current described by Equation (7.11). The bispectrum analysis, which is a nonlinear signal processing method based on high-order statistic, is commonly used to reveal the QPC from raw data; however, traditional bispectrum detectors may produce incorrect and misleading results, which interferes with the judgment made in fault diagnosis. To avoid omissions and misdiagnoses, the MSB and its normalized form are proposed in Equations (7.20) and (7.21) [54]:

$$B(f_m, f_c) = E\{X(f_c + f_m)X(f_c - f_m)X^*(f_c)X^*(f_c)\} \tag{7.20}$$

$$b(f_m, f_c) = \frac{\left|B(f_m, f_c)\right|^2}{E\left\{\left|X(f_c)X(f_c)\right|^2\right\}E\left\{\left|X(f_c + f_m)X(f_c - f_m)\right|^2\right\}} \tag{7.21}$$

where $X(f)$ is the discrete Fourier transform (DFT) of a discrete time current signal $x(k)$, $E(X)$ is the statistical expectation operator of the random variable X, superscript $*$denotes the complex conjugate, f_c and f_m are the carrier and modulation wave frequencies, respectively. $X(f)$ can be defined as

$$X(f) = \sum_{k=-\infty}^{\infty} x(k)e^{-j2\pi fk} \tag{7.22}$$

Differing from normal bispectral analysis, the MSB detector searches for the carrier frequency and its modulated frequency of $f_c \pm f_m$. To recognize the fault characteristic frequency f_{fault} (f_m), Eq. (7.20) needs to be expanded to

$$B(f_m, f_c) =$$

$$E\left\{\left|X(f_c + f_m)\right|e^{j\angle X(f_c+f_m)} \cdot \left|X(f_c - f_M)\right|e^{j\angle X(f_c-f_m)} \cdot \left|X(f_c)\right|e^{-j\angle X(f_c)} \cdot \left|X(f_c)\right|e^{-j\angle X(f_c)}\right\} \tag{7.23}$$

Then, Eq. (7.23) can be rewritten as

$$B(f_m, f_c) = E\left\{\left|X(f_c + f_m)\right|\left|X(f_c - f_m)\right|\left|X(f_c)\right|\left|X(f_c)\right|e^{j(\angle X(f_c+f_m)+\angle X(f_c-f_m)-\angle X(f_c)-\angle X(f_c))}\right\} \tag{7.24}$$

If the phase coupling occurs in sideband frequency components, the following relationship can be obtained:

$$\angle X(f_c \pm f_m) = \angle X(f_c) \pm \angle X(f_m) \tag{7.25}$$

Then, Eq. (7.23) can be simplified to

$$B(f_m, f_c) = E\left\{\left|X(f_c + f_m)\right|\left|X(f_c - f_m)\right|\left|X(f_c)\right|\left|X(f_c)\right|\right\} \tag{7.26}$$

Therefore, if a current signal is composed of components with f_c Hz and $(f_c \pm f_m)$ Hz and phases are coupled, the MSB detector will reaches its peak value in $B(f_m, f_c)$. For current components without phase coupling, the expectation in (7.24) approaches zero after taking the average of all such segmented samples. As a result, the MSB method is effective in deducing modulated components in current signals from which noise has been eliminated. By combing the physical models of the electric machine described in (7.10) and (7.11), the fault characteristic frequency of the bearing can be calculated from current signals.

To highlight the difference between faulty and healthy bearings, the MSB results should be plotted on the same graph. Thus, the 3D results are converted to 2D format by only retaining f_m when f_c is around 50 Hz. As can be seen from the comparison of MSB results shown in Figure 7.17, the magnitude of the outer-race characteristic frequency for the faulty bearing is much larger than that of the healthy bearing, meanwhile the variation of other harmonic components is small. The diagnosis of a bearing fault can rely on this contrast at a certain frequency based on MSB analysis.

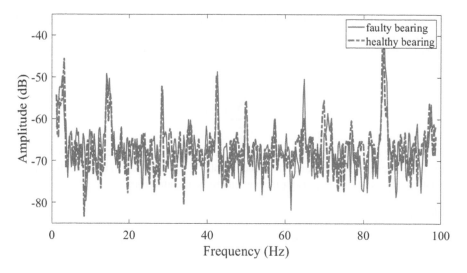

FIGURE 7.17   Comparison of current spectra by the MSB method [57].

7.2.4.3.5   Summary of the Four Data Processing Methods

From Figures 7.14 to 7.17, the amplitude for BPFO of the faulty bearing is greater than that of a healthy bearing; however, additional visible differences in the power spectrum, the EPVA spectrum, and the envelope spectrum may lead to misdiagnosis of the bearing fault. For example, the amplitude peaks at 35.9 Hz appearing in both the EPVA spectrum and the envelope spectrum are caused by the harmonic component of $50 + 35.9 = 85.9$ Hz in the power spectrum (this peak disappears in the MSB result). The reason is that the MSB detector not only analyses the amplitude of the harmonic components but also takes the phase information into account. Since the QPC in the AM cannot be identified by algorithms based on single spectrum (the second order spectrum), the algorithm, when applied to the bispectrum, will be effective, accurate, and incurs low computational cost. Based on this analysis, the results among the four data processing methods in the specific case of bearing fault diagnosis using the approximately stationary current data are compared, as shown in Table 7.4.

TABLE 7.4
Comparison of Data Processing Methods Using Current Data

Property	MSB Method	Power Spectral	EPVA	Envelope Analysis
Domain	Frequency	Frequency	Frequency	Time
Frequency extraction	Yes	Yes	Yes	Yes
Detecting the QPC	Yes	No	No	No
Complexity	High	Low	Medium	Medium
Data storage space	Low	Low	High	Low

7.2.5 Air Gap Eccentricity

Rotor eccentricity faults also form a large percentage of the faults in wind turbine generators. Moreover, it may be associated with other faults and may occur as a result of bearing failure and rotor bar broken [60], which makes the diagnosis and detection of eccentricity fault complicated. The appearance of rotor eccentricity will cause the unequal air gap between stator and rotor. As shown in Figure 7.18, static eccentricity occurs when the axis of the rotor is at a constant distance from the center of the stator meanwhile the rotor still rotates with its own axis. However, dynamic

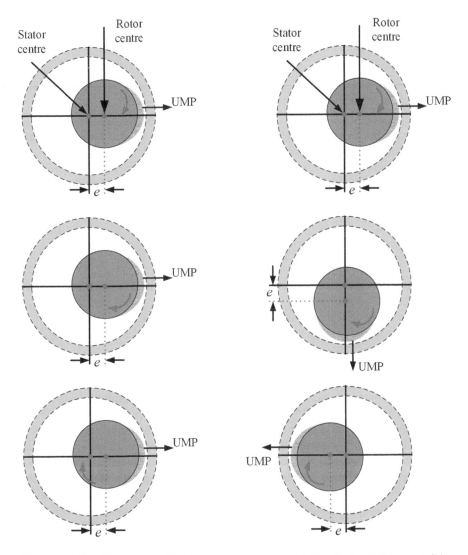

FIGURE 7.18 Illustration of different types of eccentricity: (Left) static eccentricity and (Right) dynamic eccentricity.

eccentricity occurs when the rotor center is not the true rotational axis although it still rotates with the stator axis. If the tow conditions exist together, the mixed eccentricity appears [61]. With air-gap eccentricity, circuit inductances will vary, causing asymmetrical air gap flux distribution. This imbalance produces special current waveforms and electromagnetic forces between the stator and rotor, which can be applied to monitoring the eccentricity faults of the wind power generators.

7.2.5.1   Current-Based Diagnostic Technologies for Air Gap Eccentricity Faults

Current based diagnose and monitoring technologies are widely applied in the detection of rotor eccentricity. For induction generator, the asymmetrical air gap flux distribution can produce harmonic currents with the specific frequencies as

$$f_{ec} = f_1\left(\frac{Z_r}{p}(1-s) \pm k_1\right) \pm k_2\left(f_1\frac{(1-s)}{p}\right) \tag{7.27}$$

It has been shown in [62] that only very high static eccentricity increases the principal slot harmonics in (7.27), which allows for the distinguishing between static and dynamic eccentricity-related components.

For PM generators, the impacts of eccentricity on the current spectrum can be determined by the analysis of the corresponding magnetic flux density components [63]. Therefore, the stator current can be defined as

$$i(t) = I_0\cos(\omega t - \varphi) + \sum_{k=0}^{\infty}I_k\cos\left[\left(1-\frac{k}{p}\right)\omega t - \varphi_1\right] + \sum_{k=0}^{\infty}I_k\cos\left[\left(1+\frac{k}{p}\right)\omega t - \varphi_2\right] \tag{7.28}$$

where I_0 is the amplitude of fundamental current, I_k is the amplitude of harmonic components.

According to (7.28), sideband components at frequencies $(1 \pm k/p) f_1$ ($k = 1, 2, 3, \ldots$) are observed in the distribution of stator currents, which can be utilized as a proper index for eccentricity fault diagnosis in PM generators.

As a result, monitoring the amplitude of sideband components with a particular frequency pattern which is extracted from the spectrum of stator current is a popular method [64]. Since the method utilized the current sensor in convertor, it is an economically attractive method without any extra hardware. Literature [65] presented the classical rotor slot passing frequency flux and current components with twice the supply frequency to predict the current signature pattern as a function of air-gap eccentricity. By detecting the zero-sequence current based on calculating and then summing the three-phase currents, the single-phase rotation test method and static fault diagnosis method have recently been proposed to diagnose the static eccentricity [66]. Eccentricity fault diagnosis can be done in real time by analyzing the frequency components of stator current signals with computer-aided finite element monitoring, which are discussed in [67]. According to experimental results in [68], the Park's vector approach, as illustrated from (7.15) to (7.18), can also detect the air gap eccentricity.

However, current monitoring techniques largely depend on human expertise owing to the existence of many current components in the spectrum. It has been shown in [69] that the fault signatures in time-varying operating conditions vary with a frequency proportionally to the variation of the speed. This indicates that current based diagnose technologies may not be appropriate for the speed variation case. The time-frequency analysis method, such as the short time Fourier transform, the continuous wavelet transform, and the Hilbert Huang Transform, are promising solutions to overcome the transient problem.

7.2.5.2 Torque/Power-Based Diagnostic Technologies for Air Gap Eccentricity Faults

Electromagnetic torque of the generator is achieved by multiplying the flux linkage and the stator currents. From the point of view of energy conversion, forward stator rotating field produces a steady torque, meanwhile the backward stator field interacting with the rotor field creates the oscillating torque. Frequency of the oscillating torque is defined as

$$\omega = -\omega_s - [\omega_s(1-s) + s\omega_s] = -2\omega_s \tag{7.29}$$

where ω_s is the stator current frequency of the generator. For induction generator, s is defined from 0 to 1, whereas the value of s is 0 in the synchronous generator.

Thus, the double frequency torque can be regarded as an indicator of the unbalanced rotor. According to the basic torque equation that derived from stator and rotor flux density waves, the derivation of the torque is [70]

$$T = -\frac{1}{\mu_0} l_m R_0 g_0 \int_0^{2\pi} [B_s(\theta,t) + B_r(\theta,t)] \frac{\partial}{\partial \theta_r} B_r(\theta,t) d\theta \tag{7.30}$$

where θ_r denotes the rotor position, g_0 the mean air-gap length without eccentricity, l_m the effective lamination length, B_s and B_r are stator and rotor flux density waves, respectively. Oscillating torque components are derived from stator current and air gap expressions as

$$T = -\frac{\pi l_m R_0 g_0}{\mu_0} B_s B_r' \sin(\mp \omega_r t - \varphi) \tag{7.31}$$

From (7.31), the oscillating torque components at rotor frequency ω_r are monitored by the interaction of magnetic field waves, which are obviously increased at both static eccentricity and dynamic eccentricity conditions in simulations [61]. Moreover, an efficient time-domain technique based on a modified Prony's method for the air-gap eccentricity fault detection in is presented in [71]. The stator current and apparent power are divided into short overlapped time windows and each one is analysed by the least squares Prony's method by this way, which allows tracking the frequency and amplitude of the air-gap eccentricity fault characteristic component with a high accuracy using limited data samples.

7.3   DIAGNOSTIC AND MONITORING TECHNOLOGIES OF POWER CONVERTERS FOR WIND TURBINES

Owing to its effective use in wind energy, high-power converters are in high demand. However, several statistical studies point out that power converters are a significant contributor to the overall failure rate of modern wind turbines. In order to improve the reliability and availability of the converters of wind turbines, condition monitoring and fault diagnosis are considered crucial means to achieve these goals.

Inverter faults are classified into three categories: the DC-link capacitor fault, the sensor fault and the switch device fault. The DC link capacitor fault occurs either because of an electric short circuit or an earth fault. Such a fault causes abnormal overcurrents in an input component owing to a decrease in the DC link voltage. Thus, the overcurrents may potentially destroy the diodes within the rectifier and the input source. The entire system must be immediately shut down once these faults occur.

The offset and the gain errors of the current sensor generate ripple components. The offset is signified by the fundamental stator current frequency and two times, respectively. The secondary problems can therefore occur in the generator or other parts. Thus, a proper tolerant control method is required to protect the system.

Switching device faults fall into two categories: the short-switch fault and the open-switch fault. Short-switch faults may occur as a result of alien substances within the system or the destruction of the switch devices. If this fault occurs, other switch devices, the DC link capacitor and the generator (load) will be destroyed by abnormal overcurrents. At the onset of such a fault, the system operation should be immediately halted. Another remedy to avoid switching device faults is to implement a tolerant control algorithm. An open-switch fault results from gate driver faults or the destruction of the IGBT gate. The currents will then become unbalanced, producing ripples. If the system continuously operates regardless of this fault, the second problems will be encountered in the generator or the load owing to noise and vibrations. Thus, monitoring switch device faults and being able to identify the switch in which the fault occurred is essential for reducing the cost of repairs and for improving the stability and reliability of the system.

Condition monitoring and fault diagnosis are currently considered crucial means to increase the reliability and availability of wind turbines and, consequently, to reduce the wind energy cost. In other words, reducing the cost of corrective maintenance by means of condition monitoring is a way of lowering operation and maintenance costs for wind turbine systems. The most recent techniques applied for the goal of diagnose and monitoring of the power converters in wind turbine applications can be outlined and described as follows.

7.3.1   Open-Circuit Fault Diagnostics

The direct-drive wind turbines based on permanent magnet synchronous generators (PMSGs) with full-scale power converters are an emerging and promising technology. Numerous studies show that power converters are a significant contributor to the overall failure rate of modern wind turbines. Ref. [72] addressed open-circuit fault diagnosis in the two power converters of a PMSG drive for wind turbine applications.

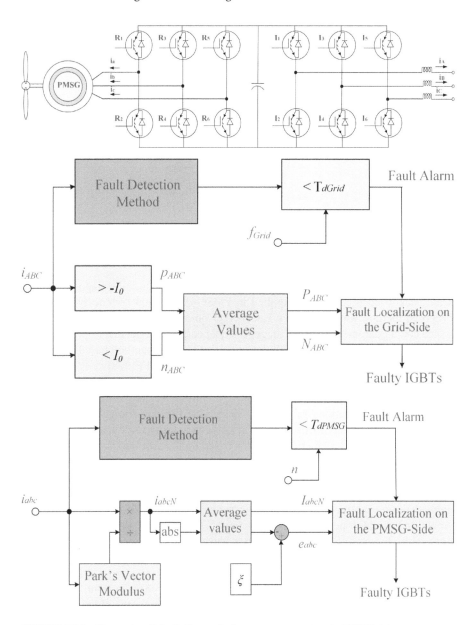

FIGURE 7.19 Open-circuit fault diagnosis for power converters in PMSG drives.

A diagnostic method has been proposed for each power converter, allowing real-time detection and localization of multiple open-circuit faults. The proposed methods in [72] are suitable for integration into the drive controller and triggering remedial actions. Figure 7.19 shows the fault diagnosis methods applied for the power converters in PMSG Drives with that for the grid-side converter in Figure 7.19 (middle) and the other presented in Figure 7.19 (bottom) for the PMSG-side converter.

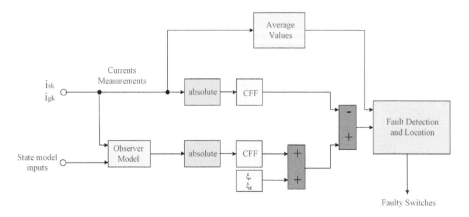

FIGURE 7.20 Block diagram of the open-circuit fault diagnostic method.

Ref. [73] presented an algorithm for multiple open-circuit faults diagnosis in full-scale back-to-back converters, applied in PMSG-drive wind turbine systems. The proposed method was based on a Luenberger observer and an adaptive threshold, which could guarantee a reliable diagnosis independently of the drive operating conditions. Figure 7.20 shows the fault detection and isolation method based on the current Luenberger observer and the Current Form Factors (CFF) [73].

Ref. [74] presented an open-circuit fault diagnostic approach for back-to-back converters of doubly fed wind turbines. Figure 7.21 gives the fault detection principle and the calculations required to determine the detection variable D [74]. The proposed fault diagnostic method in [74] can achieve fault detection and fault localization. The fault detection has been based on the absolute normalized Park's vector approach, which could detect multiple open-circuit switch faults and guarantee immunity to false alarms when the doubly fed induction generator (DFIG) operates around the synchronous speed. For fault localization, this approach has applied the normalized current average values, which could be used to identify single and double open-circuit switch faults. This method could not only diagnose multiple open-circuit faults, but that false alarms due to transients could be avoided.

In Figure 7.21, i_{jv} is the average absolute value of the phase current, I_p is the modulus of that vector of the average absolute values of each phase current, I represents the Park's vector modulus of the phase currents, and j denotes the phases a, b and c, respectively.

The authors in reference [75] presented an approach for real-time diagnostics of open-switch faults in back-to-back converters of a doubly fed wind turbine as shown in Figure 7.22. The average value of the normalized converter phase currents and the absolute normalized currents have been used as principal quantities to formulate the diagnostic variables. The proposed fault-diagnostic variables have proved to be carrying information about multiple open-circuit faults. In addition, by the combination of these variables with the average absolute values of the normalized converter phase currents, an adaptive fault-diagnostic threshold has been

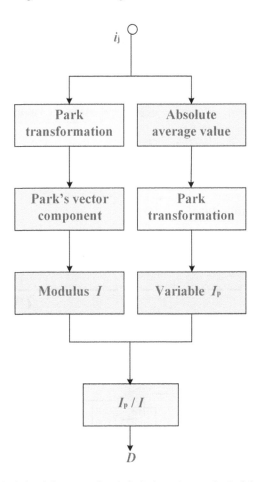

FIGURE 7.21   Principle of the open-circuit fault detection method of the back-to-back converters of doubly fed wind turbines.

proposed, which ensured the robustness of the diagnosis of single and double open-circuit faults. Finally, through the diagnostic variables and the adaptive threshold, the dynamic fault-diagnostic method for open-circuit faults in the back-to-back converter of a doubly fed wind turbine has been formed. This method could not only diagnose the multiple open-circuit faults of a back-to-back converter, but also have a better robustness.

7.3.2   MONITORING AND DIAGNOSTICS FOR SWITCH DEVICES

A converter monitoring unit (CMU), which can enable condition monitoring of wind turbine converters, is presented in [76]. The CMU is able to detect a broad range of failure modes related to IGBT power modules and associated gate drives, as shown in Figure 7.23. IGBT collector-emitter on-state voltage (vce_{on}) and current

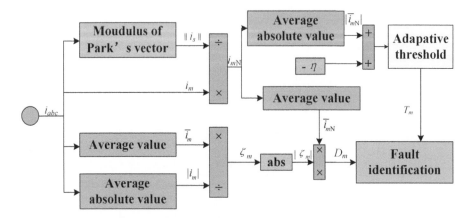

FIGURE 7.22   Open-circuit fault-diagnostic principle of the back-to-back converter for the doubly fed wind power generation system.

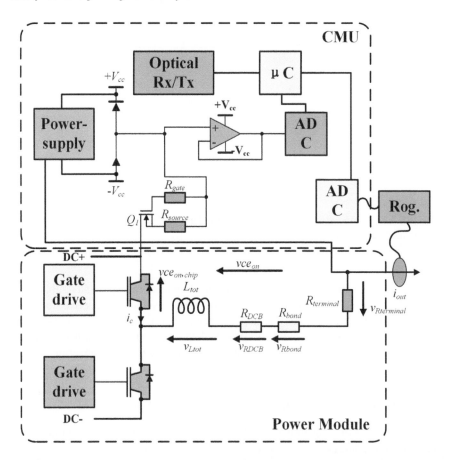

FIGURE 7.23   Condition monitoring based on converter monitoring unit for insulated gate bipolar transistor power modules and associated gate drives.

(i_c) has been sampled in the CMU and used for detection of emerging failures. A new method for compensation of unwanted inductive voltage drop in the vce_{on} measurement path has been presented in [76], enabling retrofitting of CMUs in existing wind turbines. Finally, experimental results obtained on a prototype CMU have been presented. Experimentally the vce_{on} dependency to IGBT junction temperature and deterioration of gate drive voltage has been investigated. The proposed method has shown good detectability of both thermally related degradations of the power module and detectability of deterioration of gate drive voltage. It has been also expected that the proposed method would enable detectability of other failure mechanisms.

A fault detection method for the switch devices of three-parallel power converters in a wind-turbine system has been presented in reference [77]. The proposed method has utilized the measured three phase currents that were already used for controlling the converters. Additional current and voltage sensors have not thus been necessary. Consequently, this particular feature potentially has cut costs to apply the detection algorithm. The three-phase currents have been transformed to a stationary reference frame. This reference frame has possessed specific patterns in accordance to the conditions of the switch devices in the converter. Faulty switches could be detected by analyzing the obtained pattern and pattern recognition has been achieved by a neural network that has a learning capability as shown in Figure 7.24. The proposed method was primarily able to diagnose an open fault switch in the converter.

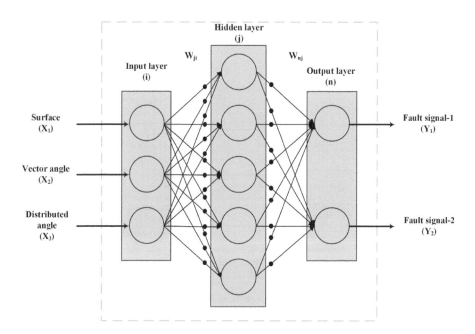

FIGURE 7.24 Neural network-based fault detection method for switch devices in three-parallel power converters.

7.3.3   MONITORING AND DIAGNOSTICS FOR DC VOLTAGE AND SENSORS

Reference [78] developed a combination of PI and predictive methods, using them simultaneously to control a three-phase grid-connected inverter, as shown in Figure 7.25. Under this control scheme, the PI controller is given a new task of monitoring and controlling the DC link voltage V_{dc}. As a result, the output current of the inverter is of high quality, and more importantly, V_{dc} can be double checked for its correctness of measurements. When the V_{dc} sensor fails or its signals are corrupted, the V_{dc} PI controller will become a V_{dc} controller, adding an extra protective function for the reliable operation of wind turbine inverters.

The authors in reference [79] presented an approach for diagnosis and mitigation of sensor malfunctioning in PMSG-based direct-drive variable speed wind energy conversion system (WECS), as shown in Figure 7.26. Malfunctioning of current sensors causes erroneous grid- and machine-side current measurements, which significantly affect the operation of grid- and machine-side controllers, and in turn, performance of the WECS system degrades. In the approach proposed in reference [79], the sliding-mode observer-based fault diagnosis theory has been used to diagnose (i.e., to detect and estimate) the error induced in the grid- and machine-side current measurements due to sensor malfunctioning. The proposed mitigation action in reference [79] has rectified the measured grid- and machine-side currents using estimated measurement errors, as soon as malfunctioning of sensors has been diagnosed, and ensured resilient operation of the WECS against sensor malfunctioning.

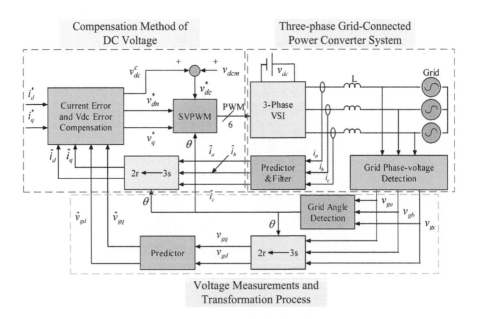

FIGURE 7.25   DC voltage monitoring and control method for three-phase grid-connected wind turbine inverters.

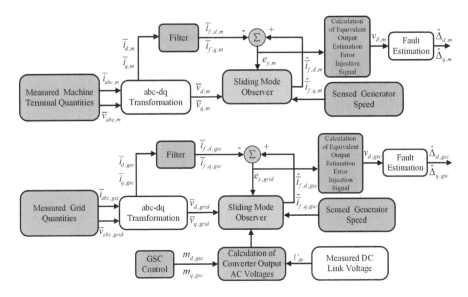

FIGURE 7.26   Sensor malfunctioning diagnosis and mitigation method of power converters in PMSG-based WECS.

7.4 DIAGNOSTIC AND MONITORING TECHNOLOGIES IN SYSTEM LEVEL FOR WIND TURBINES

7.4.1 Diagnostic and Monitoring Technology Based on Power Curve

The power curve which can describe the relationship between wind speed and electric power output, is a method widely used to evaluate the performance of wind turbines [80–83].

Adaptive neuro-fuzzy interference system (ANFIS) model has been established in [80], for wind turbine power curve detection after considering the parameters of ambient temperature and wind direction. The ANFIS model is specifically constructed as follows: (1) fuzzy interference system (FIS) structure: There are two types of common fuzzy reasoning: Mamdani and Sugeno. Because Sugeno has a fuzzy set involving only the premise part and its computational efficiency is higher, Sugeno type is used as fuzzy inference. (2) Membership functions (MF): The generalized normal distribution MF is used in the input space, and the linear membership function is used as the output space. The advantage of the generalized normal distribution MF is that it can provide a wide range of flexibility in function shape according to the parameters of the function, and can guarantee the smoothness of the transition in the input space. (3) Number of MFs/rules: The number of rules is usually determined by experts familiar with the system to be modelled. In this article, the MF was set to 3 for each input. (4) Training method: Use a mixed learning rule estimated by gradient decency and least squares.

By comparing the root mean squared (RMS), mean absolute error (MAE), mean absolute percentage error (MAPE) and standard deviation (SD) indicators

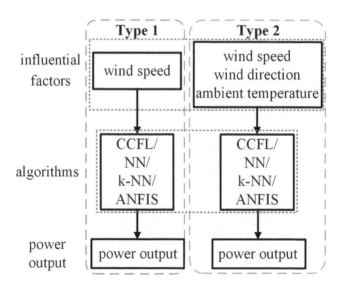

FIGURE 7.27 Model types used for power output modeling.

of ANFIS, cluster center fuzzy logic (CCFL), neural network (NN) and *k*-nearest neighbor (K-NN) in the two models, as shown in Figure 7.27, ANFIS not only shows the best performance in this indicator, but also can better detect abnormal power output faults.

In reference [81], an algorithm that can automatically calculate the limit of power curve is proposed. The algorithm automatically generates warning messages when the measured data of wind turbines deviate from the limit of the power curve or hover between the warning area and the warning area. The overall process of setting the limit value of the power curve is shown in Figure 7.28. The first stage is sorted according to the speed-power data measured by speed bin. In the second stage, the average and standard deviation of power are calculated according to each speed bin. The speed bin's width (unit: m/s) is determined by Equation (7.32).

$$B_{in} \text{ width} = \frac{1}{\text{Number of iterations of overall algorithm loop}} \tag{7.32}$$

If the b_{in}'s width is 1 m/s, it becomes "1/2 m/s, 1/3m/s ..." with each iteration of the algorithm. The third stage uses interpolation to estimate the power curves. According to the average power obtained in stage 2 as the input of this stage. The fourth stage is to eliminate the abnormal data by moving the estimated power curve left and right or up and down to get the limit of the optimal power curve. In the fifth stage, the data within the limit of the power curve will be taken as input after the abnormal data are excluded. The last stage is to determine whether to terminate the entire algorithm cycle. Calculate the average power standard deviation of each bin obtained in the second phase and compare it with the average power standard deviation of each bin in the entire algorithm cycle before.

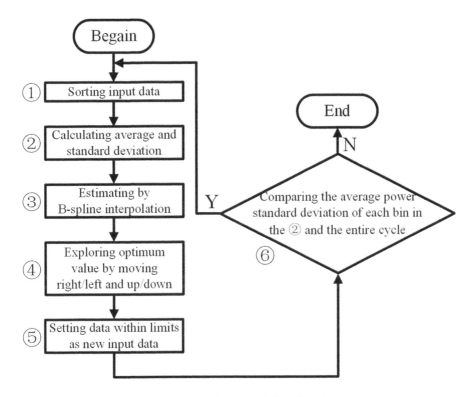

FIGURE 7.28 Automatic power curve limit calculation algorithm.

After the limit value of the power curve is obtained, the fault data queue method is used to generate the alert message. The fault data queue consists of three fault data sets. Because the queue has a FIFO data structure, when the measured data enters the warning or alert area, the data set is stored and the earliest failure data set is deleted. If the failure data set queues are all deleted, a warning or alert message for the current activity is released. If the data set in the queue is saturated, a warning or alert message is issued based on the defined conditions.

In [82], an improved Cholesky decomposition Gaussian process (GP) is used to construct a multi-variable power curve model, and longitudinal and transverse data are compared to check the operation of specific components, so as to timely detect the fan operation anomalies and degradation. According to the influence degree of different factors on the output power, wind speed and direction, pitch angle, yaw error, rotor speed and tip speed ratio are selected as the input of the model, and the output of the model is power output prediction. GP modelling is essentially a Bayesian method, which has good performance for random data. A Gaussian Process is completely specified by its mean value $m(x)$ and covariance function $k(x, x')$ and can be written as:

$$f(x) \sim GP(m(x), k(x, x')). \tag{7.33}$$

In reference [83], the squared exponential is used as the covariance function. Its matrix form is:

$$K(x_i, x_j) = \sigma_f^2 \exp\left[-\frac{1}{2}(x_i - x_j)^T D(x_i - x_j)\right] + \sigma_n^2 \delta_{ij}. \tag{7.34}$$

where σ_f^2 is the signal variance and σ_n^2 is the noise variance. $D = diag(d_1, d_2, \cdots, d_L)$ is the length scale parameter for each of the model inputs which links the time variation of the input parameter to that of the output. In this paper, a Cholesky decomposition is used to compute K^{-1} since it is fast and numerically stable. With Cholesky decomposition, the inversion of matrix K can be computed as:

$$K^{-1} = (L^T)^{-1} \times L^{-1}. \tag{7.35}$$

The predicted value of Gaussian Process becomes:

$$\bar{y}_* = K(x_*, X)K^{-1}(X, X)y = K(x_*, X)(L^T)^{-1}L^{-1}y. \tag{7.36}$$

where x^* is the input of system and y^* is unknown predicted output, c is the mean predicted of value of y^*.

When a wind turbine encounters a failure, the relationship between power and input factors will deviate from the model, which will cause the model's predicted residuals to grow, so the sequential probabilistic ratio test (SPRT) is used to detect any abnormal changes in the GP power curve residuals. SPRT consists of two possible testing hypotheses. Hypotheses H_0: the wind turbine is fault free and the model residuals have a normal distribution with mean value μ_0 and variance σ_0^2; Hypotheses H_1: the wind turbine exhibits abnormal operation with the mean value and variance of the model residuals respectively changing to μ_1 and σ_1^2 respectively. For the GP power curve model residual sequence $e_1, e_2, \cdots, e_n$, the joint probability densities for H_0 and H_1 are respectively as follows:

$$P_{0n} = \frac{1}{(\sigma_0 \sqrt{2\pi})} \exp\left(-\frac{1}{2\sigma_0^2} \sum_{i=1}^{n} (e_i - \mu_0)^2\right) \tag{7.37}$$

$$P_{1n} = \frac{1}{(\sigma_1 \sqrt{2\pi})} \exp\left(-\frac{1}{2\sigma_1^2} \sum_{i=1}^{n} (e_i - \mu_1)^2\right) \tag{7.38}$$

The SPRT ratio (or the likelihood ratio) is:

$$R_m = \frac{P_{1n}}{P_{0n}}. \tag{7.39}$$

The false alarm probability and missed alarm probability respectively are set as α and β which give a lower limit A and an upper limit B respectively as:

$$A = \frac{\beta}{1-\alpha}, B = \frac{1-\beta}{\alpha}. \tag{7.40}$$

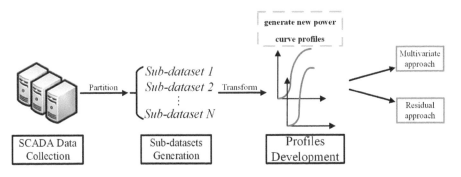

FIGURE 7.29 Online monitoring procedure of the proposed model.

If $R_m \leq A$, hypothesis H_0 should be accepted and the wind turbine is regarded as operating normally. Conversely, if $R_m > B$, hypothesis H_0 should be rejected H_1 accepted instead and the wind turbine operation is regarded as abnormal, triggering an alarm.

In [83], the turbines with weak power generation performance were evaluated by evaluating the wind curve. The least squares method fits the power curve model to the supervisory control and data acquisition (SCADA) data set to construct the curve and shape of the wind curve in continuous time intervals. In order to identify power curves with abnormal curvature and shape, in the next section, a multivariate method for monitoring the contour of the power curve and a residual method for monitoring the power curve fitting error will be introduced. The SCADA data set is divided into sub-data sets according to a fixed time interval, thereby generating a power curve profile and a power curve fitting error. Use multivariate methods and residual methods to develop control charts for monitoring. Continuously collect SCADA data, generate new power curve profiles, and apply developed control charts. The specific implementation process is shown in Figure 7.29.

Since this solution is based on the analysis of sub-data sets, the monitoring effectiveness and accuracy are high. But compared with the point-based state analysis method, this method is not timely, the reason is that it takes some time to construct a new power curve profile. Besides, the relationship between wind energy and multiple parameters was not considered in reference [83].

7.4.2 DIAGNOSTIC AND MONITORING TECHNOLOGY BASED ON EXTREME LEARNING MACHINE

Ref. [84] proposed a data-driven method based on support vector data description (SVDD) and extreme learning machine (ELM) algorithm to achieve effective monitoring of the fan's bad state. First get the wind turbine data from the sensor. Use SVDD classification algorithm to separate unhealthy data. Then, based on this balance data set, an effective classifier is constructed by ELM to monitor the unhealthy state. The specific implementation process is shown in Figure 7.30.

Use three characteristic indicators (e.g., Gini index, information value and Cramer's V) to reflect the influence of variables on pattern recognition. The

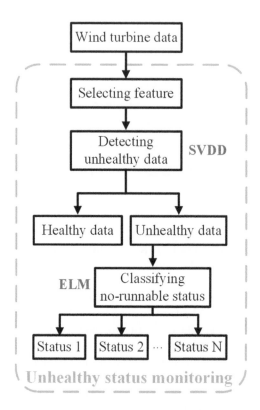

FIGURE 7.30   The framework of monitoring not-runnable status of wind turbines.

proposed method mainly includes two stages to realize the monitoring of the fan status. The first stage is to separate health data from unhealthy data. The second stage is to classify different unhealthy states, namely the various nonoperational states of the wind turbine. By comparing with the six models combining support vector machine (SVM), ELM, principal component analysis (PCA) and SVDD, this method improves the accuracy of the classification of unhealthy data and inoperable states. However, this article only considers two specific nonoperational states (WTs are stopped due to vibration in tactical air control (TAC84); WTs are stopped by remote control due to excessive power generation and reduction capacity), the use of this method has certain limitations.

In reference [85], a data-based method was proposed to estimate the health status of wind turbines, so as to improve the active power output of wind turbines. The degree of disability is estimated using an ELM algorithm combined with Bonferroni intervals. The data collected from the measurement is selected as the input of ELM model to predict the output signal of physical process. And the actual output signal obtained from the wind turbine is compared with the predicted output signal of the corresponding input signal. The Bonferroni interval method, which has more accurate confidence intervals, is applied to estimate the deviation level of each component in this study.

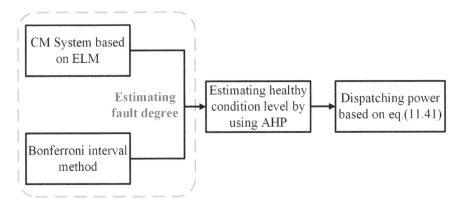

FIGURE 7.31   Flowchart of proposed method.

Next, using analytic hierarchy process (AHP), various factors such as the degree of failure, maintenance cost, and maintenance time were comprehensively examined to estimate the health condition of the turbine structure. The procedure to evaluate the health condition of the turbine using AHP is as follows: (1) Define the evaluation criteria and build a hierarchical structure; (2) compose the pairwise criterion elements; (3) establish the comparison matrix and the alternative pairwise comparison matrix; and (4) calculate the relative weights of each criterion to estimate the health of the wind turbines.

In the end, in order to reduce the fatigue load of the failed unit and improve the operating efficiency of the normal unit, power distribution control is performed according to the operating state of the unit. Therefore, the process of the power distribution scheme is defined as follows (shown in Figure 7.31): for a healthy wind turbine, they are set in normal mode, which is the maximum power under variable wind speed. The failed wind turbine is operated in the power-reduced mode according to its health condition. Therefore, the output demand P* is defined as follows:

$$P^* = \sum_{i=1}^{N1} P_{hi} + \sum_{j=1}^{N2} C_{fj} P_{fj}. \tag{7.41}$$

where N_1 and N_2 represent the number of the healthy and faulty wind turbines respectively; P_{hi} is the power output of the ith healthy turbine operating in the baseline mode, whereas P_{fj} is the power output of the jth faulty turbine working in a power-reduced mode. C_{fj} is the power distribution coefficient of the jth faulty turbine. Power was distributed according to the power distribution coefficient of the failed turbine considering the health condition.

7.4.3 DIAGNOSTIC AND MONITORING TECHNOLOGY BASED ON INTELLIGENT ALGORITHM

The least square support vector machine was used to establish the relationship function between the evaluation index and its influencing factors, which was used to determine the dynamic range of wind turbine generator (WTG) disease evaluation index. The dynamic weight of each evaluation index was used to characterize the influence of the deterioration degree of different components on WTG conditions. Then, similar cloud and fuzzy comprehensive assessment methods were used to evaluate the health status of WTG [86].

To solve the problem that the abnormal load change is not always the fault of the component itself, a WTG state assessment method is proposed. First, according to the structure and working principle of wind turbine generator unit (WTGU), WTG evaluation index system was established, and the mathematical model of each condition index was trained by using the monitoring data under normal conditions. Second, the residual and deterioration degree of each index were given, and the evaluation standard cloud was established. Then, during the period to be evaluated, one needed to select the monitoring data, standardize the data, and perform a conditional evaluation. The flow chart of condition assessment of WTG is shown in Figure 7.32.

First, the mapping relationship between input and output physical quantities is established by using the normal operation data of the pre-processed unit. Then, the prediction model of evaluation index is established to calculate the deviation between the predicted value and the measured value. The least square support vector machine (LSSVM) starts with the loss function of machine learning, optimizes the objective function with two norms, and replaces the inequality constraint in SVM with the equality constraint. And then normalization is done to get the degree of degradation. Since the sensitivity of each index to the evaluation result is different, the weight is used to represent the sensitivity degree of this index. A set of fused data is obtained through the weight fusion degradation degree, and the fused data is used to obtain the digital characteristics of the conditional cloud model to be evaluated. The advantage of this method is that it can reduce the downtime while maintaining the generator state.

Reference [87] presents an online operation optimization strategy by optimizing the operation parameters. The operation parameters related to active power are extracted by loss analysis. Then the dimension of the array of operation parameters is reduced by factor analysis, and finally the reference value of operation parameters is obtained by cluster analysis. The specific implementation process is shown in Figure 7.33. K-mean clustering algorithm is selected to obtain the reference value of operation parameters. According to the wind speed, the WT operation is divided into different working conditions. According to the clustering results, the cluster center with the maximum active power is taken as the reference value of the operation parameters. The advantage of this model is that it can accurately determine the effective values of wind speed and power during operation, and its calculation accuracy is high, avoiding the influence of high-dimensional data. However, this paper does not give the operating parameters and reasons that lead to the decrease of operating efficiency.

Because wind turbines undergo a variety of state changes during operation, including normal operation of the turbine, idling, maintenance/repair mode, failure mode, weather downtime, etc. Therefore, state-based maintenance tools are needed to predict failure patterns in the system. In reference [88], a method for

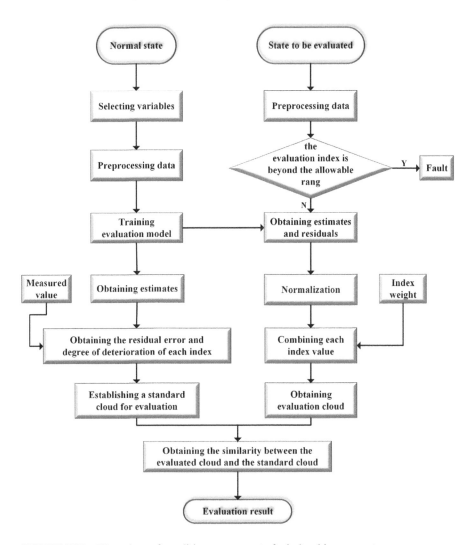

FIGURE 7.32   Flow chart of condition assessment of wind turbine generator.

predicting the state of wind turbines is proposed with three typical steps, as shown in Figure 7.34.

Step 1: Abstraction of turbine state. The possible states of the turbine are classified according to the different stages of prediction. In the first stage, the turbine can be divided into four states: normal turbine, failure, weather shutdown and maintenance shutdown. Where the weather shutdown category corresponds to a turbine shutdown due to adverse weather conditions, and any other shutdown is considered a maintenance shutdown. The second stage is classified by replacing the faults in stage I with the actual fault types, which are pitch overrun 0°, pitch thyristor 2 fault, axle 1 fault pitch

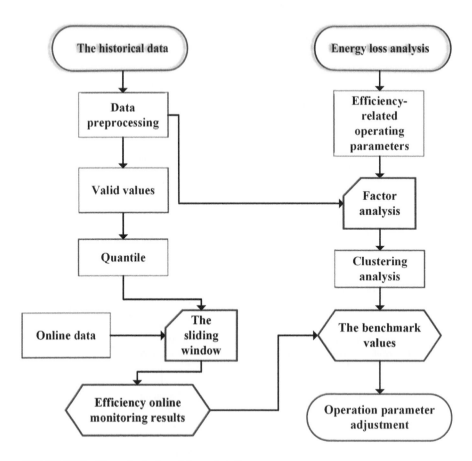

FIGURE 7.33   Flow chart of model construction.

controller and pulse sensor motor defect. The third stage also added practical failures such as out of control yaw, brush wear warning, blade Angle unreliability and advanced response generator.

Step 2: Learn Strategy. Data mining algorithm is used to establish wind turbine failure prediction model, through the use of five data algorithm, NN, SVM, random forest algorithm (RFA), boosting tree algorithm (BTA) and general chi-square automatic interaction detector (CHAID) to training, two-thirds of the primitive data types of the remaining one third of the original data for testing. The geometric mean (gmean) of the output category is used as the evaluation index, and the expression is:

$$gmean = \sqrt[n]{\prod_{i=1}^{n} acc_i} \tag{7.42}$$

where acc_i is the accuracy of class i, n is the total number of output classes.

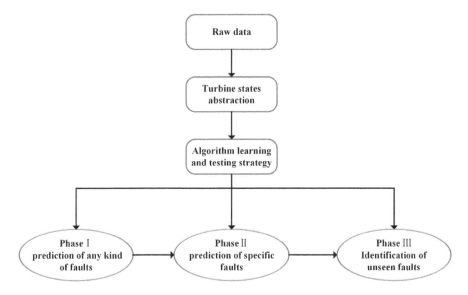

FIGURE 7.34   Framework of the proposed approach.

Step 3: State prediction. The primary goal of the first phase is to predict any type of failure. The second stage predicts specific failures. Replace the type of fault with the output type in phase I with the actual fault type. The third stage predicts invisible failures in different wind turbines.

7.4.4   OTHER DIAGNOSTIC AND MONITORING METHODS

An overall monitoring system based on field programmable gate array CPU has been proposed [89]. The system can not only provide status and subsynchronous control interactions (SSCI) monitoring functions at the same time, but also record the data required for analysis after the event. A monitoring system based on the overall internet of things (IoT) was developed for wind turbines. The system used a field programmable gate array (FPGA)-CPU hybrid controller to continuously complete data acquisition, data processing, and data recording to improve system performance. In this monitoring system, as long as the predefined detection mechanism is triggered on the FPGA target, the vibration and voltage signals were recorded on the CPU target. The recorded data can be sent from the FPGA target to the CPU target and stored to the network-attached storage (NAS). Through the monitoring and detection of SSCI events and the status of wind turbines, this method can provide detailed information for the system operators to avoid unnecessary coordination between the two independent monitoring systems.

Reference [90] proposed a method for tracking the power signal using an adaptive filter based on continuous wavelet transform (CWT) to monitor the state of the wind turbine. The technology can track the energy in the power signal at the specified fault-related frequency band rather than at all frequencies of the monitoring signal, and can display the results graphically. The energy A in the fault-related frequency band at each time interval can be calculated by using Equations (7.43) and (7.44). These calculations were repeated until the curve of the energy change in the fault-related frequency band was obtained, and then the change in the operating condition of the WT can be evaluated.

$$CWT_{local}(b,a) = \frac{1}{\sqrt{|a|}} \int_{-\infty}^{\infty} x(t)\psi^* \left(\frac{t-b}{a}\right) dt \qquad (7.43)$$

$$A(t_0 + T/2) = \max(|CWT_{local}|(b,a)|) \quad \begin{cases} a \in [a_{\min} \quad a_{\max}] \\ b \in [t_0 \quad t_0 + T] \end{cases} \qquad (7.44)$$

where CWT_{local} is the matrix of wavelet coefficients, a is the wavelet scale and $a = \omega_0/\omega$, ω represents frequency, and T is the time interval.

This method has been proven to detect two types of faults, including electrical and mechanical faults. The advantage of this technique is to reduce the calculation time of feature extraction, low cost and strong versatility.

In reference [91], an overview on grid-friendly wind turbines (WTs) and relevant technologies for control and monitoring was presented. In this paper, the aerodynamic conversion system, generator and front-end speed regulation (FESR) system module composed of FESR are established, which is similar to double-fed induction generator (DFIG) system. Thus, a friendly grid connection is achieved, with better stability and low-voltage ride-through (LVRT) capability. As shown in Figure 7.35, a "virtual generator" is introduced to achieve the torques of the low-speed shaft virtually. Then, a PI-based virtual controller can be developed, and the design interface for FESR is achieved in Bladed. As FESR contains two transmission modules, mechanical strength and fatigue tests should be performed.

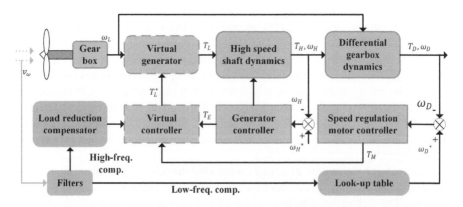

FIGURE 7.35 Modeling and control scheme for FESR WT in reference [91].

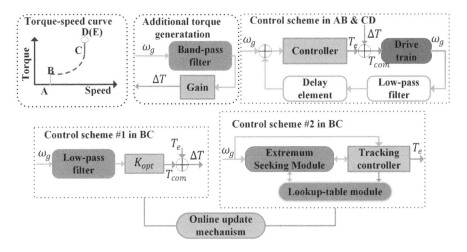

FIGURE 7.36   Suggested torque control scheme for load reduction in reference [91].

The improvement methods for load reduction of the system are as follows: (1) IPC should solve various influences caused by coupling nonlinearity and uncertainty of pitch blade, so as to realize accurate system model of IPC; (2) Proper torque control is also helpful to reduce load. One suggested approach is shown in Figure 7.36. The schematic curve of generator torque–speed depicts the four generator operation areas, as AB, BC, CD, and DE. The torque control command is compensated with the additional torque derived from the band-pass filter and gain block, so that the compound command is achieved.

Similarly, grid-friendly WTs capabilities can be enhanced in power electronics technologies, such as novel configurations or converter topologies. From the point of view of the monitoring scheme based on SCADA, a comprehensive scheme consisting of "data preprocess," "regular data process," and "fault data process" is proposed to get a comprehensive monitoring mechanism for the system.

7.5   SUMMARY

The generator and converter are main components of an electric drive train in the wind turbine. Therefore, diagnose and monitoring technologies for generators and converters are the focus of this chapter. Since the converter contains voltage and current signals of both converter and generator, converter-based advanced diagnose and monitoring technologies for wind turbines are very promising. For wind power generators, converter-based diagnose and monitoring technologies for winding, bearing, and eccentricity faults are introduced. For converters in wind power systems, converter-based diagnose and monitoring technologies for three typical faults, i.e., open-circuit fault, switch device faults, and DC voltage and sensor faults are summarized. Finally, the mainstream converter-based diagnose and monitoring technologies for the whole wind turbine system have also been reviewed in this chapter.

REFERENCES

1. Maekawa, N., Riahy, G. H., Hosseinian, S. H., et al. 2011. Wind power optimal capacity allocation to remote areas taking into account transmission connection requirements. *IET Renewable Power Generation* 5(5): 347–355.
2. Qiao, W. and Lu, D. 2015. A survey on wind turbine condition monitoring and fault diagnosis—part I: components and subsystems. *IEEE Transactions on industrial electronics* 62(10): 6536–6545.
3. Amirat, Y., Benbouzid, M. E. H., Bensaker, B., et al. 2007. Condition monitoring and fault diagnosis in wind energy conversion systems: a review. *Proceedings of 2007 Electric Machines & Drives Conference*: 1434–1439.
4. Malekpour, M., Phung, B.T. and Ambikairajah, E. 2017. Online technique for insulation assessment of induction motor stator windings under different load conditions. *IEEE Transactions on Dielectrics and Electrical Insulation* 24(1): 349–358.
5. Stavrou, A., Sedding, H. G. and Penman, J. 2001. Current monitoring for detecting inter-turn short circuits in induction motors. *IEEE Transactions on Energy Conversion* 16(1): 32–37.
6. Yang, S., Hsu, Y., Chou, P., et al. 2018. Fault detection and tolerant capability of parallel-connected permanent magnet machines under stator turn fault. *IEEE Transactions on Industry Applications* 54(5): 4447–4456.
7. Mohammed, O. A., Abed, N. Y. and Ganu, S. 2006. Modeling and characterization of induction motor internal faults using finite-element and discrete wavelet transforms. *IEEE Transactions on Magnetics* 42(10): 3434–3436.
8. Yagami, Y., Araki, C., Mizuno, Y., et al. 2015. Turn-to-turn insulation failure diagnosis of stator winding of low voltage induction motor with the aid of support vector machine. *IEEE Transactions on Dielectrics and Electrical Insulation* 22(6): 3099–3106.
9. Cruz, S. M. A. and Cardoso, A. J. M. 2005. Multiple reference frames theory: a new method for the diagnosis of stator faults in three-phase induction motors. *IEEE Transactions on Energy Conversion* 20(3): 611–619.
10. Tallam, R. M., Habetler, T. G. and Harley, R. G. 2002. Self-commissioning training algorithms for neural networks with applications to electric machine fault diagnostics. *IEEE Transactions on Power Electronics* 17(6): 1089–1095.
11. Lee, S. B., Younsi, K. and Kliman, G. B. 2005. An online technique for monitoring the insulation condition of AC machine stator windings. *IEEE Transactions on Energy Conversion* 20, (4): 737–745.
12. Cash, M. A., Habetler, T. G. and Kliman, G. B. 1998. Insulation failure prediction in AC machines using line-neutral voltages. *IEEE Transactions on Industry Applications* 34(6): 1234–1239.
13. Zamudio-Ramirez, I., Antonino-Daviu, J. A., Osornio-Rios, R. A., et al. 2020. Detection of winding asymmetries in wound-rotor induction motors via transient analysis of the external magnetic field. *IEEE Transactions on Industrial Electronics* 67(6):5050–5059.
14. Sapena-Bano, A., Riera-Guasp, M., Puche-Panadero, R., et al. 2015. Harmonic order tracking analysis: a speed-sensorless method for condition monitoring of wound rotor induction generators. *IEEE Transactions on Energy Conversion* 30(3):833–841.
15. Vedreño-Santos, F., Riera-Guasp, M., Henao H., et al. 2014. Diagnosis of rotor and stator asymmetries in wound-rotor induction machines under nonstationary operation through the instantaneous frequency. *IEEE Transactions on Industrial Electronics* 61(9):4947–4959.
16. Zaggout, M., Tavner, P., Crabtree, C., et al. 2014. Detection of rotor electrical asymmetry in wind turbine doubly-fed induction generators. *IET Renewable Power Generation* 8(8):878–886.
17. Djurovic, S., Crabtree, C. J., Tavner, P. J., et al. 2012. Condition monitoring of wind turbine induction generators with rotor electrical asymmetry. *IET Renewable Power Generation* 6(4):207–216.

18. Ibrahim, R. K., Watson, S. J., Djurović, S., et al. 2018. An effective approach for rotor electrical asymmetry detection in wind turbine DFIGs. *IEEE Transactions on Industrial Electronics* 65(11):8872–8881.

19. Wu, Y. C. and Li, Y. G. 2015. Diagnosis of rotor winding interturn short-circuit in turbine generators using virtual power. *IEEE Transactions on Energy Conversion.* 30(1):183–188.

20. Bellini, A., Filippetti, F., Tassoni, C., et al. 2008. Advances in diagnostic techniques for induction machines. *IEEE Transactions on Industry Electronics* 55(12): 4109–4126.

21. Plazenet, T., Boileau, T., Caironi, C., et al. 2018. A comprehensive study on shaft voltages and bearing currents in rotating machines. *IEEE Transactions on Industry Applications* 54(4): 3749–3759.

22. Immovilli, F., Bianchini, C., Cocconcelli, M., et al. 2013. Bearing fault model for induction motor with externally induced vibration. *IEEE Transactions on Industry Electronics* 60(8): 3408–3418.

23. Ojaghi, M., Sabouri, M. and Faiz, J. 2017. Analytic model for induction motors under localized bearing faults. *IEEE Transactions on energy conversion* 33(2): 617–626.

24. Zika, T., Gebeshuber, I. C., Buschbeck, F., et al. 2009. Surface analysis on rolling bearings after exposure to defined electric stress. Proceedings of the Institution of Mechanical Engineers, *Part J: Journal of Engineering Tribology* 223(5): 787–797.

25. Chen, S., Lipo, T. A. and Fitzgerald, D. 1996. Modeling of motor bearing currents in PWM inverter drives. *IEEE Transactions on Industry Applications* 32(6): 1365–1370.

26. K Kalaiselvi, J. and Srinivas, S. 2014. Bearing currents and shaft voltage reduction in dual-inverter-fed open-end winding induction motor with reduced CMV PWM methods. *IEEE Transactions on Industrial Electronics* 62(1): 144–152.

27. Zitzelsberger, J., Hofmann, W., Wiese, A., et al. 2005. Bearing currents in doubly-fed induction generators. Proceeding of 2005 European Conference on Power Electronics and Applications: 9.

28. Muetze, A. and Binder, A. 2007. Practical rules for assessment of inverter-induced bearing currents in inverter-fed AC motors up to 500 kW. *IEEE Transactions on Industrial Electronics* 54(3): 1614–1622.

29. Buckley, G. W., Corkins, R. J. and Stephens, R. N. 1988. The importance of grounding brushes to the safe operation of large turbine generators. *IEEE Transactions on Energy Conversion* 3(3): 607–612.

30. Bo, L., M, C., Y, T., et al. 2000. Neural-network-based motor rolling bearing fault diagnosis. *IEEE Transactions Industrial Electronics*, 2000, 47(5): 1060–1069.

31. Gong, X. and Qiao, W. 2013. Bearing fault diagnosis for direct-drive wind turbines via current-demodulated signals. *IEEE Transactions on Industry Electronics* 60(8): 3419–3428.

32. Blodt, M., Regnier, J. and Faucher, J. 2009. Distinguishing load torque oscillations and eccentricity faults in induction motors using stator current wigner distributions. *IEEE Transactions on Industry Applications* 45(6): 1991–2000.

33. Qiao, W. and Lu, D. 2015. A survey on wind turbine condition monitoring and fault diagnosis—part II: signals and signal processing methods. *IEEE Transactions on Industry Electronics* 62(10): 6546–6557.

34. Zhang, P. and Neti, P. 2015. Detection of gearbox bearing defects using electrical signature analysis for doubly fed wind generators. *IEEE Transactions on Industry Applications* 51(3): 2195–2200.

35. Blodt, M., Granjon, P., Raison, B., et al. 2008. Models for bearing damage detection in induction motors using stator current monitoring. *IEEE Transactions on Industry Electronics* 55(4): 1813–1822.eite, V. C. M. N., Borges, D. S. J. G., Veloso, G. F. C.,

et al. 2015. Detection of Localized Bearing Faults in Induction Machines by Spectral Kurtosis and Envelope Analysis of Stator Current. IEEE Transactions on Industrial Electronics 62(3): 1855–1865.

36. An, X., Jiang, D., Li, S., et al. 2011. Application of the ensemble empirical mode decomposition and Hilbert transform to pedestal looseness study of direct-drive wind turbine. *Energy* 36(9): 5508–5520.

37. Immovilli, F., Bellini, A., Rubini, R., et al. 2010. Diagnosis of bearing faults in induction machines by vibration or current signals: a critical comparison. *IEEE Transactions on Industry Applications* 46(4): 1350–1359.

38. Elbouchikhi, E., Choqueuse, V., Amirat, Y., et al. 2017. An efficient Hilbert–Huang transform-based bearing faults detection in induction machines. *IEEE Transactions on Energy Conversion* 32(2): 01–413.

39. Pandarakone, S. E., Mizuno, Y. and Nakamura, H. 2017. Distinct fault analysis of induction motor bearing using frequency spectrum determination and support vector machine. *IEEE Transactions on Industry Applications* 53(3): 3049–3056.

40. Silva, J. L. H. and Cardoso, A. J. M. 2005. Bearing failures diagnosis in three-phase induction motors by extended Park's vector approach. Proceedings of 31st Annual Conference of IEEE Industrial Electronics Society: 6.

41. Ahmadi, H. and Moosavian, A. 2011. Fault diagnosis of journal-bearing of generator using power spectral density and fault probability distribution function. Proceedings of International Conference on Innovative Computing Technology: 30–36.

42. Mbo'o, C. P. and Hameyer, K. 2016. Fault diagnosis of bearing damage by means of the linear discriminant analysis of stator current features from the frequency selection. *IEEE Transactions on Industry Applications* 52(5): 3861–3868.

43. Boudinar, A. H., Benouzza, N., Bendiabdellah, A., et al. 2016. Induction motor bearing fault analysis using a Root-MUSIC method. *IEEE Transactions on Industry Applications* 52(5): 3851–3860.

44. Dalvand, F., Kang, M., Dalvand, S., et al. 2018. Detection of generalized-roughness and single-point bearing faults using linear prediction-based current noise cancellation. *IEEE Transactions on Industrial Electronics* 65(12): 9728–9738.

45. Tao, R., Li, Y. L. and Wang, Y. 2009. Short-time fractional fourier transform and its applications. *IEEE Transactions on Signal Processing* 58(5): 2568–2580.

46. Singh, S. and Kumar, N. 2016. Detection of bearing faults in mechanical systems using stator current monitoring. *IEEE Transactions on Industrial Informatics* 13(3): 1341–1349.

47. Abid, F. B., Zgarni, S. and Braham, A. 2018. Distinct bearing faults detection in induction motor by a hybrid optimized SWPT and aiNet-DAG SVM. *IEEE Transactions on Energy Conversion* 33(4): 1692–1699.

48. Schlechtingen, M., Santos, I. F. and Achiche, S. 2013. Wind turbine condition monitoring based on SCADA data using normal behavior models. Part 1: System description. *Applied Soft Computing Journal* 13(1): 259–270.

49. Li, G. and Shi, J. 2012. Applications of Bayesian methods in wind energy conversion systems. *Renewable Energy* 43: 1–8.

50. Refaat, S. S., Abu-Rub, H., Saad, M. S., et al. 2013. ANN-based for detection, diagnosis the bearing fault for three phase induction motors using current signal. Proceedings of 2013 IEEE International Conference on Industrial Technology: 253–258.

51. Liu, T. I., Singonahalli, J. H. and Iyer, N. R. 1996. Detection of roller bearing defects using expert system and fuzzy logic. *Mechanical systems and signal processing* 10(5): 595–614.

52. Stack, J. R., Habetler, T. G. and Harley, R. G. 2003. Fault classification and fault signature production for rolling element bearings in electric machines. *IEEE Transactions on Industry Applications* 40(3): 735–739.

53. Nikias, C. L. and Mendel, J. M. 1993. Signal processing with higher-order spectra. *IEEE Signal processing magazine* 10(3): 10–37.
54. Stack, J. R. Harley, R.G. and Habetler, T. G. 2004. An amplitude modulation detector for fault diagnosis in rolling element bearings. *IEEE Transactions on Industry Electronics* 51(5): 1097–1102.
55. Hamomd, O., Alabied, S., Xu, Y., et al. 2017. Vibration based centrifugal pump fault diagnosis based on modulation signal bispectrum analysis. Proceedings of 2017 23rd International Conference on Automation and Computing: 1–5.
56. Alwodai, A., Yuan, X., Shao, Y., et al. 2012. Modulation signal bispectrum analysis of motor current signals for stator fault diagnosis. Proceedings of 18th International Conference on Automation and Computing: 1–6.
57. Chen, X., Xu, W., Liu, Y., et al. 2020. Bearing corrosion failure diagnosis of doubly fed induction generator in wind turbines based on stator current analysis. *IEEE Transactions on Industry Electronics* 67(5): 3419–3430.
58. Cardoso, A. J. M., Cruz, S. M. A. and Fonseca, D. S. B. 1999. Inter-turn stator winding fault diagnosis in three-phase induction motors, by Park's vector approach. *IEEE Transactions on Energy Conversion* 14(3): 595–598.
59. Cruz, S. M. A. and Cardoso, A. J. M. 2001. Stator winding fault diagnosis in three-phase synchronous and asynchronous motors, by the extended Park's vector approach. *IEEE Transactions on industry applications* 37(5): 1227–1233.
60. Blodt, M., Granjon, P., Raison, B., et al. 2008. Models for bearing damage detection in induction motors using stator current monitoring. *IEEE Transactions on Industry Electronics* 55(4): 1813–1822.
61. Salah, A. A., Dorrell, D.G. and Guo, Y. 2019. A review of the monitoring and damping unbalanced magnetic pull in induction machines due to rotor eccentricity. *IEEE Transactions on Industry Applications* 55(3): 2569–2580.
62. Nandi, S. Ilamparithi, T. C., Lee, S. B., et al. 2011. Detection of eccentricity faults in induction machines based on nameplate parameters. *IEEE Transactions on Industry Electronics* 58(5): 1673–1683.
63. Ebrahimi, B. M., Faiz, J. and Roshtkhari, M. J. 2009. Static-, dynamic-, and mixed-eccentricity fault diagnoses in permanent-magnet synchronous motors. *IEEE Transactions on Industry Electronics* 56(11): 4727–4739.
64. Thomson, W. T., Fenger, M. 2001. Current signature analysis to detect induction motor faults. *IEEE Industry Applications Magazine* 7(4): 26–34.
65. Thomson, W. T. 2001. On-line MCSA to diagnose shorted turns in low voltage stator windings of 3-phase induction motors prior to failure. Proceedings of 2001 IEEE International Electric Machines and Drives Conference: 891–898.
66. Gyftakis, K. N. and Kappatou, J.C. 2013. A novel and effective method of static eccentricity diagnosis in three-phase PSH induction motors. *IEEE Transactions on Energy Conversion* 28(2): 405–412.
67. Thomson, W. T. and Barbour, A. 1998. On-line current monitoring and application of a finite element method to predict the level of static airgap eccentricity in three-phase induction motors. *IEEE Transactions on Energy Conversion* 13(4): 347–357.
68. Cardoso, A. J. M. and Saraiva, E. S. 1991. Computer aided detection of airgap eccentricity in operating three-phase induction motors, by Park's vector approach. Proceedings of 1991 IEEE Industry Applications Society Annual Meeting: 94–98.
69. Gritli, Y., Bellini, A., Rossi, C., et al. 2017. Condition monitoring of mechanical faults in induction machines from electrical signatures: Review of different technique. Proceedings of 2017 IEEE 11th International Symposium on Diagnostics for Electrical Machines, Power Electronics and drives: 77–84.

70. Blodt, M., Regnier, J. and Faucher, J. 2009. Distinguishing load torque oscillations and eccentricity faults in induction motors using stator current Wigner distributions. *IEEE Transactions on Industry Applications* 45(6): 1991–2000.

71. Yahia, K. Sahraoui, M., Cardoso, A. J. M., et al. 2016. The use of a modified prony's method to detect the airgap-eccentricity occurrence in induction motors. *IEEE Transactions on Industry Applications* 52(5): 3869–3877.

72. Freire, N. M. A., Estima, J. O. and Cardoso, A. J. M. 2013. Open-circuit fault diagnosis in PMSG drives for wind turbine applications. *IEEE Transactions on Industrial Electronics* 60(9): 3957–3967.

73. Jlassi, I., Estima, J. O., El Khil, S. K., Bellaaj, N. M. and Cardoso, A. J. M. 2015. Multiple open-circuit faults diagnosis in back-to-back converters of PMSG drives for wind turbine systems. *IEEE Transactions on Power Electronics* 30(5): 2689–2702.

74. Zhao, H. and Cheng, L. 2017. Open-circuit faults diagnosis in back-to-back converters of DF wind turbine. *IET Renew. Power Generation* 11(4): 417–424.

75. Zhao, H. and Cheng, L. 2018. Open-switch fault-diagnostic method for back-to-back converters of a doubly fed wind power generation system. *IEEE Transactions on Power Electronics* 33(4): 3452–3461.

76. Rannestad, B., Maarbjerg, A. E., Frederiksen, K., Munk-Nielsen, S. and Gadgaard, K. 2018. Converter monitoring unit for retrofit of wind power converters *IEEE Transactions on Power Electronics*, 33(5): 4342–4351.

77. Ko, Y.-J., Lee, K.-B., Lee, D.-C., Kim, J.-M. 2012. Fault diagnosis of three-parallel voltage-source converter for a high-power wind turbine. *IET Power Electronics*, 5(7):1058–1067.

78. Wang, Z. and Chang, L. 2008. A DC voltage monitoring and control method for three-phase grid-connected wind turbine inverters. *IEEE Transactions on Power Electronics* 23(3): 1118–1125.

79. Saha, S., Haque, M. E. and Mahmud, M. A. 2018. Diagnosis and mitigation of sensor malfunctioning in a permanent magnet synchronous generator based wind energy conversion system. *IEEE Transactions on Energy Conversion* 33(3): 938–948.

80. Schlechtingen, M., Santos, I. F. and Achiche, S. 2013. Using data-mining approaches for wind turbine power curve monitoring: a comparative study. *IEEE Transactions on Sustainable Energy* 4(3): 671–679.

81. Park, J., Lee, J., Oh, K., et al. 2014. Development of a novel power curve monitoring method for wind turbines and its field tests. *IEEE Transactions on Energy Conversion* 29(1): 119–128.

82. Guo, P. and Infield, D. 2020. Wind turbine power curve modelling and monitoring with gaussian process and SPRT. *IEEE Transactions on Sustainable Energy* 11(1): 107–115.

83. Long, H., Wang, L., Zhang, Z., et al. 2015. Data-driven wind turbine power generation performance monitoring. *IEEE Transactions on Industrial Electronics* 62(10): 6627–6635.

84. Ouyang, T., He, Y. and Huang, H. 2019. Monitoring wind turbines' unhealthy status: a data-driven approach. *IEEE Transactions on Emerging Topics in Computational Intelligence* 3(2): 163–172.

85. Qian, P., Ma, X., Zhang, D., et al. 2019. Data-driven condition monitoring approaches to improving power output of wind turbines. *IEEE Transactions on Industrial Electronics* 66(8): 6012–6020.

86. Li, J., Li, Q. and Zhu, J. 2019. Health condition assessment of wind turbine generators based on supervisory control and data acquisition data. *IET Renewable Power Generation* 13(8): 1343–1350.

87. Gu, Y. and Xing, Y. 2019. Online monitoring of wind turbine operation efficiency and optimization based on benchmark values. *IEEE Access* 7: 132193–132204.

88. Kusiak, A. and Verma, A. 2012. A data-mining approach to monitoring wind turbines. *IEEE Transactions on Sustainable Energy* 3(1): 150–157.

89. Zhao, L., Zhou, Y., Matsuo, I. B. M., et al. 2020. The design of a remote online holistic monitoring system for a wind turbine. *IEEE Transactions on Industry Applications* 56(1): 14–21.

90. Yang, W., Tavner, P. J., Crabtree, C. J., et al. 2010. Cost-effective condition monitoring for wind turbines. *IEEE Transactions on Industrial Electronics* 57(1): 263–271.

91. Li, P., Song, Y., Li, D., et al. 2015. Control and monitoring for grid-friendly wind turbines: research overview and suggested approach. *IEEE Transactions on Power Electronics* 30(4): 1979–1986.

8 A Comprehensive Stability Analysis of Multi-Converter-Based DC Microgrids

Mohammad Habibullah, N. Mithulananthan, and Rahul Sharma
School of Information Technology and Electrical Engineering, The University of Queensland

Md. Rakibuzzaman Shah
School of Engineering, Information Technology and Physical Sciences, Federation University

CONTENTS

8.1   INTRODUCTION

Direct current microgrid (DCMG) has emerged as an efficient next-generation electrical energy delivery system. DCMGs enable smooth and cost-effective integration of distributed energy sources (DESs) [1–4] with the various end-users. Among the DESs, photovoltaic (PV) system is the widely utilized, which produces power with DC output. Besides, the widespread deployment of electric vehicles (EVs) into the market is expected within a couple of years. The DC-powered fast-charging stations will be necessary for these EVs in both urban and rural areas. Furthermore, a number of equipment used in residential, commercial, and industrial buildings can be powered by DC sources/devices, which require multiple conversion stages for AC grid connection. These numerous conversion stages not only increase the overall operational cost but also reduce the reliability and efficiency of systems. Given these future scenarios, efficacy and cost-effectiveness of the electricity delivery system could be improved by introducing DCMG with highly efficient DC-DC converters.

A DCMG consists of various sources, namely, photovoltaics, battery energy storage, AC grid interface, AC or DC generator, wind turbine, and loads. DCMG can be used in commercial buildings, military centers, EV charging stations, educational institutes, ships, remote rural areas, and others. Besides, it can be a feasible solution for electrifying an area following a natural disaster or even after blackout [5]. The advantages of DCMGs over the conventional AC microgrids are enormous. Reactive power and grid synchronizing issues are absent in DCMG. Moreover, the DCMG is not susceptible to skin effect, angular, and frequency instability problems [5]. Further, there is no issue of inrush current as there is no traditional transformer in the DCMG [6].

Nevertheless, despite numerous advantages, there are technical challenges associated with the DCMG that need to be addressed. For instance, designing a suitable converter along with its appropriate controller for an industrial-scale DCMG is very challenging. Besides, stability issues are very significant for the DC bus. The stability of the DC system deteriorates due to external disturbances and resonances in the

system [7, 8]. Therefore, a comprehensive modelling of microgrid with an appropriate mathematical model and effective control is sought to ensure the stable operation.
Despite the substantial efforts, minimizing the oscillation still remains a key challenge for DCMG [9, 10].

A thorough stability assessment using three different topologies of DCMG is
presented in this chapter. First, the DC-DC converters (both buck and boost types)
are designed in a step-by-step manner. Second, these converters are interconnected
to the energy resources and load under three different topologies. Third, various
disturbances are applied in DCMG to investigate the voltage oscillation, resonance
issues, and the entire stability of the microgrid. Subsequently, impedance analysis,
time-domain simulation, and fast Fourier transform (FFT) analysis are concurrently
conducted to analyze and understand the stability issues within DCMG. Finally,
based on the stability performances, the most appropriate topology is recommended
for further examination.

8.2 DESIGN OF DC MICROGRID

To design a DCMG, different sources such as DC–DC converters and DC link
capacitance are essential. The following subsections discuss the design procedure of DC-DC buck converters, boost converters, and DC link capacitance in
detail.

8.2.1 Design of DC–DC Buck Converter

A buck converter is usually used to step down the voltage level. A few steps are
considered to design a DC–DC buck converter. These are: (a) appropriate parameter
search; (b) controller development; and (c) performance verification.

8.2.1.1 Basic Configuration and Parameter Selection of a Buck Converter

Figure 8.1 shows the schematic diagram and control of the buck converter. To design
a buck converter, it is necessary to determine the appropriate value of duty cycle,

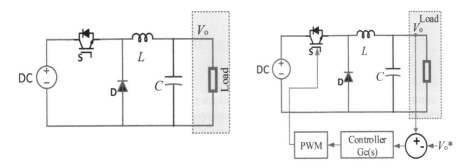

FIGURE 8.1 Schematic diagram: (Left) buck converter, (Right) buck converter with proportional, and integral (PI) controller.

inductor, capacitor, resistor, and switching frequency. Duty cycle can be defined by Equation (8.1).

$$D = 1 - \frac{V_{g(min)} \times \eta}{V_d} \qquad (8.1)$$

In Equation (8.1), $V_{g(min)}$, V_d, and η are, respectively, minimum input voltage, desired output voltage, and efficiency of the converters. Then, an inductor can be estimated by Equation (8.2).

$$L = \frac{V_{out} \times (V_{in} - V_{out})}{\Delta I_L \times f_s \times V_{in}} \qquad (8.2)$$

In Equation (8.2), V_g, V_d, ΔI_L, and f_s are known as typical input voltage, desired output voltage, estimated inductor ripple current, and switching frequency, respectively. Inductor ripple current can be obtained by Equation (8.3).

$$\Delta I_L = \frac{(V_{in(max)} - V_{out}) \times D}{f_s \times L} \qquad (8.3)$$

If the inductor value is recommended by the datasheet, then the inductor ripple current can be calculated by Equation (8.4).

$$\Delta I_L = (0.2 - 0.4) \times I_{out(max)} \qquad (8.4)$$

A capacitor can be obtained by Equation (8.5).

$$C_{min} = \frac{\Delta I_L}{8 \times f_s \times \Delta V_{out}} \qquad (8.5)$$

In Equation (8.5), ΔV_{out} is the desired output voltage ripple, which can be found by Equation (8.6). The estimated parameters of the buck converter are given in Table 8.1.

$$\Delta V_{out} = ESR \times \Delta I_L \qquad (8.6)$$

8.2.1.2 Conventional PI Controller

A controller is usually used to regulate and fix the output voltage. In this chapter, a conventional PI controller has been considered to regulate the output voltage of buck and boost converter. Four design criteria have been considered to design the PI controller for the studied buck and boost converter, i.e., root locus, settling time, phase margin (PM), and gain margin (GM). A conventional proportional and integral gain is then selected on the basis of a root locus, PM, step response, and GM. To ensure the system stability, PM should be higher than 60 degrees, while GM must be greater than 6 dB [11]. The $K_p = 0.9$ pu and $K_i = 0.005$ pu have

TABLE 8.1

Detailed Parameter Calculation for Buck Converter

10 kW Buck Converter	5 kW Buck Converter	1 kW Buck Converter
$P = V \times I = 10 \text{ kW}$	$P = V \times I = 5 \text{ kW}$	$P = V \times I = 1 \text{ kW}$
$R = \dfrac{V_d}{I} = 16\Omega$	$R = \dfrac{V_d}{I} = 32\Omega$	$R = \dfrac{V_d}{I} = 145\Omega$
$D = \dfrac{V_{out} \times \eta}{V_{in(max)}} = 0.5333$	$D = \dfrac{V_{out} \times \eta}{V_{in(max)}} = 0.5333$	$D = \dfrac{V_{out} \times \eta}{V_{in(max)}} = 0.51$
$\Delta I_L = \dfrac{\left(V_{in(max)} - V_{out}\right) \times D}{f_s \times L}$	$\Delta I_L = \dfrac{\left(V_{in(max)} - V_{out}\right) \times D}{f_s \times L}$	$\Delta I_L = \dfrac{\left(V_{in(max)} - V_{out}\right) \times D}{f_s \times L}$
$= 21.78 \text{ A}$	$= 2.52 \text{ A}$	$= 2.14 \text{ A}$
$L = \dfrac{V_{out} \times \left(V_{in} - V_{out}\right)}{\Delta I_L \times f_s \times V_{in}}$	$L = \dfrac{V_{out} \times \left(V_{in} - V_{out}\right)}{\Delta I_L \times f_s \times V_{in}}$	$L = \dfrac{V_{out} \times \left(V_{in} - V_{out}\right)}{\Delta I_L \times f_s \times V_{in}}$
$= 0.00043 \text{ H}$	$= 0.0037 \text{ H}$	$= 0.0044 \text{ H}$
$C_{min} = \dfrac{\Delta I_L}{8 \times f_s \times \Delta V_{out}}$	$C_{min} = \dfrac{\Delta I_L}{8 \times f_s \times \Delta V_{out}} = 0.0006$	$C_{min} = \dfrac{\Delta I_L}{8 \times f_s \times \Delta V_{out}}$
$= 0.00006 \text{ F}$	$= 0.00063 \text{ F}$	$= 0.00006 \text{ F}$
$\Delta V_{out} = ESR \times \Delta I_L = 2.17 \text{ V}$	$\Delta V_{out} = ESR \times \Delta I_L = 0.25 \text{ V}$	$\Delta V_{out} = ESR \times \Delta I_L = 0.21 \text{ V}$

been selected based on the PM of 90 degrees and GM of 195 dB for the load-side converter. The root locus, bode plot, and the step response of the conventional PI controller are given in Figure 8.2.

8.2.1.3   Verification Using TIme-Domain Simulation

The converter along with the controller has been designed in MATLAB® Simulink® platform, and the time-domain simulations results are given in Figure 8.3.

From Figures 8.3 (a)–(b), it is clear that both converters provide oscillation-free and stable output voltage. The design aspects of the boost converter are discussed next.

8.3   DESIGN OF DC–DC BOOST CONVERTER

A boost converter is employed to scale up the voltage level in the DC system. Using the similar approach as stated in Section 8.2, there are three steps to design a DC-DC boost converter: (a) appropriate parameter selection; (b) controller design; and (c) performance testing by time-domain simulation. Figure 8.4 shows the schematic diagram and control of the boost converter.

8.3.1   Basic Configuration and Parameter Selection

To design a boost converter, it is necessary to determine the value of duty cycle, inductor, capacitor, resistor, and switching frequency. Duty cycle can be found by as given in Equation (8.7).

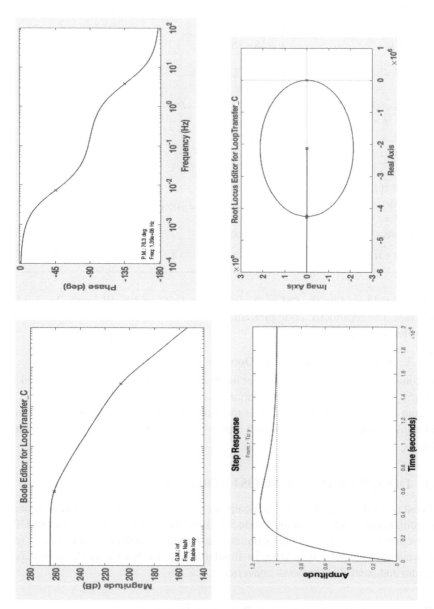

FIGURE 8.2 The root locus, step response, phase, and gain of the overall system.

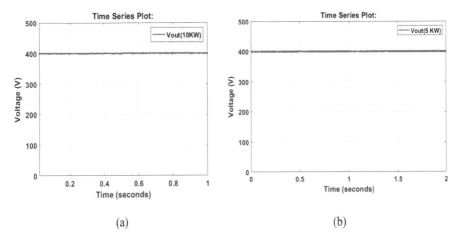

(a)   (b)

FIGURE 8.3   Output voltage: (a) 10 kW buck converter, (b) 5 kW buck converter.

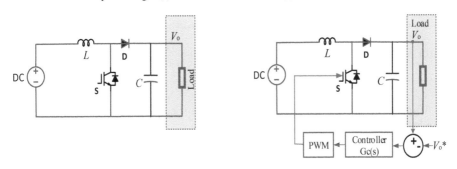

FIGURE 8.4   Schematic diagram: (Left) boost converter, (Right) boost converter with PI controller.

$$D = 1 - \frac{V_{g(\min)} \times \eta}{V_d} \tag{8.7}$$

Here, $V_{g(\min)}$ is the minimum input voltage, V_d is the desired output voltage, and η is the efficiency of the converters. Then, an inductor can be obtained by as given in Equation (8.8).

$$L = \frac{V_g \times \left(V_d - V_g\right)}{\Delta I_L \times f_s \times V_d} \tag{8.8}$$

Here, V_g is the typical input voltage, ΔI_L is the estimated inductor ripple current, and f_s is the switching frequency. Inductor ripple current can be found by Equation (8.9).

$$\Delta I_L = (0.2 - 0.4) \times I_{out(\max)} \times \frac{V_d}{V_g} \tag{8.9}$$

TABLE 8.2

Detailed Parameter Calculation for Boost Converter

20 kW Boost Converter	10 kW Boost Converter
$P = V \times I = 750*26.66 = 20$ kW	$P = V \times I = 750*13.333 = 10$ kW
$R = \dfrac{V_d}{I} = 750/26.66 = 28.13\ \Omega$	$R = \dfrac{V_d}{I} = 750/13.333 = 56.264\ \Omega$
$D = 1 - \dfrac{V_{in}}{V_{out}} = 1 - 380/750 = 0.49333$	$D = 1 - \dfrac{V_{in}}{V_{out}} = 1 - 38/750 = 0.49333$
The inductor ripple current is	The inductor ripple current is
$= 0.2 \times I_{out(max)} \times \dfrac{V_{out}}{V_{in}} = 10.52$ A	$= 0.2 \times I_{out(max)} \times \dfrac{V_{out}}{V_{in}} = 5.263$ A
$L = \dfrac{V_g \times (V_d - V_g)}{\Delta I_L \times f_s \times V_d} = 0.003562$ H	$L = \dfrac{V_g \times (V_d - V_g)}{\Delta I_L \times f_s \times V_d} = 0.003562$ H
$C = \dfrac{I_{out(max)} \times D}{f_s \times \Delta V_d} = 0.00227$ F	$C = \dfrac{I_{out(max)} \times D}{f_s \times \Delta V_d} = 0.00095$ F
$\Delta V_{out} = ESR \times \left(\dfrac{I_{out(max)}}{1-D} + \dfrac{\Delta I_L}{2} \right)$	$\Delta V_{out} = ESR \times \left(\dfrac{I_{out(max)}}{1-D} + \dfrac{\Delta I_L}{2} \right)$
$= 0.01 \times \left(\dfrac{26.66}{1-0.49333} + \dfrac{10.52}{2} \right) = 0.58$ V	$= 0.01 \times \left(\dfrac{13.333}{1-0.49333} + \dfrac{5.263}{2} \right) = 0.69248$ V

If inductor value is recommended by the data sheet, then inductor ripple current can be obtained using Equation (8.10).

$$\Delta I_L = \frac{V_g \times D}{f_s \times L} \tag{8.10}$$

The appropriate capacitor can be obtained using Equation (8.11).

$$C = \frac{I_{out(max)} \times D}{f_s \times \Delta V_d} \tag{8.11}$$

In Equation (8.11), ΔV_d is the desired output voltage ripple, which can be found as given in Equation (8.12).

$$\Delta V_{out} = ESR \times \left(\frac{I_{out(max)}}{1-D} + \frac{\Delta I_L}{2} \right) \tag{8.12}$$

The calculation of the studied boost converters are given in Table 8.2.

8.3.2 PI CONTROLLER DESIGN AND VERIFICATION

A PI controller is designed by using the MATLAB control toolbox. The gain is selected considering root locus, settling time, phase margin, and gain margin. A time-domain simulation is conducted to verify the controller performance, and the results are presented in Figure 8.5.

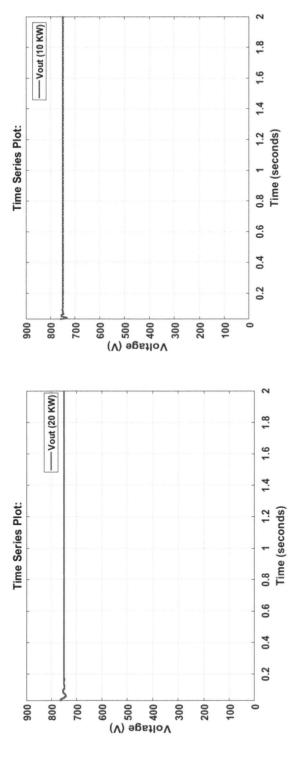

FIGURE 8.5 Output voltage: (Left) 10 kW boost converter, (Right) 20 kW boost converter.

8.4   DC LINK CAPACITOR DESIGN

The DC bus is commonly formed by electromagnetic interference (EMI) filter (capacitance), which helps in maintaining the DC bus voltage and provides a filtering function to DCMG. A DC-link capacitance for a small system (with two units) can be expressed by Equation (8.13).

$$C_{dc} = \left(\frac{V_s}{Z_s} - \frac{V_{out}}{Z_L} \right) \tag{8.13}$$

Consequently, the DC-link capacitance for n number of parallel units of systems can be found by Equation (8.14).

$$C_{dc} = \left(\frac{V_s}{Z_{sn}} - \frac{V_{out}}{Z_{Ln}} \right) \tag{8.14}$$

where, Z_{sn} is the total impedance at the source side converters and Z_{Ln} is the total impedance of load side converters, and can be expressed as Equations (8.15) and (8.16).

$$Z_{sn} = Z_1 \parallel Z_2 \parallel Z_3 .... Z_n$$

$$= \left(\frac{1}{Z_1} + \frac{1}{Z_2} + \frac{1}{Z_3} + ... \frac{1}{Z_n} \right)^{-1} \tag{8.15}$$

$$Z_{Ln} = Z_{L1} + Z_{L2} + Z_{L3} + ... Z_{Ln}$$

$$= \left(\frac{1}{Z_{L1}} + \frac{1}{Z_{L2}} + \frac{1}{Z_{L3}} + ... \frac{1}{Z_{Ln}} \right)^{-1} \tag{8.16}$$

In addition, an industrial scale of DC grids, loads and sources are connected via long DC cable with higher inductance values. The characteristic impedance of the DC line can be obtained as Equation (8.17) [12].

$$Z_{link} = \sqrt{\frac{R + JwL}{JwC}} \tag{8.17}$$

where, R is the resistance, L is the inductance, and C is the capacitance of the DC link. This cable inductance induces resonance with DC link capacitance, and the resonance could propagate towards the power electronic devices or network or in both [13].

8.5   METHODOLOGIES FOR STABILITY ASSESSMENT

Three methods, namely, impedance analysis, time-domain simulation, and FFT analysis can be used for DC grid stability assessment; each of them has its own merits and demerits [7]. This section will briefly discuss the pros and cons of these three methodologies.

8.5.1 Impedance Analysis

Middlebrook et al. [14] introduced the theory of impedance analysis in 1976. This theory has been extensively used to assess the small-signal stability (SSS) of an interconnected system since the mid 90s [15]. This method is also useful to check the SSS of power electronic (PE)-based power systems, and it could examine the whole system in accordance with the input and output features of its each and every subsystem [7]. In impedance analysis, a system splits into two sides first. Then, the output and input impedances are extracted, respectively, from the source and load subsystem of the particular system. Finally, the ratio of these two impedances is applied to find out the system stability. According to Nyquist theory [14], the stability of an interconnected system can be determined by the proportion of the output and input impedances (Z_s / Z_L). If the curve is far away from (–1,0) point, then it is considered as a stable system. The system is deemed unstable when the curve cross or touch (-1,0) point. In a nutshell, there are some specific advantages of impedance analysis over the other methods. Studied analytical expression of the large and multi-converter DCMG could be obtained by impedance analysis. Moreover, it is possible to detect harmonic oscillation (which is potentially originated from PE devices) by impedance analysis.

In this chapter, three different topologies have been considered, and under all topologies, the boost converter has been placed in the source side, whereas buck converter has been placed in the load side. Hence, the output impedances from the boost converter and the input impedances from the buck converter have been estimated. The output impedance of the investigated boost converter can be expressed as Equation (8.18).

$$Z_{ocl}(s) = \frac{Z_o(s)}{1 + T(s)} \tag{8.18}$$

Here, $Z_o(s)$ is the open-loop output impedance of boost converter, and $T(s)$ is the loop gain. Open-loop output impedance and loop gain can be expressed as Equations (8.19) and (8.20).

$$Z_o(s) = \frac{s^2 + s\left(C_1 \times R_1\left(1 + D^2\right) + L_1 + \left(R_1 \times D^2\right)\right)}{L_1 \times C_1 \times R_1} \tag{8.19}$$

$$T(s) = \left(K_p + \frac{K_i}{s}\right) \times \frac{\left(\dfrac{V_{in}}{(1-D)^2}\right) \times \left(1 - s\dfrac{L_1}{(1-D)^2 \times R_1}\right)}{\left(\dfrac{L_1 \times C_1}{(1-D)^2}\right)\left(s^2 + s\left(\dfrac{1}{C_1 \times R_1} + \dfrac{(1-D)^2}{L_1}\right)\right) + \dfrac{(1-D)^2}{L_1 \times C_1}} \tag{8.20}$$

where C_1, L_1, R_1, and D is the capacitor, inductor, resistor, and duty cycle in the boost converter. Z_o is the open-loop output impedance of the boost converter. The input impedance of the investigated buck converter can be written as Equation (8.21).

$$Zicl(s) = \frac{Z_i(s) \times R_2 \times \left(1 + T(s)\right)}{R_2 + Z_i(s) - D^2 \times T(s)} \tag{8.21}$$

where $Z_i(s)$ and $T(s)$ are the open-loop input impedance and loop gain, respectively, and they can be expressed as Equations (8.22) and (8.23).

$$T(s) = \left(K_p + \frac{K_i}{s}\right) \times \left[\left(\frac{V_{in}}{L_2 \times C_2}\right) \times \frac{(1 + s \times C_2)}{s^2 + s\left(\frac{1}{C_2 \times R_2} + \frac{1}{L_2}\right) + \frac{1}{L_2 \times C_2}}\right] \tag{8.22}$$

$$Z_i(s) = \frac{(s^2 L_2 C_2) + R_2}{D^2 \times (1 + sR_2 C_2)} \tag{8.23}$$

Here, three individual converters having the power rating of 1 kW, 5 kW, and 10 kW, respectively, have been used in load side, and input impedances of the individual unit can be expressed as Equations (8.24), (8.25), and (8.26), respectively.

$$Z_{i2}(s) = \frac{(s^2 L_2 C_2) + R_2}{D^2 \times (1 + sR_2 C_2)} \tag{8.24}$$

$$Z_{i3}(s) = \frac{(s^2 L_3 C_3) + R_3}{D^2 \times (1 + sR_3 C_3)} \tag{8.25}$$

$$Z_{i4}(s) = \frac{(s^2 L_4 C_4) + R_4}{D^2 \times (1 + sR_4 C_4)} \tag{8.26}$$

Since all converters have been connected in parallel, therefore the total input impedance can be written as Equation (8.27).

$$Z_{tin} = \frac{Z_{ti2} \times Z_{ti3} \times Z_{ti4}}{Z_{ti2} + Z_{ti3} + Z_{ti4}} \tag{8.27}$$

Eventually, DC bus impedance can be expressed as Equation (8.28).

$$Z_{bus} = \frac{Z_0 \times Z_{tin}}{Z_0 + Z_{tin}} \tag{8.28}$$

8.5.2   TIME-DOMAIN SIMULATION

In this chapter, after the impedance assessment, time-domain simulation has been taken into consideration, precisely to study the power oscillations [16]. In the time-domain analysis, a mode is disturbed first. Then, the behavior of state variables is studied by resolving differential and algebraic equations by numerical integration techniques (i.e., backward Euler and trapezoidal rule), and can be written as Equation (8.29).

$$\dot{x} = f(x, u, m, n)$$
$$0 = g(x, y, m, n) \tag{8.29}$$

where, x and u are the vectors of state variables and input variables, respectively; m and n are control and uncontrollable variables, respectively. In summary, it can be said that apart from the experimental study, the time domain simulation is the most accurate way to study power oscillations or any instability problems. The demerit of the time-domain analysis is doing a critical analysis, for instance, various weak modes, the dominant states variable, and the sensitivity of weak modes with respect to the parameter variation cannot be captured. Therefore, impedance scanning, and time-domain simulation, along with FFT analysis, can be used as complementary solutions to get a complete understanding of the high-frequency oscillation.

8.5.3 FAST FOURIER TRANSFORM ASSESSMENT

The FFT analysis is usually conducted to see the frequency domain response in DCMG. In this study, FFT analysis is conducted to find out the resonance frequency in DC bus for the studied DCMG. The multi-converter-based DC microgrid and stability study are followed in Section 8.6.

8.6 NUMERICAL RESULTS AND DISCUSSION

A multi-converter-based DC microgrid could be composed of highly nonlinear components, and it should be operated under all operating conditions. Hence, the stability of the system should be ensured at every moment to avoid any adverse consequences. Three different topologies are considered to design a multi-converter-based DC microgrid. Some of the exciting results from Topology I are illustrated in Section 8.6.1.

8.6.1 MICROGRID TOPOLOGY I (ONE SOURCE AND THREE LOADS)

In Topology I, a PV generator along with a single unit of the converter on the generation side and three groups of the converter on the load side are considered (See Figure 8.6). A conventional PI controller is considered on both source and load sides to regulate the output voltage of the respective converters. The parameter of the converters of DCMG is shown in Table 8.3. Then, an analytical expression is

---

TABLE 8.3
Parameter of Investigated DC Microgrid

Boost Converter		Buck Converter 1		Buck Converter 2		Buck Converter 3	
$V_s =$ 375V	$L_1 =$ 156$e-6H$	$V_s = 750V$	$L_2 =$ 4.2857$e-4H$	$V_s = 750V$	$L_3 =$ 0.0044H	$V_s = 750V$	$L_4 =$ 0.0037333H
$V_o =$ 750V	$R_1 = 29\Omega$	$V_o = 400V$	$R_2 = 16\Omega$	$V_o = 380V$	$R_3 = 145\Omega$	$V_o = 400V$	$R_4 = 32\Omega$
$D =$ 0.5	$C_1 =$ 1000$e-6F$	$D = 0.5$	$C_2 =$ 470$e-6F$	$D = 0.5$	$C_3 =$ 500$e-6F$	$D = 0.5$	$C_4 =$ 900$e-6F$

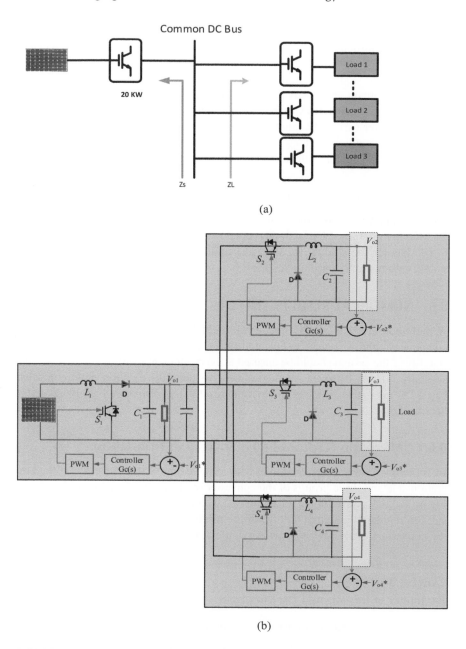

FIGURE 8.6   Topology I; Small-scale DCMG: (a) schematic diagram, (b) detail circuit diagram.

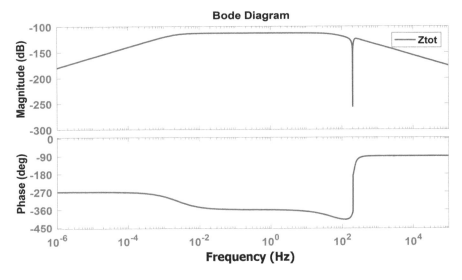

FIGURE 8.7   Frequency response of impedance seen from the DC link.

developed, and the total bus impedance is calculated and plotted in MATLAB, as shown in Figure 8.7.

Figure 8.7 shows the impedance frequency response Z_t, which is seen from the DC link. As can be seen, there has been a steep decline at 119 Hz with a magnitude of −121 dB in DC bus. Apart from this, overall, the total impedance magnitude is steady and under satisfactory state. Hence, a sensitive mode at 120 Hz is identified, which is the critical focal points of this investigation.

As the output power of the renewable resources is entirely dependent on the weather and fluctuates within a short period, the uncertainty of the distributed resources and demand are taken into account in this study. Therefore, both load and source sides disturbances are simultaneously applied to see the impact of such incidents on DC bus of the DCMG; which are shown in Figure 8.8 (a)–(d).

8.6.1.1   Impact of Voltage Disturbances

In scenario I, a DC microgrid is investigated considering a single unit of a DC/DC step-up converter along with the traditional single-variable controller on the source side, and three units of DC/DC step-down converters along with the traditional PI controller on the load side, as shown in Figure 8.6. The parameters of the converters are given in Table 8.3.

In this study, a series of voltage disturbances are applied on the source side, and the responses are recorded and presented in Figures 8.8. Figure 8.8 (a) represents a disturbance-free and ideal DC bus voltage. It is observed that the DC bus voltage is aligned with the reference voltage. On the other hand, Figures 8.8 (b)–(c), presents the impact of 15% and 25% voltage disturbance on source side, respectively. Almost similar scenarios are observed for these two cases except for the frequency of oscillation. A 100 Hz frequency of oscillation is observed

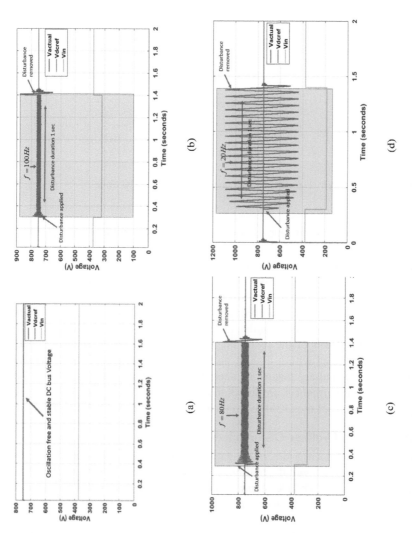

FIGURE 8.8 Common DC bus voltage with a single-variable classical PI controller: (a) disturbance-free (an ideal 20 kW DC microgrid system), (b) impact of 15% voltage disturbance, (c) impact of 25% voltage disturbance, and (d) impact of 50% voltage disturbance.

during the 15% voltage disturbances, whereas 80 Hz frequency of oscillation is noticed during 25% voltage disturbances. Besides, an overshoot of around 900 voltage peak and power oscillation are observed during the voltage disturbances. Figure 8.8 (d) presents the DC bus voltage with respect to the 50% voltage disturbance. It is noticed that during the time of disturbances, voltage fluctuates between 500 V and around 1100 V. A 20 Hz frequency of oscillation is observed during the disturbance time. It is also noticed that after removing disturbance, DC bus voltage comes to steady-state stable condition without having any voltage oscillation in all the cases.

8.6.1.2   Impact of Load Disturbances

In this scenario, a series of load disturbances have been applied, and the results are shown in Figure 8.9. Three loads having the capacity of 1 kW, 5kW, and 10 kW have been considered for this investigation. Figure 8.9 (a)–(b) presents the DC bus voltage with respect to individual 5 kW and 10 kW load changes, respectively. It is observed that with respect to load change, DC bus voltage starts to oscillate, but on a small scale. Besides, similar consequences are observed during and after the disturbances in both cases. A continuous power oscillation with 120 Hz frequency of oscillation is observed in DC bus, after removing the load disturbance.

8.6.1.3   Impact of Both Sides Disturbances

In this case, disturbances are applied concurrently on both load and source sides to investigate the most significant impact on DC bus in the DC microgrid. It is assumed that the voltage disturbances first initiate on source side and then propagate onto load side. Four scenarios are presented in Figure 8.10.

Figure 8.10 (a) presents the impact of 15% voltage and 5 kW load changes between 0.3 and 1.5 s on DC bus in the DCMG. A continuous power oscillation is observed after removing the disturbances. Besides, two modes, one at 70 Hz and another at 120 Hz, are noticed during and after the event. Figure 8.10 (b) presents the impact of 25% voltage and 5 kW load changes in the investigated DCMG. A significant power oscillation is observed during the disturbances, and it remains continued even after removing the disturbance (although the amplitude of oscillation is lower). In addition, two different modes at 80 Hz and 120 Hz are observed. The 80 Hz oscillation is observed during the disturbances, whereas 120 Hz oscillation is observed after the disturbances. A similar scenario is noticed in Figure 8.10 (c), which presents the impact of 10 kW load and 25% voltage disturbances when applied concurrently. Two modes, one at 60 Hz and another at 120 Hz, are noticed during and after the incident. In contrast, completely different scenarios are observed in Figure 8.10 (d), where 50% voltage and 10 kW load disturbance on the DC bus are applied concurrently. It is noticed that during the time of disturbances, voltage fluctuates between 500 V and around 1150 V. Two different modes at 20 Hz and 120 Hz are observed during this interference. It is also noticed that after removing disturbance, DC bus voltage gets returned to the steady state, with a continuous power oscillation at 120 Hz.

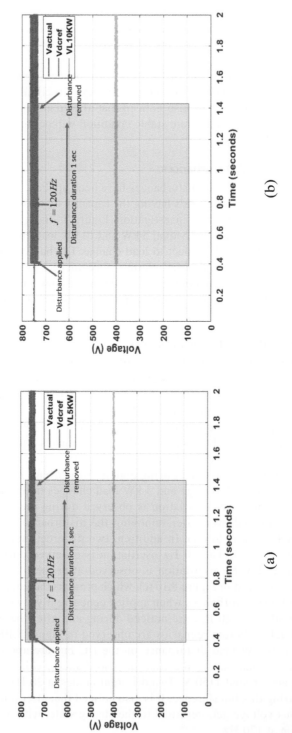

FIGURE 8.9 Impact of load disturbances on DC bus in DCMG: (a) 5 kW load disturbance, (b) 10 kW load disturbances.

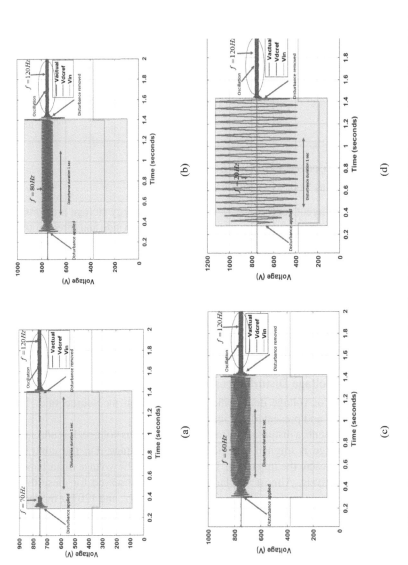

FIGURE 8.10 Common DC bus voltage with a single variable classical PI controller: (a) impact of 5 kW load and 15% voltage disturbance (0.3–1.5 s), (b) impact of 5 kW load and 25% voltage disturbance (0.3–1.5 s), (c) impact of 10 kW load and 25% voltage disturbance (0.3–1.5 s), and (d) impact of 10 kW load and 50% voltage disturbance (0.3–1.5 s).

TABLE 8.4

DC Microgrid Parameters—Topology II

Boost Converter 1 and 2		Buck Converter 1		Buck Converter 2		Buck Converter 3	
$V_s =$ 375V	$L_1 =$ 156e−6H	$V_s =$ 750V	$L_2 =$ 4.2857e−4H	$V_s =$ 750V	$L_3 =$ 0.0044H	$V_s =$ 750V	$L_4 =$ 0.0037333H
$V_o =$ 750V	$R_1 = 56\Omega$	$V_o = 400V$	$R_2 = 16\Omega$	$V_o = 380V$	$R_3 = 145\Omega$	$V_o = 400V$	$R_4 = 32\Omega$
$D = 0.5$	$C_1 =$ 470e−6F	$D = 0.5$	$C_2 =$ 470e−6F	$D = 0.5$	$C_3 =$ 470e−6F	$D = 0.5$	$C_4 =$ 900e−6F

8.6.2 MICROGRID TOPOLOGY II (TWO SOURCES AND THREE LOADS IN SERIES CONFIGURATION)

In Topology II, two different sources, i.e., PV and battery energy storage are considered. Consequently, two units of the converter on the generation side and three units of the converter on the load sides are considered, as shown in Figure 8.11. The parameter of the converters are presented in Table 8.4. All the sources and loads are connected in parallel to each other. Then, an analytical expression for impedance analysis is developed and plotted in MATLAB, as shown in Figure 8.12. As can be seen, a susceptible frequency mode of oscillation is observed at around 120 Hz on DC bus. The uncertainty of the distributed resources and loads are taken into account for this study. The source, load, and disturbances on both sides imposed to see the impact of such events on DC bus of the DC microgrid; are given in Figure 8.13 (a–d).

8.6.2.1 Impact of Voltage Disturbances

In this study, a series of voltage disturbance are applied on the source side. The responses are recorded and presented in Figure 8.13.

Figure 8.13 (a) represents a disturbance-free and ideal DC bus voltage. It is observed that the DC bus voltage is aligned with the reference voltage. On the contrary, Figures 8.13 (b)–(d), present the impacts of 15%, 25%, and 50% voltage disturbances on source side, respectively. Almost similar scenarios are observed for these three cases except for the frequency of oscillation. The 60 Hz, 110 Hz, and 120 Hz modes of the frequency of oscillation are noticed at DC bus voltage due to small to large source-side voltage disturbances. It is also noticed that after removing disturbance, DC bus voltage gets returned to steady state without having any significant voltage oscillation.

From the results, it is observed that overshoot with a power oscillation could be seen in DC bus with respect to source side voltage change. However, it decays to the pre-disturbance value after clearing the disturbance.

8.6.2.2 Impact of Load Disturbances

In this scenario, a series of load disturbances are applied; the results are given in Figure 8.14. The loads in the system are perturbed for 1 second for this investigation.

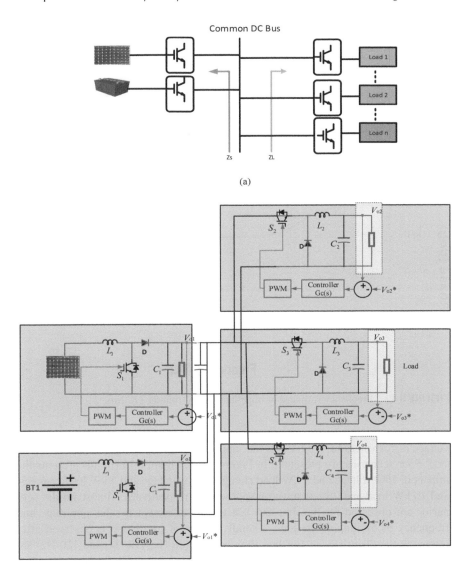

FIGURE 8.11 Topology II: (a) schematic diagram, (b) circuit diagram.

Figure 8.14 (a) and (b) presents the DC bus voltage with respect to 5 kW and 10 kW load changes, respectively. It is observed that DC bus voltage starts to oscillate with small amplitude and it continues even after removing the disturbances. Similar consequences are observed during and after the disturbances in both cases. A continuous power oscillation with 80 Hz and 160 Hz frequency of oscillation are observed in DC bus.

8.6.2.3 Impact of Both Side Disturbances

In this case, disturbances are applied on both sides simultaneously to investigate the most significant disturbances on DC bus in the DC microgrid. It is assumed that the

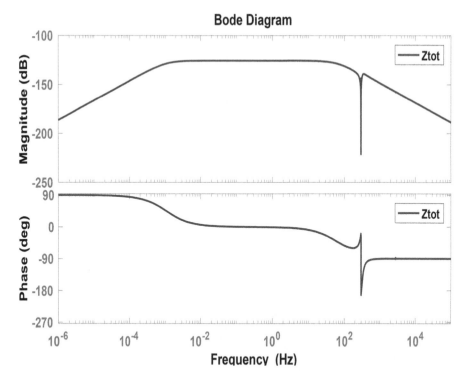

FIGURE 8.12   Frequency response of impedance seen from the DC link.

voltage disturbances first initiate on the source side and then propagate onto the load side. Four scenarios are presented in Figures 8.15. Figure 8.15 (b)–(d) presents the impact of 15% voltage and 5 kW load changes, 25% voltage and 5 kW load changes, and 10 kW load and 25% voltage on source side, respectively. Almost similar scenarios are observed for the first and last two cases except for the amplitude and frequency of oscillation. Besides, a small amount of oscillation is observed after removing the disturbances. Further, three modes (80 Hz, 110 Hz, and 120 Hz) are noticed during and after the incident. It is also noticed that in all the four cases, during the disturbances, time-frequency of oscillation is 120 Hz, which is very similar to the impedance scanning result as presented previously.

8.6.3   MICROGRID TOPOLOGY III (TWO SOURCES AND THREE LOADS IN PARALLEL CONFIGURATION)

In Topology III, two different sources PV and synchronous generator (SG) are considered and connected in parallel to each other. The schematic diagram of the system is given in Figure 8.16. The parameters of the converters are presented in Table 8.5. Then, an analytical expression for impedance analysis is developed and plotted in MATLAB, as shown in Figure 8.17.

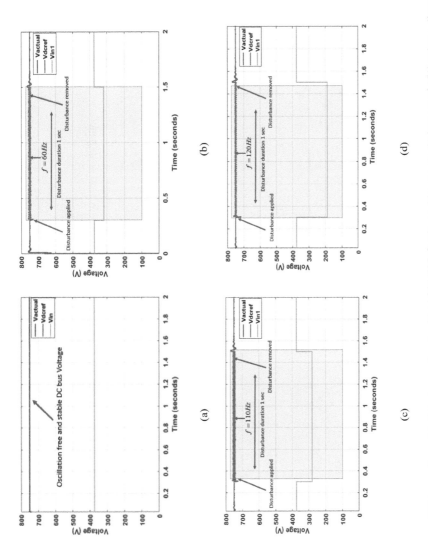

FIGURE 8.13   Common DC bus voltage: (a) disturbance-free (an ideal 20 kW DC microgrid system), (b) impact of 15% voltage disturbance, (c) impact of 25% voltage disturbance, and (d) impact of 50% voltage disturbance.

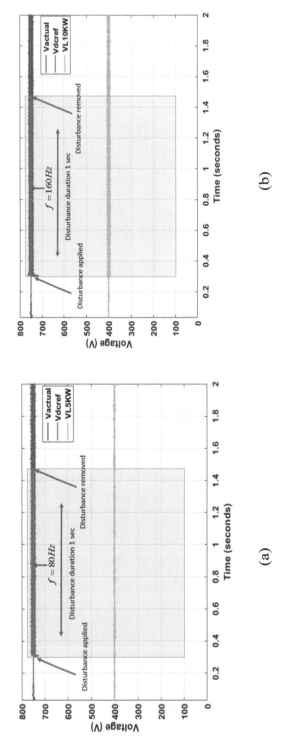

FIGURE 8.14 Impact of load disturbances on DC bus in DCMG: (a) 5 kW load, (b) 10 kW load disturbances.

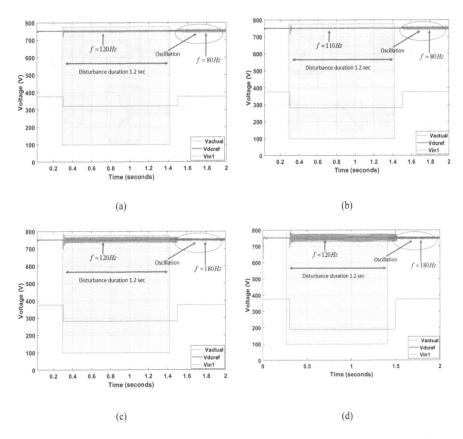

(a)   (b)

(c)   (d)

FIGURE 8.15   Common DC bus voltage: (a) impact of 5 kW load and 15% voltage disturbance between 0.3 and 1.5 s, (b) impact of 5 kW load and 25% voltage disturbance between
0.3 and 1.5 s, (c) impact of 10 kW load and 25% voltage disturbance between 0.3 and 1.5 s,
and (d) impact of 10 kW load and 50% voltage disturbance between 0.3 and 1.5 s.

As can be seen, a susceptible frequency mode of oscillation is observed at around
120 Hz on DC bus. Similar to prior topologies, the uncertainty of the distributed
resources and loads are taken into account for this study.

8.6.3.1   Impact of Voltage Disturbances

In this study, a series of voltage disturbances are applied on the source side. The
responses are recorded and presented in Figure 8.18. Figure 8.18 (a) represents a
disturbance-free and ideal DC bus voltage. It is observed that the DC bus voltage
is aligned with the reference voltage. On the other hand, Figure 8.18 (b)–(c) presents the impact of 15% and 25% voltage disturbance on source side, respectively.
Almost similar scenarios are observed for these two cases other than the frequency
of oscillation. An oscillation mode of 120 Hz is observed during the 15% voltage
disturbances, whereas a 60 Hz frequency of oscillation is noticed during 25% voltage disturbances. Figure 8.18 (d) presents the DC bus voltage with respect to the

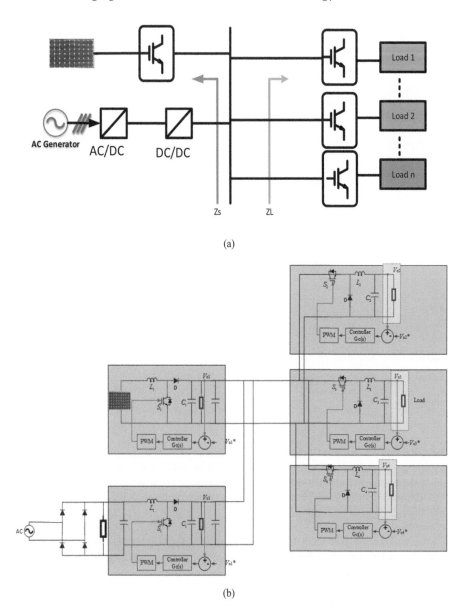

(a)

(b)

FIGURE 8.16   Topology III: (a) schematic diagram, (b) circuit diagram.

50% voltage disturbance. It is noticed that during the time of disturbances, voltage fluctuates between 500 V and around 1100 V. A 20 Hz frequency of oscillation has been observed during the disturbance time. It is also noticed that after clearance of the disturbance, DC bus voltage gets returned to steady state without having any noticeable voltage oscillation.

8.6.3.2   Impact of Load Disturbances

In this scenario, a series of load disturbances are applied. The results are shown in Figure 8.19. Figure 8.19 (a)–(b) presents the DC bus voltage with respect to 5 kW and 10 kW load changes, respectively. It is observed that with respect to load change, DC bus voltage starts to oscillate, but with small amplitude. Besides, in both cases, similar consequences are observed during and after the disturbances except for the frequency of oscillation. A continuous power of oscillation with 60 Hz and 120 Hz frequency is observed in the DC bus.

8.6.3.3   Impact of Disturbances on Both Sides

In this case, disturbances are applied on both sides concurrently to investigate the most significant impact on the DC bus. It is assumed that the voltage disturbances first originate on the source side and then propagate onto the load side. Four scenarios are presented in Figure 8.20. Figure 8.20 (a)–(d) presents the impact of 15% voltage and 5 kW load changes, 25% voltage and 5 kW load changes, 10 kW load and 25% voltage disturbances, 50% voltage and 10 kW load disturbance, respectively, between 0.3 and 1.5 s on a DC bus in the DCMG. A similar consequence is observed in all the four cases except for the amplitude and frequency of oscillation at DC bus. During the disturbances, 40 Hz frequency mode is observed, while 60 Hz and 120 Hz frequency modes are observed after the disturbances. Besides, a continuous power oscillation is observed after removing the disturbances.

8.7   DISCUSSION

The impact of the possible uncertainties on common DC bus in DCMG has been analysed in three different topologies. Throughout the study, a classical conventional single variable PI controller is considered. It is noticed that in Topology I, the modes at 20 Hz, 80 Hz, and 100 Hz appear at DC bus voltage due to small to large source-side voltage disturbances. In contrast, due to small to significant load disturbances, only 120 Hz modes of the frequency of oscillation is observed at DC bus voltage. Finally, 60 Hz, 70 Hz, 80 Hz, and 120 Hz modes of the frequency of oscillation are observed for simultaneous disturbances at voltage, load and both sides. In the case of

---

TABLE 8.5
Parameters of the Investigated DC Microgrid

Boost Converter 1 and 2		Buck Converter 1		Buck Converter 2		Buck Converter 3	
$V_s =$ 375V	$L_1 =$ 156e – 6H	$V_s = 750V$	$L_2 =$ 4.2857e – 4H	$V_s =$ 750V	$L_3 =$ 0.0044H	$V_s =$ 750V	$L_4 =$ 0.0037333H
$V_o =$ 750V	$R_1 = 56\Omega$	$V_o = 400V$	$R_2 = 16\Omega$	$V_o =$ 380V	$R_3 = 145\Omega$	$V_o =$ 400V	$R_4 = 32\Omega$
$D = 0.5$	$C_1 =$ 470e – 6F	$D = 0.5$	$C_2 = 470e – 6F$	$D = 0.5$	$C_3 =$ 470e – 6F	$D = 0.5$	$C_4 =$ 900e – 6F

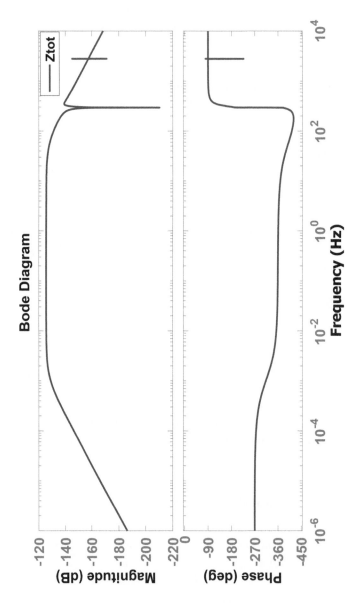

FIGURE 8.17   Frequency response of impedance seen from the DC link.

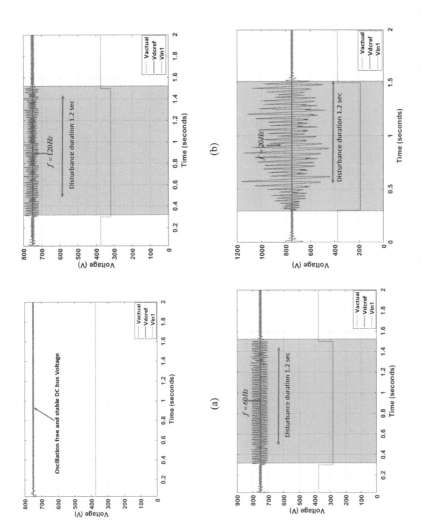

FIGURE 8.18   Common DC bus voltage: (a) disturbance-free (an ideal 20 kW DC microgrid system), (b) impact of 15% voltage disturbance, (c) impact of 25% voltage disturbance, and (d) impact of 50% voltage disturbance.

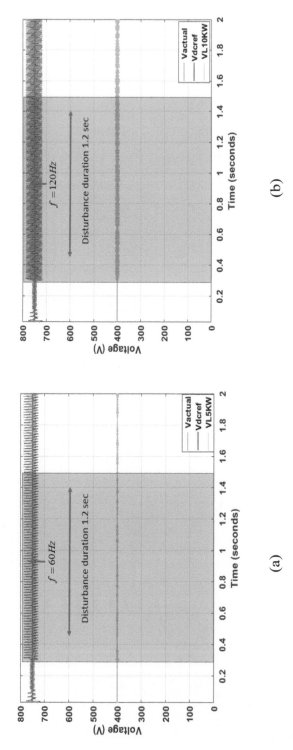

FIGURE 8.19 Impact of load side disturbances: (a) 5 kW load disturbance, (b) 10 kW load disturbance.

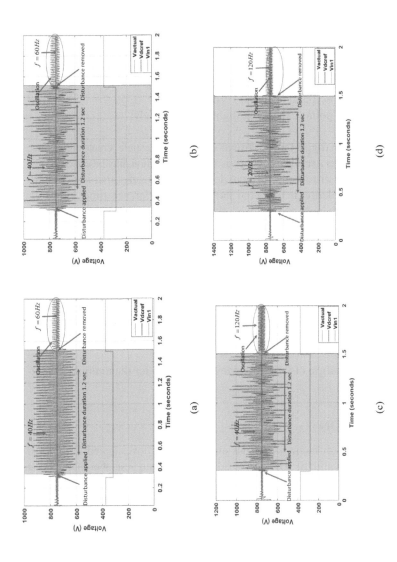

FIGURE 8.20   Common DC bus voltage: (a) impact of 5 kW load and 15% voltage disturbance between 0.3 and 1.5 s, (b) impact of 5 kW load and 25% voltage disturbance between 0.3 and 1.5 s, (c) impact of 10 kW load and 25% voltage disturbance between 0.3 and 1.5 s, and (d) impact of 10 kW load and 50% voltage disturbance between 0.3 and 1.5 s.

Topology II, two dominant modes, one at 60 Hz and another at 120 Hz, are observed due to small to large source-side voltage disturbances. In contrast, due to small to significant load disturbances, only 120 Hz mode is noticed at DC bus. Finally, due to simultaneous disturbances at both sides, 80 Hz, 120 Hz, and 180 Hz modes are observed. In Topology III, three dominants modes (20 Hz, 60 Hz, and 120 Hz) are observed due to small to large source-side voltage disturbances. In contrast, due to small to significant load disturbances, only 120 Hz mode is noticed at DC bus voltage. Finally, due to concurrent disturbances at both sides, 40 Hz, 60 Hz, and 120 Hz modes are observed. As can be seen with respect to various disturbances, different modes of oscillation could have been ejected, which potentially could interact with other integrated sources. However, after removing disturbances, in all four cases, the similar frequency of oscillation (120 Hz) is observed.

The following key points can be highlighted in this study:

- Disturbances can trigger the sensitive mode, which potentially creates resonance on the DC bus. As a result, a continuous voltage oscillation may be observed
- A simultaneous disturbance is more destructive than any others. A continuous power oscillation is observed during the interference time. However, when the disturbances are removed, the bus voltage returns to the predetermined value with a negligible amount of power oscillation
- Source voltage changes can create a significant impact on the DC bus. A significant amount of voltage distortion and a noticeable power oscillation are observed in the DC bus. Likewise, arbitrary load changes can cause substantial influence on the DC bus. Disturbances can trigger the sensitive mode, which potentially creates resonance on the DC bus. As a result, a significant amount of voltage oscillation could have been seen
- Resonance frequency is a significant contributor to voltage oscillation in DC bus in a multi-converter-based DC microgrid
- The stability margin of the system could be higher for the multiple generation system

8.8 CONCLUSION

In this chapter, three different topologies are extensively investigated to understand instability problems within the DC microgrids. The study suggests that there are many challenges in designing and operating an industrial-scale DC microgrid. The results show that disturbances can potentially deteriorate controller performance and contribute to oscillation and resonance at the DC bus. Further, it is found that disturbances cause power oscillation in the distribution system and make the system oscillatory and unstable. It is also found that the disturbances at both sides are more significant than any others. Overshoot along with continuous power oscillation has been observed during and after the disturbances. To find out the modes of the oscillation, FFT analysis and impedance scanning are conducted, and different modes of the frequency of oscillation are observed. For instance, the 20 Hz, 80 Hz, and 120 Hz modes of the frequency of oscillation are observed at DC bus voltage due to small to large source-side voltage disturbances. In contrast, due to small to significant load disturbances, only

120 Hz mode of the frequency of oscillation is observed at DC bus voltage. Finally, due to concurrent both sides disturbances, 20 Hz, 80 Hz, and 120 Hz modes of the frequency of oscillation are observed. Based on the voltage oscillation with respect to uncertainties, topology II is proposed for further study and examination.

Overall, this chapter focuses on power oscillation in an industrial DC microgrid by following literature review to provide information on the significance of industrial-scale DC microgrid research. Then, this research zooms into the pros and cons of the existing methodologies, which have recently been used for stability analysis. The research focuses on the modeling of industrial-scale DC grid components and demonstrates a few case studies. The studies demonstrate and discuss the presence of resonance and power oscillation at the DC bus. Above all, the outcome of this chapter leads to a better understanding of voltage and power oscillations in DC microgrids, and modeling and optimization of industrial DC microgrid along with EV power and components.

REFERENCES

1. D. Kumar, F. Zare, and A. Ghosh, "DC microgrid technology: System architectures, AC grid interfaces, grounding schemes, power quality, communication networks, applications, and standardizations aspects", *IEEE Access*, vol. 5, pp. 12230–12256, 2017.
2. S. K. Sahoo, A. K. Sinha, and N. Kishore, "Control techniques in AC, DC, and hybrid AC–DC microgrid: A review", *IEEE Journal of Emerging and Selected Topics in Power Electronics*, vol. 6, no. 2, pp. 738–759, 2017.
3. E. Planas, J. Andreu, J. I. Gárate, I. M. De Alegría, and E. Ibarra, "AC and DC technology in microgrids: A review", *Renewable and Sustainable Energy Reviews*, vol. 43, pp. 726–749, 2015.
4. J. J. Justo, F. Mwasilu, J. Lee, and J.-W. Jung, "AC-microgrids versus DC-microgrids with distributed energy resources: A review", *Renewable and Sustainable Energy Reviews*, vol. 24, pp. 387–405, 2013.
5. S. Augustine, J. E. Quiroz, M. J. Reno, and S. Brahma, "DC microgrid protection: Review and challenges", Sandia National Lab.(SNL-NM), Albuquerque, NM (United States), 2018.
6. S. Beheshtaein, R. M. Cuzner, M. Forouzesh, M. Savaghebi, and J. M. Guerrero, "DC microgrid protection: A comprehensive review", *IEEE Journal of Emerging and Selected Topics in Power Electronics*, 2019.
7. M. Amin and M. Molinas, "Small-signal stability assessment of power electronics based power systems: A discussion of impedance-and eigenvalue-based methods", *IEEE Transactions on Industry Applications*, vol. 53, no. 5, pp. 5014–5030, 2017.
8. M. Amin, M. Molinas, and J. Lyu, "Oscillatory phenomena between wind farms and HVDC systems: The impact of control," *2015 IEEE 16th Workshop on Control and Modeling for Power Electronics (COMPEL)*, 2015, pp. 1–8.
9. M. Habibullah, N. Mithulananthan, F. Zare, and D. S. Alkaran, "Investigation of power oscillation at common DC bus in DC grid," *2019 IEEE International Conference on Industrial Technology (ICIT)*, 2019, pp. 1695–1700.
10. M. Habibullah, N. Mithulananthan, F. Zare, and R. Sharma, "Impact of control systems on power quality at common DC Bus in DC grid," *2019 IEEE PES GTD Grand International Conference and Exposition Asia (GTD Asia)*, 2019, pp. 411–416.
11. X. Feng, J. Liu, and F. C. Lee, "Impedance specifications for stable DC distributed power systems", *IEEE Transactions on Power Electronics*, vol. 17, no. 2, pp. 157–162, 2002.

12. G. Vitale, "Characterization of a DC grid for power line communications in smart grids," *Control and Modeling for Power Electronics (COMPEL), 2014 IEEE 15th Workshop*, 2014, pp. 1–10.
13. H. Liu, H. Guo, J. Liang, and L. Qi, "Impedance-based stability analysis of MVDC systems using generator-thyristor units and DTC motor drives", *IEEE Journal of Emerging and Selected Topics in Power Electronics*, vol. 5, no. 1, pp. 5–13, 2017.
14. A. Riccobono and E. Santi, "Comprehensive review of stability criteria for DC power distribution systems", *IEEE Transactions on Industry Applications*, vol. 50, no. 5, pp. 3525–3535, 2014.
15. M. Amin and M. Molinas, "Non-parametric impedance based stability and controller bandwidth extraction from impedance measurements of HVDC-connected wind farms," *arXiv preprint arXiv:1704.04800*, 2017.
16. K. Prasertwong, N. Mithulananthan, and D. Thakur, "Understanding low-frequency oscillation in power systems", *International Journal of Electrical Engineering Education*, vol. 47, no. 3, pp. 248–262, 2010.

9 Stability of Remote Microgrids: Control of Power Converters

Mohd. Hasan Ali
University of Memphis

Sagnika Ghosh
Tennessee State University

CONTENTS

9.1 INTRODUCTION

The flexible AC transmission system (FACTS) devices are one of the most well-known and reliable solutions to tackle the stability issues of power grids. FACTS devices include the static VAR compensator (SVC), static synchronous compensator (STATCOM), thyristor controlled braking resistor (TCBR), TSC, static synchronous series compensator (SSSC), thyristor controlled series capacitor (TCSC), unified power flow controller (UPFC), superconducting magnetic energy storage (SMES), etc. Their benefits include improvement of the stability of the grid, reactive power control, control of active and reactive power flows on the grid, loss minimization, and increased grid efficiency. However, in all these methods, typically conventional proportional-integral (PI) control techniques are used. Moreover, since the existing power grids are being transitioned into the smart grids, with integration of renewables and sophisticated power electronic controller, stability enhancement devices should be equipped with intelligent based controllers such as a fuzzy logic controller (FLC), neural network, etc. In addition, power networks are highly nonlinear. Therefore, it is appropriate to employ nonlinear based intelligent control techniques for smart power grids.

Based on these backgrounds, this chapter first deals with the nonlinearity that arises due to the balanced and unbalanced faults occurring at different points in the microgrid system. To improve the stability due to nonlinearity issues, intelligent controller-based FACTs devices have been used and explained. Detailed description on testbed models of the remote microgrid samples and their control systems developed in MATLAB®/Simulink® environment along with the obtained system responses have been provided in this chapter.

9.2 BACKGROUND AND MOTIVATION

Microgrids are distributed networks comprised of various distributed generators, storage devices, and controllable loads that can operate either interconnected or isolated from the main distribution grid as a controlled entity [1]. To the utility, the microgrid can be thought of as a single controllable load that can respond in seconds to meet the needs of the transmission system. To the customer, the microgrid can meet their special needs; such as, enhancing local reliability, reducing feeder losses, supporting local voltages, providing increased efficiency through the use of waste heat, voltage sag [2], correction or providing uninterruptible power supply functions to name a few. The focus is on systems of distributed resources that can switch from grid connection to island operation without causing problems for critical loads. Different microgrid control strategy and power management techniques are discussed in [3]. Premium power is a concept based on the use of power electronic equipment (such as custom power devices and active filters), multi utility feeders and uninterruptible power supplies to provide power to users having sensitive loads. This power must have a higher level of reliability and power quality than normally supplied by the utility. These technologies require power electronics to interface with the power network and its loads.

The stability issues in a microgrid can be small signal, transient and voltage stability. Small signal stability issues could arise from sudden load changes, feedback controller, power limit of distributed generators [4]. Voltage stability is created by dynamic load, tap changers and reactive power limits in a microgrid. Temporary, permanent faults lead to transient stability issues. To reduce stability issues, there are number of ways like reactive compensation, voltage regulation with distributed generations, load controller and current limiters of microgrid.

A fault within microgrid creates the transient stability problems as well as voltage stability issues. The voltage stability in a remote microgrid is related to the reactive compensation of the network but in a utility microgrid the main source of the voltage stability problems is the tap changers. With few sources and confined loads, limiters in the micro sources and under voltage load shedding create most of the voltage stability problems in a facility microgrid.

One of the most important microgrid design procedure is to ensure constant voltage and frequency within the standard permissible levels [5]. There are various microgrid controls which have been developed to maintain the voltage and frequency of the microgrid within permissible levels. Some of these control approaches are: (1) master/slave control [6]; (2) droop-based control [7]; and (3) the modified active-power droop control. In [8], the transition from the grid-connected mode to autonomous operation is explored. In most cases, when there is a fault in the network, the microgrid operates in islanded mode [9]. In these cases, the microgrid becomes more vulnerable to stability issues even when there are conventional fault clearing mechanism are in place.

In microgrid there are droop-controlled inverters (DC to AC) which help transfer DC power to AC power. To have a proper functioning microgrid, inverters are needed [10]. AC to DC and DC to DC converters play vital role in power system applications [11]. When converters are operating individually, they are stable. But in a microgrid, several converters, inverters function in together which can cause instability [12]. These stability issues [13] also arise due to multiple parallel connected converters with droop control [14–15]. In such system models, converters are modeled as voltage sources with control loop dynamics and output filter dynamics while the switching dynamics are ignored [16]. Based on the stability analysis, several methods have been proposed in the previous literature to enhance system stability such as a controller based on the second derivative of the output capacitor voltage [14], arctan gradient algorithm concept [17], an adaptive decentralized droop controller [18], virtual complex impedance [19], and high gain angle droop control. Both eigenvalue and impedance analyses can both tell if the system is stable in the current state. However, they are not reliable if model analysis and stability of each DG has to be obtained. For that, a nonlinear theory is developed. Since the microgrid is a complex structure, it can be described by a nonlinear dynamic model. With the theory, the characteristics of each DG can be ascertained [4]. Uncertainty is a major hindrance in small signal study. Many of those factors impact the damping ratio and frequency, which in turn, affect the stability of small signals.

Power sharing among different DGs in an isolated microgrid [18] is possible by employing droop control or by using some centralized communication [19]. Traditionally active power-frequency and reactive power voltage droop is implemented to control frequency and voltage in DGs having a power-electronic interface [20]. However, these droop control laws are applicable for high-voltage (HV) microgrids where tie-line inductance is greater than its resistance and droop control laws improve power sharing [21] and are shown to improve efficiency for low-voltage (LV) resistive microgrids. In the last two decades (approximately), the traditional droop control laws have been modified to improve power sharing [19] and/or stability of microgrid. Small-signal stability of microgrid deteriorates at higher droop gains while it is immune to other parameters, such as controller gains and tie-line impedance [22].

Just as FACTS devices improve the reliability and quality of power transmission by simultaneously enhancing both power transfer volume and stability, the power electronic controllers are reliable for the distribution systems. Some of these devices are voltage regulator, dynamic voltage restorer, line reactor, surge suppressors, power conditioners, isolation transformers, uninterrupted power supply, proper wiring and grounding, filters, etc. [23].

One important method to improve the stability of microgrid system is reactive power compensation. The TSC is a device to compensate for the reactive power [24–25]. The TSC is used because it provides instantaneous response to the changes in the system parameters. The TSC also has another advantage in that it generates no harmonics and requires no filtering. The TSC is also used for power factor improvement, to mitigate harmonics, active and reactive power control, voltage stability. Typically, conventional PI control techniques are used [26] for TSC. However, since the existing power grids are being transitioned into the smart grids, with integration of renewables [27] and sophisticated power electronic controllers, stability enhancement devices should be equipped with intelligent based controllers, such as an FLC, neuroFLCs. Moreover, microgrid networks are highly nonlinear. Therefore, it is appropriate to employ nonlinear based intelligent control techniques for microgrids [28]. These facts motivate to deal with the subject matter of this chapter.

9.3   STABILITY ISSUES DUE TO NONLINEARITY

A remote microgrid includes a variety of operational and energy measures including smart meters, smart appliances, renewable energy resources, and energy efficient resources. The important features of a modern power grid are reliability, flexibility in network topology, efficiency, load balancing, peak curtailment, sustainability, market-enabling, demand response support, etc. However, with respect to modern power grids, there are a number of major challenges such as nonlinearity that have serious impact on system stability.

When renewable resources, smart meters, etc., are connected or integrated to the existing grid, it is known as the modern grid. The integration of renewables help with the demand response. The ongoing change of energy supply from large, centralized power plants based on nuclear or fossil fuels to smaller, decentralized sources based on renewable energies poses an enormous challenge for design and stable operation of the grid. The structure of the power grid has to be optimized to increase its stability against fluctuations and robustness against failures. However, large-scale failures are consequences of the collective dynamics of the power grid and are often caused by nonlocal mechanisms, one being the nonlinearity issue that deteriorates the power quality of the system [29]. As the modern grid consists of various renewables resources, converters, inverters, controllers, switching of the inverters and convert-ers, etc., it adds more nonlinearity to the dynamic nature of the network. However, handling such nonlinearities and stochastic nature of the system is very important. Controller parameters that are optimal for one set of operating conditions might be ineffective for the other sets.

To overcome such challenges, the nonlinear dynamics of the power system is con-sidered in the controller design technique. The solution to this nonlinear control is the application of different types of nonlinear controllers like the exact linearization design method for scalar nonlinear control systems, FLCs, discrete-time predictive control, feedback linearizing control, Oscillator-based nonlinear controller, bounded integral controller (BIC), ANFIS [30–31], decentralized nonlinear controller [32], static nonlinear control, etc.

FLC tolerates the uncertainty, imprecision, or fluctuation of the input param-eters as well as provides an opportunity to introduce expert knowledge in control rules. This type of controller yields good outcomes under changing operating conditions and time-varying input signals. The FLC has been utilized as an effi-cient tool to stabilize power systems in a wide range of operating conditions and various devices such as power system stabilizer (PSS) and FACTS. One of the powerful tools is the artificial neural network (ANN), which has applica-tions in embedded control systems and information processing. The ANFIS is an intelligent technique that can be obtained by the combination of fuzzy infer-ence system (FIS) and ANN [33]. The imprecision and uncertainty of the system modeling is taken into consideration by the fuzzy logic, while the neural network gives it a sense of adaptability [34] and rapid learning capacity. The ANFIS model has the advantage of having both numerical and linguistic knowledge [35]. The neural network helps in back-propagation to structured network to automate the fuzzy parametric tuning. ANFIS models are also able to explain past data and predict future behaviour.

9.4   THYRISTOR SWITCHED CAPACITOR

A TSC is a FACTS device used for reactive power compensating in electrical power systems. A TSC normally comprises three main items of equipment: The main capacitor bank, the thyristor valve and a current-limiting reactor, which is usually air-cored. The largest item of equipment in a TSC, the capacitor bank is constructed from rack-mounted outdoor capacitor units, each unit typically having a rating in the range 500–1000 kilovars (kVAR). The power capacitor is connected in series with an anti-parallel thyristor [24–25]. To protect the thyristors, a current limiting inductor (reactor) is used which limits the peak current and the rate of rise of current (di/dt) when the TSC turns on at an incorrect time. In TSC, thyristors are connected in anti-parallel pairs and these pairs are connected in series. The anti-parallel connection of the thyristor makes the flow of current in both directions unlike the commercially available thyristors that can conduct only in one direction.

In this study, the TSC is connected in shunt to the line (shown at bus 8 of the hybrid power grid model of Figure 9.1) to function as a reactive power compensation device during faults [36]. It consists of capacitors which are usually switched with the

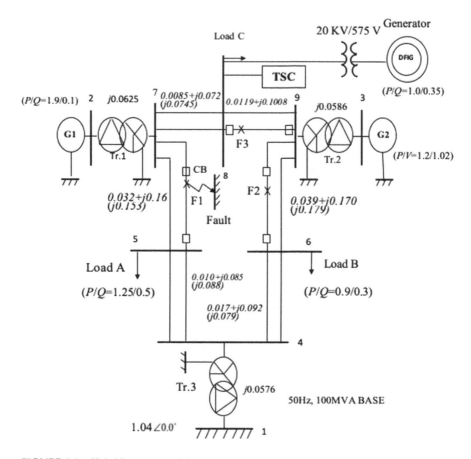

FIGURE 9.1   Hybrid system model.

help of back to back thyristors as shown in Figure 9.2. Thyristors work as a switching device for controlling the switching of the capacitor. Alpha (α) is the switching pulse of the thyristor as shown in Figure 9.3. The thyristors are controlled such that the current through the thyristor, I_{TSC} is a function of the firing angle, α. The value of I_{TSC} is obtained using Equation (9.1). The reactive power that would be injected into the system is given by Equation (9.2). Figure 9.4 shows the graphical representation of the relationship between the TSC reactive power and the firing angle. As the firing angle increases, the delivered reactive power decreases. Therefore, the reactive power diminishes at high firing angles. The thyristors work in full conduction mode when the firing angle, α is 0° and the reactive power is injected into the system and when α is 180°, the TSC is turned off. The capacity of the capacitor, C, used in this work is 30 MVAR.

$$I_{TSC} = \frac{2}{\pi} \frac{V_{PCC}}{X_C} \left[\frac{\sin 2\alpha}{2} + \pi - \alpha \right] \qquad (9.1)$$

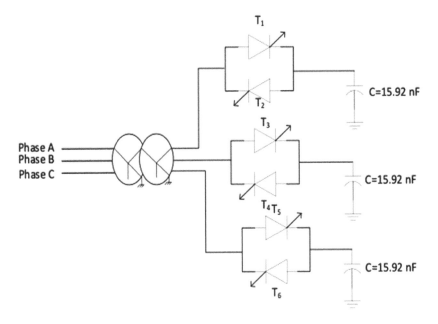

FIGURE 9.2   Single line diagram of thyristor switched capacitor model.

The reactive power is given by,

$$Q = I_{TSC} * V_{PCC} = \frac{2V_{PCC}^2}{\pi X_C}\left[\frac{\sin 2\alpha}{2} + \pi - \alpha\right] \tag{9.2}$$

9.5 FAULTS

Under normal operating conditions of a power network, current flows through all elements which are designed according to their ratings. Any power system parameters can be analyzed by calculating system voltages and currents. When faults occur, due

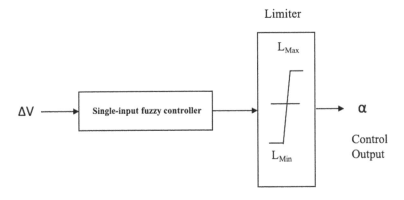

FIGURE 9.3   A control block for switching of thyristor switched capacitor.

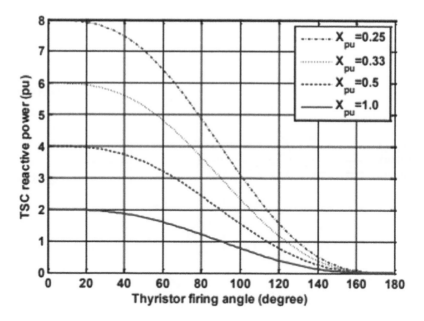

FIGURE 9.4   TSC reactive power versus firing angle.

to natural events or accidents, a connection between a phase and another phase is established or between the ground and the phase or in some cases both. Faults can also be defined as a flow of excessive current through improper path that can lead to interruption of power, equipment damage, personal injury, or in worst case, death.

There are two types of faults which can occur on any transmission lines; balanced faults & unbalanced faults. Unbalanced faults can be classified into single line-to-ground (1LG) faults, line-to-line (2LL) faults and double line-to-ground (2LG) faults. The 1LG faults occurs when one phase of the transmission lines makes a contact with ground by either ice, wind, falling tree or any other incidents. About 70% of all transmission faults accounts under this category. 2LL faults results due to one phase touching another. Of all faults, 15% occur due to 2LL faults. 2LG faults occur when two phases come in contact with the ground and contributes to 10%. Balanced faults or three phase-to-ground (3LG) faults occur when all the three phases touch the ground due to falling tower, failure of equipment or even a line breaking and touching the remaining phases.

9.6   STABILITY IMPROVEMENT BY NONLINEAR CONTROLLERS OF THE REMOTE MICROGRID

In this chapter, three controllers such as the FLC, ANFIS and static nonlinear controller have been considered for stability improvement of remote microgrid. Their performance has been compared with that of a conventional proportional-integral-derivative (PID) controller. The designs of all these controllers and simulation results are described below.

9.6.1   Fuzzy Controller vs. proportional-integral-derivative Controller

For case I, the FLC-based TSC is proposed for power stability improvement, and the performance of the proposed FLC-based TSC is compared with that of the PID controller-based TSC.

9.6.1.1   Fuzzy Logic Controller

An FLC is a simple controller which is based on IF-THEN rules. It is mostly used when the mathematical modeling is difficult to implement and there is significant nonlinearity in the system [37]. Fuzzy logic applications generate precise solutions from certain or approximate data which is much like human decision making. Fuzzy logic is also used in cases where the information is nonlinear.

FLC appears very useful when linearity and time invariance of the controlled process cannot be assumed, when the process lacks a well posed mathematical model, or when human understanding of the process is very different from its model. FLC provides a formal methodology for representing, manipulating and implementing a human experience-based knowledge about how to control a system. Fuzzy logic uses human knowledge and expertise to deal with uncertainties in the process of control.

The FLC design procedure is described below.

    a. Fuzzification: For the design of the FLC, the total voltage deviation at the PCC and the firing angle of thyristor, are selected as the input and output, respectively. The membership functions for the input and output variables are shown in Figure 9.5 and 9.6, respectively. Through a process of trial and error with various functions, it was found that the triangular membership function leads to better power stability enhancement, thus it is adopted for both input and output variables. In Figure 9.5, the symbols are defined as N: negative, Z: zero, and P: positive. In Figure 9.6, 1, 2 and 3 are the membership functions of the firing-angle of the thyristors.

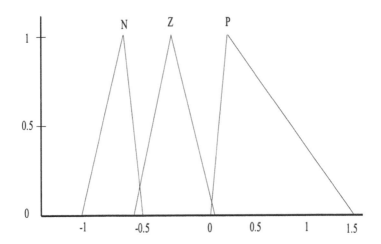

FIGURE 9.5   Membership function of TVD (pu).

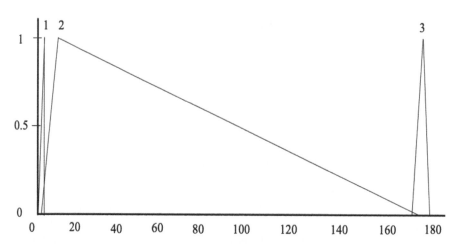

FIGURE 9.6   Membership function of firing angle, α (degrees).

  b. The equation of the triangular membership functions used in this work is as follows [36]:

$$\mu_{Ai}(x) = \frac{1}{b}(b - 2|x - a|) \qquad (9.3)$$

where $\mu_{Ai}(x)$ is the value of the grade of membership, "b" is the width, "a" is the coordinate of the point at which the grade of membership is 1, and "x" is the value of the input variable (TVD: Total voltage deviation for this case).

  c. Fuzzy rule base: The proposed fuzzy control strategy is very simple because it has only three control rules. It is important to note that the control rules have been developed from the viewpoint of practical system operation and by trial and error. The rules given in terms of the input variable, TVD, and the output variable, α, are outlined as follows:
- If TVD is P, then the output, α is $0°$ (1).
- If TVD is N, then the output, α is $100°$ (2).
- If TVD is Z, then the output, α is $180°$ (3).

  d. Fuzzy inference: For the inference mechanism of the FLC, Mamdani's method [36] has been utilized. According to Mamdani, the degree of conformity W_i of each fuzzy rule is as follows:

$$W_i = \mu_{Ai}(x) \qquad (9.4)$$

where $\mu_{Ai}(x)$ is the value of grade of membership, and i is the rule number.

  e. Defuzzification: The center-of-area method is the most well-known and rather simple defuzzification method [36], which is implemented to determine the output crispy value (i.e., the firing angle of the thyristor, α). Once the input variable is fuzzified and sent to the fuzzy rule base, the output of the rule base

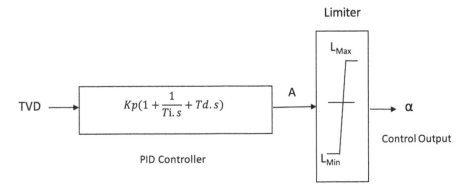

FIGURE 9.7   Proportional-integral-derivative controller.

is then aggregated and defuzzied. In aggregation, all the output fuzzy sets are added in a logical way, which produces a crisp control signal.

$$\alpha = \frac{\sum W_i C_i}{\sum W_i} \tag{9.5}$$

where, C_i is the value of α in the fuzzy rule base.

9.6.1.2 PID Controller

PID controllers are proportional-integral-derivative controllers. More than half of the industrial controllers in use today are PID controllers. Most control systems use PID controllers because of their general applicability like, when the mathematical model of a plant is not known and the analytical design methods cannot be used, PID controllers are proved effective.

In this section, in order to see how much effective the proposed fuzzy controlled TSC is, its performance has been compared with that of the PID controlled TSC. The PID control block diagram for the TSC is shown in Figure 9.7. The controller takes the TVD at the PCC as an input and feeds its output to a limiter block. The limiter is used to limit the output of PID controller within the range L_{Min} and L_{Max} as required. The PID controller parameters shown in Table 9.1 are determined by trial and error method for optimizing the best system performance. The control output is the switching pulse of the thyristors shown in Figure 9.2.

TABLE 9.1

Proportional-Integral-Derivative Parameters

			Limiter	
K_p	T_i	T_d	L_{max}	L_{min}
100	0.01	10	180	0

9.6.2 ANFIS CONTROLLER

The ANFIS is a simple data learning technique that uses fuzzy logic and ANN to transform given inputs into a desired output through highly interconnected neural network processing elements and information connections, which are weighted to map the numerical inputs into an output. The ANFIS has several features [37] such as: It refines fuzzy IF-THEN rules to describe the behavior of a complex system, it does not require prior human expertise, it is easy to implement and enables fast and accurate learning. It also offers desired data set, greater choice of membership functions to use, strong generalization abilities, and excellent explanation facilities through fuzzy rules.

The algorithms used in this section is the Sugeno-type ANFIS learning and structure. The ANFIS structure is divided into two parts, namely the predecessor and the conclusion part. In fuzzy logic, the two parts are related to each other by rules. For the controller, the Sugeno-type inference system is used, and the rules are given by,

$$\text{If } \left(x_1 = A_i \right) \text{ and } \left(x_2 = B_i \right) \text{ then } f_i = a_i x_1 + b_i x_2 + c_i \tag{9.6}$$

where, x_1 and x_2 are the inputs to the controller. In this section, x_1 indicates the PCC voltage deviation (ΔV_{PCC}) and x_2 represents the time derivative value of ΔV_{PCC} ($d\Delta V_{PCC}/dt$). Ai and Bi are the fuzzy sets. The output within the fuzzy region is f_i; a_i, b_i and c_i are the controller design parameter r_s, i is the number of Membership Functions (MFs) of each input.

Figure 9.8 shows the five layers architecture of ANFIS in which a circle indicates a fixed node and a square indicates an adaptive node. The structure of ANFIS network is composed of five layers.

Layer 1: This layer consists of inputs variables (ΔV_{PCC} and $d\Delta V_{PCC}/dt$). The membership function used is triangular. The output is developed by sampling the

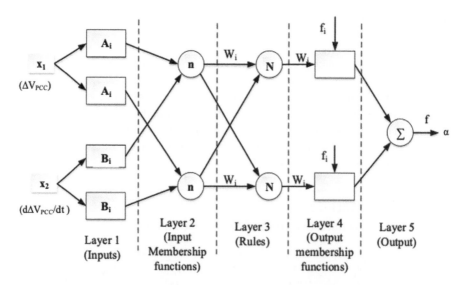

FIGURE 9.8 ANFIS architecture.

one-dimensional input variable ΔV_{PCC} and $d\Delta V_{PCC}/dt$ uniformly and estimating firing angle (α) for each sampled point.

Layer 2: As shown in Figure 9.8, the input layer is multiplied with nodes and the products of the second layer is the input of the third layer, which is the firing strength of a rule [37].

Layer 3: The third layer is known as the rule layer. The nodes in the rule, define the ratio (w_i) of the ith rule's firing strength to the sum of all rules firing strengths. Initial rules are valued after generating the training data. A hybrid leaning algorithm is used to optimize the initial rules by Sugeno-type fuzzy inference system (FIS). The previous membership functions are found out by the method of back propagation, which is an iterative process. A linear regression analysis is employed for the parameter optimization [37].

Layer 4: This layer (output MF) is called the defuzzification layer. In this layer, the output, i.e., the firing strength and the rule is generated.

Layer 5: The final layer represents the overall structure, which is the summation of all the output from layer 4. The result is then transformed into a crisp output.

In this section, triangular membership functions have been used, where the epochs are 30. The data required for the ANFIS controller is generated from the model with fault in it. After the training is done, the parameters determined are as the following: The nodes in the second layer are 75, linear parameters are 75, nonlinear parameters are 30, training data pairs are 320, and the fuzzy rules are 25.

9.7   STATIC NONLINEAR CONTROLLER

In this section, a static nonlinear controller is implemented [37] for evaluating the performance of the TSC in power stability improvement in more detail. It is noteworthy that, since the integration of the wind generator to a synchronous generator-based power network adds nonlinearity, a nonlinear controller is incorporated to generate the firing angle of the thyristor. However, for the fault scenarios considered, the performance of the cubic nonlinear function is better compared to that of other functions. The equation used for the nonlinear controller to generate the firing angle (α) in this work is given by (9.7),

$$\alpha = K * TVD^3 \tag{9.7}$$

where TVD indicates the PCC voltage deviation. To make the controller work better and perform well, the parameter K can be tuned. The block diagram of the static nonlinear controller is shown in Figure 9.9. Through a trial-and-error approach, K has been set to 0.1 to improve the power stability of the microgrid.

9.8   SIMULATION RESULTS AND DISCUSSION

9.8.1   Fuzzy Logic Controller and PID Controller

In this section, both balanced and unbalanced types of permanent faults are considered. It is assumed that the fault occurs at 0.6 second, the circuit breaker opens at

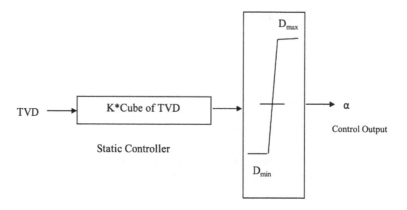

FIGURE 9.9   A control block of controller for switching of TSC.

0.7 second, the circuit breaker recloses after 0.7 second, and reopens again after 0.1 second of the reclosing instance.

Figures 9.10–9.14 show the voltage response at PCC, voltage responses for synchronous generator GI and G2, voltage response of wind generator, and the dc link voltage of the wind generator, due to a 3LG permanent fault at F1 point of Figure 9.1. From the responses, it is seen that the fuzzy logic controlled TSC works well to improve the stability of the system. Also, the performance of the fuzzy controlled TSC is better than that of the PID controlled TSC.

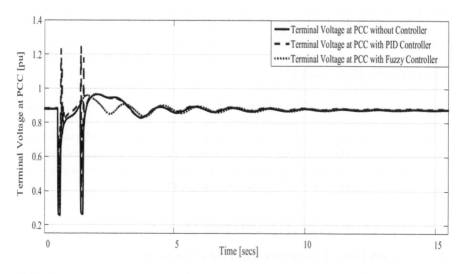

FIGURE 9.10   Voltage response at PCC.

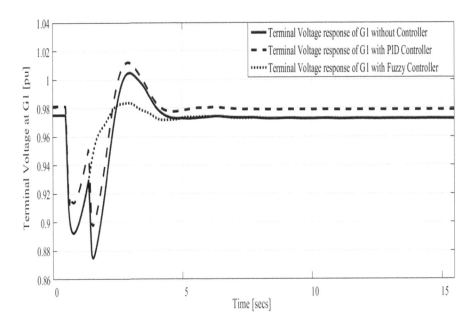

FIGURE 9.11   Voltage response of G1 synchronous generator.

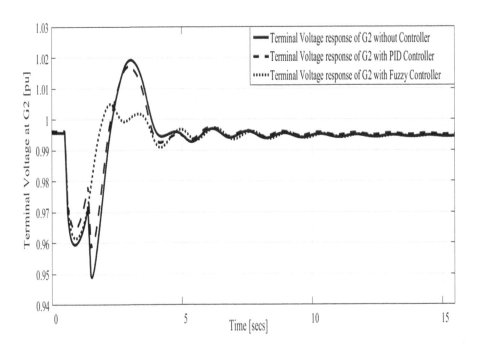

FIGURE 9.12   Voltage response of G2 synchronous generator.

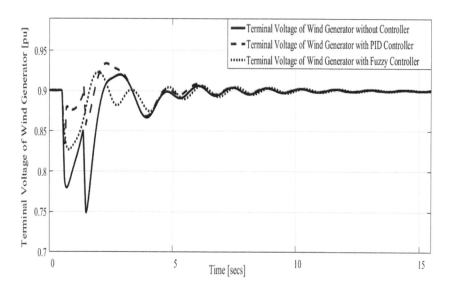

FIGURE 9.13   Voltage of wind generator.

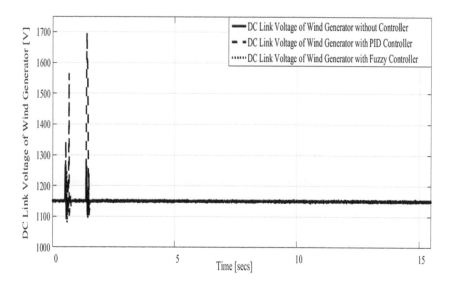

FIGURE 9.14   DC link voltage of wind generator.

9.8.1.1   Voltage Index

To evaluate the effectiveness of the proposed TSC methods in more detail, the voltage index, V_{index} shown below in Equation (9.8) is considered.

$$V_{index} = \int_{0}^{T} |\Delta V| \, dt \qquad (9.8)$$

TABLE 9.2

Voltage Indices for PCC Voltage for Permanent Faults

		Voltage Index PCC		
Fault Type	Fault Point	Without Control	With PID-TSC	With FLC-TSC
3LG	F1	0.3769	0.2808	0.2469
	F2	0.1774	0.1665	0.1572
	F3	0.6373	0.5016	0.2270
1LG	F1	0.1754	0.1568	0.1502
	F2	0.1036	0.0896	0.0792
	F3	0.6373	0.5016	0.2612

where ΔV indicates the total voltage deviation at the PCC and T is the simulation time of 15 secs. The lower the value of the index, the better the system performance is.

Tables 9.2–9.4 show the values of voltage indices at PCC, voltage indices of synchronous and wind generators for permanent 3LG faults and 1LG faults at points F1, F2 and F3 of the power system with and without the TSC control method. From the indices it is evident that the fuzzy controlled TSC works better as compared to the PID controlled TSC in improving the stability.

9.8.1.2 Total Harmonic Deviation

THD is defined as the measure of the effective value of the harmonic components of a distorted fundamental waveform. In other words, the THD is the summation of all harmonic components of the voltage or current waveform compared against the fundamental component of the voltage or current waveform [38–39], as shown in Equation (9.9). Harmonic distortion is usually caused by nonlinear loads connected to the grid by customers. Harmonics affects the shape and characteristics of a voltage and

TABLE 9.3

Voltage Indices for Synchronous Generator (G1 and G2) Permanent Faults

		Voltage Index G1 and G2		
Fault Type	Fault Point	Without Control	With PID-TSC	With FLC-TSC
3LG	F1	0.2871	0.2682	0.1928
	F2	0.2232	0.1870	0.1560
	F3	0.3290	0.3371	0.3056
1LG	F1	0.976	0.1440	0.1052
	F2	0.1276	0.0983	0.0122
	F3	0.3290	0.3371	0.0690

TABLE 9.4

Voltage Indices for Wind Generator for Permanent Faults

Fault Type	Fault Point	Voltage Index Wind		
		Without Control	With PID-TSC	With FLC-TSC
3LG	F1	0.3269	0.2625	0.2291
	F2	0.1874	0.1706	0.1564
	F3	0.5326	0.4106	0.3124
1LG	F1	0.1602	0.1861	0.940
	F2	0.0786	0.0704	0.0694
	F3	0.5326	0.4106	0.1883

current waveform relative to fundamental frequency. These current or voltage harmonics can disrupt the quality and stability of the system. These harmonic currents can result in equipment failure, conductor failure and fires. Voltage distortion can lead to overheating of the equipment, electronic failures and maintenance difficulties.

THD has been calculated by using the Equation (9.9).

$$THD = \frac{\sqrt{V_2^2 + V_3^2 + V_4^2 + \cdots + V_n^2}}{V_1} * 100\% \qquad (9.9)$$

where, V_1 is the fundamental voltage, and V_2, V_3... V_n are the higher order harmonic components of the PCC voltage. The higher the THD, the more the distortion on the fundamental signal. The limits of the THD are around 5%.

Table 9.5 shows the values of THD for permanent 3LG and 1LG faults at F1, F2 and F3 with fuzzy controlled TSC, PID controlled TSC and without any control method. It is evident from the THD values that the effects of harmonics are reduced when the controllers are used. However, the performance of the fuzzy controlled TSC is better than that of the PID controlled TSC.

TABLE 9.5

Total Harmonic Distortion at PCC for Permanent Faults

Fault Type	Fault Point	Total Harmonic Distortion at PCC		
		Without Control	With PID-TSC	With FLC-TSC
3LG	F1	4.479	4.059	3.552
	F2	3.295	3.93	1.820
	F3	3.775	3.361	1.058
1LG	F1	5.710	5.547	4.973
	F2	3.295	2.59	1.951
	F3	3.775	3.361	1.707

9.8.2 Fuzzy Logic Controller, ANFIS Controller and Static Controller

In this section, the simulations are executed through the MATLAB/Simulink software. Both balanced and unbalanced types of temporary and permanent faults have been considered. It is assumed that the fault occurs at 0.1 second, the circuit breaker opens at 0.2 second, and the circuit breaker recloses after 0.7 second and reopens again after 0.1 second of the reclosing instance in case of permanent fault.

Figures 9.15–9.16 show the voltage responses at PCC, wind generator terminal voltage, and the DC link voltage of the wind generator, due to a 3LG temporary and permanent fault at F3 points in Figure 9.1. From the responses, it is seen that the ANFIS controlled TSC and static nonlinear controlled TSC work well to improve the stability of the microgrid system. Also, the performance of the controlled TSC is better than that of the fuzzy controlled TSC. The responses also show that the ANFIS controlled TSC performs somewhat better than the static nonlinear controlled TSC.

Figures 9.17–9.18 show the voltage responses at the PCC, wind generator terminal voltage, and the DC link voltage of the wind generator, due to a 1LG temporary and permanent fault at F3 points in Figure 9.1. From the responses, it is seen that the ANFIS controlled TSC and static nonlinear controlled TSC work well to improve the microgrid stability. Also, the performance of the ANFIS controlled TSC is better

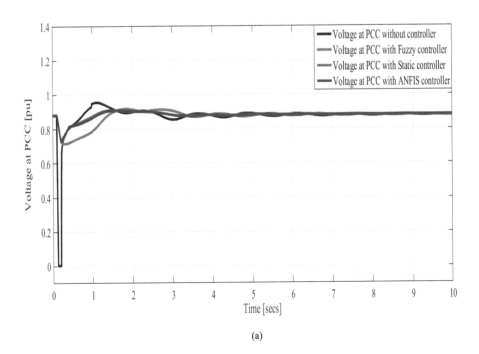

(a)

FIGURE 9.15 Voltage responses when the system is subjected to 3LG temporary fault at F3: (a) voltage at PCC. (Continued)

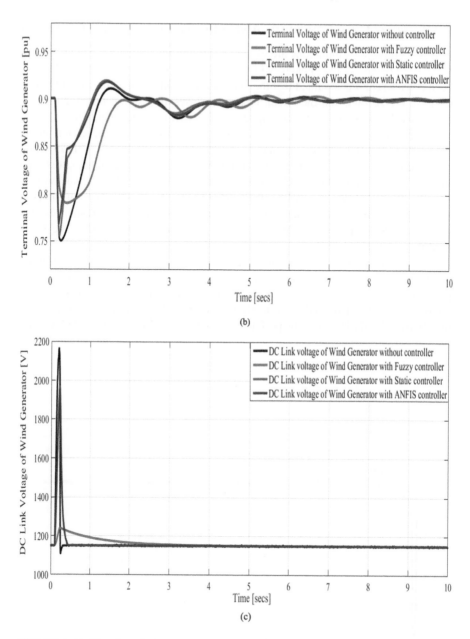

FIGURE 9.15 (Continued) Voltage responses when the system is subjected to 3LG temporary fault at F3: (b) terminal voltage of wind generator and (c) DC link voltage of the wind generator.

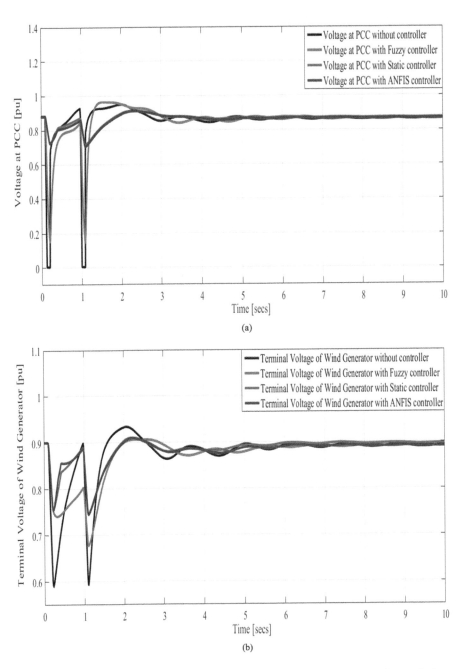

FIGURE 9.16 Voltage responses when the system is subjected to 3LG permanent fault at F3: (a) voltage at PCC and (b) terminal voltage of wind generator. (Continued)

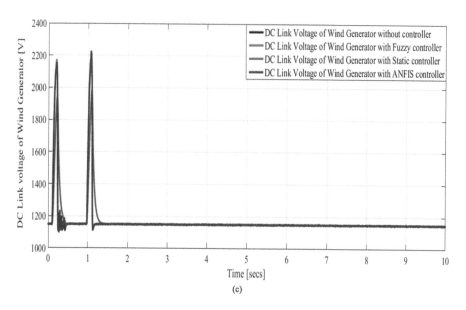

FIGURE 9.16 (Continued)   Voltage responses when the system is subjected to 3LG permanent fault at F3 (c) DC link voltage of the wind generator.

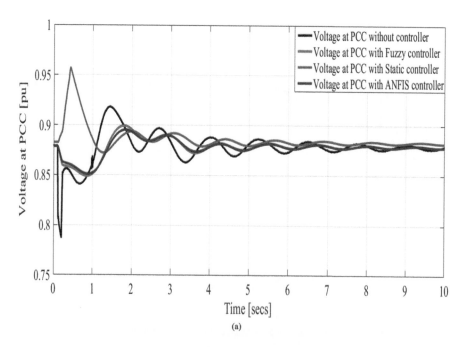

FIGURE 9.17   Voltage responses when the system is subjected to 1LG temporary fault at F3: (a) voltage at PCC. (Continued)

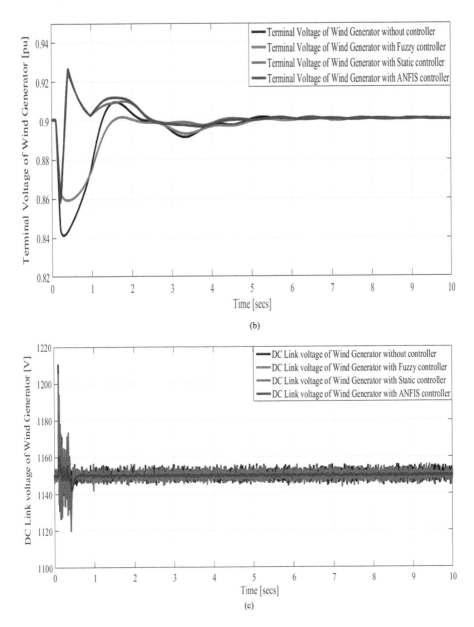

FIGURE 9.17 (Continued)   Voltage responses when the system is subjected to 1LG temporary fault at F3: (b) terminal voltage of wind generator and (c) DC link voltage of the wind generator.

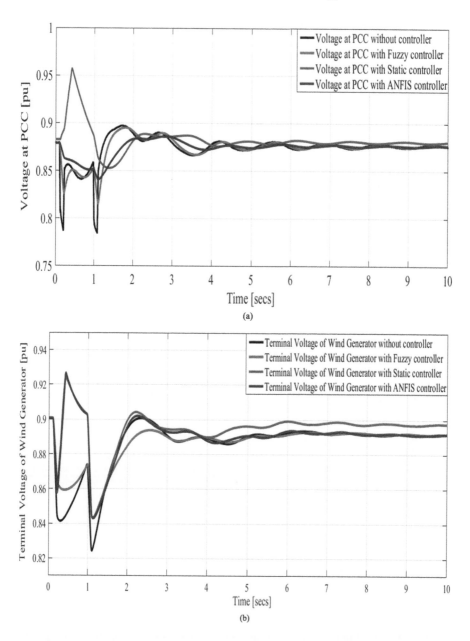

FIGURE 9.18   Voltage responses when the system is subjected to 1LG permanent fault at F3: (a) voltage at PCC, (b) terminal voltage of wind generator. (Continued)

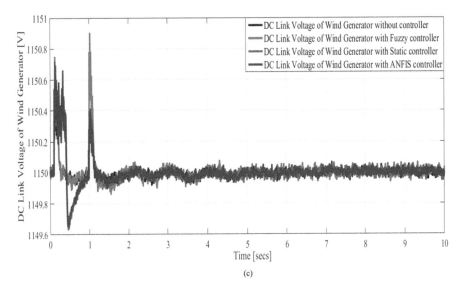

FIGURE 9.18 (Continued)   Voltage responses when the system is subjected to 1LG permanent fault at F3: (c) DC link voltage of the wind generator.

than that of the fuzzy controlled TSC. The responses also show that the ANFIS
controlled TSC performs somewhat better than the static nonlinear controlled TSC.

9.8.2.1   Voltage Index

Voltage indices are shown in Tables 9.6–9.11. It shows the values of voltage indices at PCC, voltage indices of synchronous and wind generators for temporary and

TABLE 9.6

Voltage Indices for PCC Voltage for Temporary Faults

		Voltage Index PCC (Temporary)			
Fault Type	**Fault Point**	**Without Control**	**With Fuzzy- TSC**	**With SNC-TSC**	**With ANFIS-TSC**
3LG	F1	0.3474	0.3417	0.3242	0.3120
	F2	0.1506	0.1489	0.946	0.987
	F3	0.3777	0.3659	0.3534	0.3492
1LG	F1	0.1942	0.1676	0.1808	0.1620
	F2	0.0901	0.0719	0.1225	0.0898
	F3	0.2895	0.2715	0.2292	0.2675
2LG	F1	0.1943	0.1670	0.1661	0.179
	F2	0.1193	0.1011	0.1055	0.1062
	F3	0.1885	0.1754	0.1729	0.1681
2LS	F1	0.2418	0.926	0.1661	0.1642
	F2	0.1193	0.1011	0.1055	0.1069
	F3	0.1885	0.1754	0.1729	0.1681

TABLE 9.7
Voltage Indices for PCC Voltage for Permanent Faults

Fault Type	Fault Point	Voltage Index PCC			
		Without Control	With Fuzzy- TSC	With SNC-TSC	With ANFIS-TSC
3LG	F1	0.3474	0.3417	0.3242	0.3120
	F2	0.1506	0.1489	0.946	0.987
	F3	0.3777	0.3659	0.3534	0.3492
1LG	F1	0.1942	0.1676	0.1808	0.1620
	F2	0.0901	0.0719	0.1225	0.0898
	F3	0.2895	0.2715	0.2292	0.2675
2LG	F1	0.2935	0.2644	0.2458	0.26
	F2	0.958	0.09916	0.1158	0.1411
	F3	0.3667	0.3199	0.3262	0.1573
2LS	F1	0.3491	0.1676	0.302	0.192
	F2	0.09008	0.1240	0.1225	0.0898
	F3	0.4031	0.1465	0.2292	0.2678

permanent 3LG, 1LG, 2LG and 2LS faults at points F1, F2, and F3 of the power system with and without the TSC control methods. The voltage index is calculated using the Equation (9.8). From the indices it is shown that the ANFIS controlled TSC and static nonlinear controlled TSC perform better as compared to the fuzzy controlled TSC in improving the stability. It is also observed from the responses that

TABLE 9.8
Voltage Indices for Synchronous Generator (G1 and G2) During Temporary Faults

Fault Type	Fault Point	Voltage of G1 & G2 Voltage Index			
		Without Control	With Fuzzy- TSC	With SNC-TSC	With ANFIS-TSC
3LG	F1	0.2939	0.2154	0.2210	0.2164
	F2	0.2541	0.1851	0.2006	0.1763
	F3	0.2601	0.1986	0.2061	0.1980
1LG	F1	0.1940	0.975	0.129	0.1647
	F2	0.1860	0.1221	0.1686	0.1037
	F3	0.1729	0.1121	0.995	0.1059
2LG	F1	0.1442	0.1699	0.1878	0.1918
	F2	0.1229	0.1586	0.1766	0.1496
	F3	0.1206	0.1601	0.1720	0.1505
2LS	F1	0.1940	0.975	0.1878	0.1647
	F2	0.1229	0.1586	0.1766	0.1509
	F3	0.1206	0.1601	0.1720	0.1505

TABLE 9.9

Voltage Indices for Synchronous Generator (G1 and G2) Permanent Faults

Fault Type	Fault Point	Voltage of G1 and G2 Voltage Index (Perm)			
		Without Control	With Fuzzy-TSC	With SNC-TSC	With ANFIS-TSC
3LG	F1	0.2769	0.2593	0.2691	0.2588
	F2	0.2615	0.2142	0.2165	0.2125
	F3	0.2569	0.2426	0.2040	0.2140
1LG	F1	0.909	0.1211	0.1051	0.1020
	F2	0.1164	0.099	0.1088	0.1017
	F3	0.1673	0.1590	0.1542	0.1584
2LG	F1	0.1970	0.1875	0.1081	0.1754
	F2	0.1640	0.1428	0.1488	0.1216
	F3	0.1877	0.1576	0.1441	0.927
2LS	F1	0.277	0.911	0.1672	0.1200
	F2	0.2164	0.1863	0.1588	0.917
	F3	0.2569	0.1124	0.1842	0.1548

the performance of ANFIS controlled TSC is somewhat better than that of the static nonlinear controlled TSC.

9.8.2.2 Total Harmonic Deviation

Tables 9.12–9.13 show the THD values for temporary and permanent 3LG, 1LG, 2LG, and 2LS faults at points F1, F2, and F3 with the ANFIS controlled TSC, static

TABLE 9.10

Voltage Indices for Wind Generator for Temporary Faults

Fault Type	Fault Point	Wind Generator Voltage Index (Temp)			
		Without Control	With Fuzzy-TSC	With SNC-TSC	With ANFIS-TSC
3LG	F1	0.2230	0.1117	0.0842	0.0983
	F2	0.915	0.0538	0.0487	0.0484
	F3	0.2026	0.1078	0.0862	0.0836
1LG	F1	0.2226	0.0518	0.0835	0.0533
	F2	0.0865	0.0549	0.0551	0.0429
	F3	0.0907	0.0475	0.0299	0.0322
2LG	F1	0.1671	0.07504	0.05321	0.05741
	F2	0.1065	0.03922	0.0506	0.04522
	F3	0.1519	0.08064	0.0533	0.05577
2LS	F1	0.2226	0.05179	0.05321	0.05327
	F2	0.1065	0.03922	0.0506	0.0456
	F3	0.1519	0.08064	0.0533	0.05059

TABLE 9.11

Voltage Indices for Wind Generator for Permanent Faults

Fault Type	Fault Point	Wind generator Voltage Index			
		Without Control	With Fuzzy-TSC	With SNC-TSC	With ANFIS-TSC
3LG	F1	0.2457	0.1893	0.968	0.975
	F2	0.989	0.0841	0.0634	0.0703
	F3	0.2371	0.2102	0.1742	0.2030
1LG	F1	0.1498	0.0627	0.0625	0.0615
	F2	0.0732	0.0472	0.0622	0.0444
	F3	0.1673	0.0886	0.0718	0.0593
2LG	F1	0.2253	0.1062	0.0777	0.08651
	F2	0.102	0.04754	0.0687	0.06258
	F3	0.2651	0.1503	0.1200	0.01475
2LS	F1	0.3099	0.06267	0.939	0.06795
	F2	0.07321	0.07256	0.06223	0.04442
	F3	0.3371	0.05451	0.07181	0.1107

nonlinear controlled TSC, fuzzy controlled TSC and without any control method. The THD index is calculated using Equation (9.9). From the THD values, it is observed that the proposed controllers reduced the effects of harmonics. However, the performance of the ANFIS controlled TSC and static nonlinear controlled TSC is better than that of the fuzzy controlled TSC. It is also clear from the indices that the ANFIS controlled TSC is somewhat better than the static nonlinear controlled TSC.

TABLE 9.12

Total Harmonic Distortion at PCC for Temporary Faults

Fault Type	Fault Point	THD (Permanent)			
		Without Control	With Fuzzy-TSC	With SNC-TSC	With ANFIS-TSC
3LG	F1	2.400	0.892	0.8042	0.8068
	F2	1.382	0.8267	0.8180	0.899
	F3	2.482	0.8457	1.290	0.6469
1LG	F1	1.289	0.9869	0.7905	0.8169
	F2	1.215	0.8651	0.8317	0.7875
	F3	2.915	0.1049	0.8773	0.7340
2LG	F1	2.188	0.8156	0.7741	0.0002
	F2	3.005	0.7372	0.7	0.8229
	F3	4.762	1.019	0.7379	0.3174
2LS	F1	1.171	0.1869	0.8209	0.8169
	F2	4.215	0.7694	0.8317	0.7875
	F3	4.303	0.6745	0.8773	0.7427

TABLE 9.13

Total Harmonic Distortion at PCC for Permanent Faults

Fault Type	Fault Point	THD (Temporary)			
		Without Control	With Fuzzy-TSC	With SNC-TSC	With ANFIS-TSC
3LG	F1	2.025	0.7933	0.8062	0.799
	F2	2.686	0.7948	0.8168	0.7793
	F3	1.047	0.9177	0.8900	0.7465
1LG	F1	2.677	0.8258	0.7915	0.6922
	F2	2.053	0.8020	0.4290	0.7846
	F3	2.963	1.6503	1.352	1.112
2LG	F1	1.004	0.8102	0.7939	0.8353
	F2	5.618	0.8304	0.8757	0.8547
	F3	2.41	0.7664	0.9458	0.6943
2LS	F1	2.677	0.8258	0.7939	0.9922
	F2	5.618	0.8304	0.8757	0.6910
	F3	2.41	0.7664	0.9458	0.6943

9.9   CHAPTER SUMMARY

This chapter deals with the stability issues due to nonlinearity in a remote microgrid system. Different nonlinear controller-based TSC such as the ANFIS controlled TSC, Static nonlinear control-based TSC, and FLC TSC are designed and used to improve the system stability. Both balanced and unbalanced faults at different points are considered to analyze the stability of the microgrid system. Quantitative analysis is done in terms of voltage index and THD to evaluate the performance of the system and observe the effectiveness of the nonlinear controllers. From the results, it is shown that the FLC TSC is effective to enhance the stability of the hybrid system. Also, the performance of the FLC TSC is better than that of the PID controlled TSC. Moreover, the ANFIS and static nonlinear control based TSC methods are effective to enhance the stability of the hybrid grid system. Also, the nonlinear controlled TSC performs better than the fuzzy controlled TSC.

REFERENCES

1. G. Antonis, St. Tsikalakis, and N. D. Hatziargyriou, "Centralized control for optimizing microgrids operation," *IEEE*, 2011, doi: 978-1-4577-1002-5/11/.
2. J. A. Martinez and J. M. Arnedo, "Voltage sag studies in distribution networks-part T: System modeling," *IEEE Trans. Power Deliv.*, vol. 21, no. 3, pp. 338–345, 2012.
3. C. K. Subasri, R. S. Charles, and P. Venkatesh, "Power quality improvement in a wind farm connected to grid using FACTS device," *Power Electron Renew. Energy Syst.*, vol. 326, no. 4, pp. 1203–1212, 2012.
4. R. Majumder, "Some aspects of stability in microgrids," *IEEE Trans. Power Sys.*, vol. 28, no. 3, Aug. 2009.

5. H. Kasem Alaboudy and H. H. Zeineldin, "Microgrid stability characterization subsequent to fault-triggered Islanding incidents," *IEEE Trans. Power Del.*, vol. 27, no. 2, Apr. 2012.

6. J. A. Lopes, C. Moreira, and A. Madureira, "Defining control strategies for microgrids islanded operation," *IEEE Trans. Power Syst.*, vol. 21, no. 2, pp. 916–924, May 2006.

7. K. De Brabandere, B. Bolsens, J. Van den Keybus, A. Woyte, J. Driesen, and R. Belmans, "A voltage and frequency droop control method for parallel inverters," *IEEE Trans. Power Electron.*, vol. 22, no. 4, pp. 1107–1115, Jul. 2007.

8. H. Karimi, H. Nikkhajoei, and M. R. Iravani, "Control of an electronically-coupled distributed resource unit subsequent to an islanding event," *IEEE Trans. Power Del.*, vol. 23, no. 1, pp. 493–501, Jan. 2008.

9. F. Katiraei, M. R. Iravani, and P. W. Lehn, "Micro-grid autonomous operation during and subsequent to islanding process," *IEEE Trans. Power Del.*, vol. 20, no. 1, pp. 248–257, Jan. 2005.

10. M. Kabalan, P. Singh, and D. Niebur, "Large signal lyapunov-based stability studies in microgrids: A review," *IEEE Trans. Smart Grid*, to be published, 2019.

11. K. De Brabandere *et al.*, "A voltage and frequency droop control method for parallel inverters," *IEEE Trans. Power Electron.*, vol. 22, no. 4, pp. 1107–1115, Jul. 2007.

12. S. V. Iyer, M. N. Belur, and C. Chandorkar, "Analysis and mitigation of voltage offsets in multi-inverter microgrids," *IEEE Trans. Energy Convers.*, vol. 26, no. 1, pp. 354–363, Mar. 2011.

13. Y. Li and L. Fan, "Stability analysis of two parallel converters with voltage–current droop control," *IEEE Trans. Power Del.*, vol. 32, no. 6, pp. 2389–2397, Dec. 2017.

14. W. Yao, M. Chen, J. Matas, J. M. Guerrero, and Z. M. Qian, "Design and analysis of the droop control method for parallel inverters considering the impact of the complex impedance on the power sharing," *IEEE Trans. Ind. Electron.*, vol. 58, no. 2, pp. 576–588, Feb. 2011.

15. W. R. Issa, M. A. Abusara and S. M. Sharkh, "Impedance interaction between islanded parallel voltage source inverters and the distribution network", In *Proceedings of 7th IET International Conference on Power Electronic Machine Drives*, pp. 1–6, Apr. 2014.

16. C. C. Chang, D. Gorinevsky, and S. Lall, "Stability analysis of distributed power generation with droop inverters," *IEEE Trans. Power Syst.*, vol. 30, no. 6, pp. 3295–3303, Nov. 2015.

17. F. Cavazzana, P. Mattavelli, M. Corradin and I. Toigo, "Grid sensitivity considerations on multiple parallel inverters systems", In *Proceedings of 2016 IEEE 8th International Conference of Power Electronic Motion Control*, pp. 993–999, May 2016.

18. C. N. Rowe, T. J. Summers, R. E. Betz, D. J. Cornforth, and T. G. Moore, "Arctan power 209; Frequency droop for improved microgrid stability," *IEEE Trans. Power Electron.*, vol. 28, no. 8, pp. 3747–3759, Aug. 2013.

19. T. L. Vandoorn, J. D. M. D. Kooning, B. Meersman, J. M. Guerrero, and L. Vandevelde, "Automatic power-sharing modification of P/V droop controllers in low-voltage resistive microgrids," *IEEE Trans. Power Del.*, vol. 27, no. 4, pp. 2318–2325, Oct. 2012.

20. Nimish Soni, Suryanarayana Doolla, and Mukul C. Chandorkar, "Improvement of transient response in microgrids using virtual inertia," *IEEE Trans. Power Del.*, vol. 28, no. 3, July 2013.

21. S.-J. Ahn, J.-W. Park, I.-Y. Chung, S.-I. Moon, S.-H. Kang, and S.-R. Nam, "Power-sharing method of multiple distributed generators considering control modes and configurations of a microgrid," *IEEE Trans. Power Del.*, vol. 25, no. 3, pp. 2007–2016, Jul. 2010.

22. T. L. Vandoorn, B. Meersman, L. Degroote, B. Renders, and L. Vandevelde, "A control strategy for islanded microgrids with DC-link voltage control," *IEEE Trans. Power Del.*, vol. 26, no. 2, pp. 703–79, Apr. 2011.

23. J. C. Vasquez, J. M. Guerrero, A. Luna, P. Rodriguez, and R. Teodorescu, "Adaptive droop control applied to voltage-source inverters operating in grid-connected and islanded modes," *IEEE Trans. Ind. Electron.*, vol. 56, no. 10, pp. 4088–4096, Oct. 2009.

24. S. D. Swain, P. K. Ray, and A. K. B. Mohanty, "Improvement of power quality using a robust hybrid series active power filter," *IEEE Trans. Power Electron.*, vol. 32, no. 5, pp. 3490–3498, 2017.

25. A. D. Baing and A. J. G. Jamnani, "Closed loop control of thyristor switched capaciotr (TSC) for instantaneous reactive power compensation," *Int. J. Eng. Develop. Res. (IJEDR)*, pp. 84–87, 2014.

26. P. B. and B. V. Sumangala, "Implementation of thyristor switched capacitor for reactive power compensation at secondary of distribution level feeders for voltage stability," *Int. J. Eng. Res. Technol. (IJERT)*, vol. 2, no. 5, 2013.

27. R. Sharma, A. Singh and A. N. Jha, "Performance evaluation of tuned PI controller for power quality enhancement for linear and non linear loads", In *Recent Advances and Innovations in Engineering (ICRAIE)*, 2014.

28. K. Sharma and S. S. Sudhir, "Effects on major power quality issues due to incoming induction generators in power system," *ARPN J. Eng. Appl. Sci.*, vol. 5, no. 2, p. 9, 2010.

29. M. Zarghami, M. L. Crow and A. S. Jagannathan, "Nonlinear control of FACTS controllers for damping inter-area oscillations in power systems," *IEEE Trans. Power Del.*, vol. 25, no. 4, pp. 319–3121, 2010.

30. N. D. G. MPI for Dynamics and Self-Organization Göttingen, "Dynamics of Modern Power Grids," 2018.

31. N. Walia, H. Singh, and A. Sharma, "ANFIS: Adaptive neuro-fuzzy inference system-A survey," *Int. J. Comput. Appl.*, vol. 123, no. 9, pp. 32–38, 2015.

32. S. R., Panda and S. Khuntia, "ANFIS approach for SSSC controller design for the improvement of transient stability performance," *Math. Comput. Model*, vol. 57, no. 1-2, pp. 289–300, 2013.

33. A. R. Roosta, D. Georges and A. N. Hadj-Said, "Decentralized Nonlinear Controller Design for Multimachine Power Systems Via Backstepping", *2003 European Control Conference (ECC)*, Cambridge, UK, 2003, pp. 246–251, doi: 10.23919/ECC.2003.7084962.

34. A. Kusagur, S. F. Kodad, and S. Ram, "Modelling & simulation of an ANFIS controller for an AC drive," *World J. Model. Sim.*, vol. 8, no. 1, pp. 36–49, 2012.

35. M. Mahdavi, L. Lil, J. Zhu1 and S. Mekhilef, "An adaptive Neuro-Fuzzy controller for maximum power point tracking of photovoltaic systems", In TENCON 2015–2015 IEEE Region 10 Conference, 2015.

36. S. Ghosh and M.H. Ali, "Augmentation of power quality of grid-connected wind generator by Fuzzy Logic controlled TSC," In *Proceedings of the IEEE T&D Conference & Exposition, Denver*, CO, USA, April 16–19, 2018.

37. M. K. H and M. H. Ali, "Transient stability augmentation of PV/DFIG/SG-based hybrid power system by nonlinear control-based variable resistive FCL," *IEEE Trans. Sustain. Energy*, vol. 6, no. 4, 2015.

38. N. Rao, "Harmonic analysis of small-scale industrial loads and harmonic mitigation techniques in industrial distribution system," *IJERA*, vol. 3, no. 4, pp. 1511–1540, 2013.

39. S. Saha and C. Nandi, "Modelling and harmonic analysis of domestic/industrial loads", In *National Seminar on Energy Science and Enngineering (NSESE)*, 2013.

10 Mixed AC/DC System Stability under Uncertainty

Mohd Hasan Ali and Morteza Daviran Keshavarzi
Department of Electrical & Computer
Engineering, University of Memphis

CONTENTS

10.1 INTRODUCTION

Mixed AC/DC power system, especially a hybrid AC/DC microgrid (HMG) is getting more nad more attention from the power community worldwide [1–2]. HMGs are electrical distribution networks that serve both AC and DC power systems, ensuring high reliability and efficiency levels and incorporating autonomous control systems [3]. The purpose of operating these grids is to supply energy from distributed energy resources (DERs), e.g., wind turbines, photovoltaic (PV) arrays, diesel generators, etc., to the loads independent of main utility grid in a relatively small area of coverage such as remote towns, university campuses, hospitals, military bases, ships, etc. However, they can operate in grid-connected mode to exchange power with the main grid. In an HMG system, operational functionalities of both AC and DC currents are combined to avoid frequent conversions from AC to DC and DC to AC to minimize energy losses and directly feed DC loads from DC current primary sources [4]. However, the uncertain or intermittent nature of renewable resources constitutes the risk of a power outage for sensitive loads inside the microgrid. The technical development has made it feasible to collocate energy storage (ES) systems to enhance power quality as well as grid reliability. In HMGs, an ES is an essential component that is used mainly for peak shaving and minimization of power and frequency fluctuations resulting from solar irradiance variation or wind speed change. Improving the resiliency and transient stability during large/small-signal disturbances in an islanded HMG comprising different DERs exhibiting diverse dynamic characteristics is a challenging task [5].

The purpose of this chapter is to discuss new methods that provide overall stability and control improvements in hybrid AC/DC microgrids. Typically, auxiliary control devices are used to improve the transient stability of the microgrid, incurring extra cost. However, it is interesting to explore whether an existing component of the microgrid such as the battery energy storage system (BESS) or a hybrid energy storage system (HESS) can be utilized for resiliency enhancement [5] during large-signal disturbances in an HMG system. If an existing component, i.e., the BESS can be used to improve the resiliency of the HMG, then the cost of many auxiliary devices can be saved. Based on this fact, in this chapter, the impact of BESS dynamic control on hybrid AC/DC microgrid's transient stability during disturbed conditions has been analyzed. The primary contribution of this chapter is to design control systems for the hybrid AC/DC microgrid, including different kinds of DERs, BESS, and static/dynamic loads. The control philosophy has different roles in islanded and grid-connected modes of microgrids. To regulate the bus voltage and AC microgrid frequency during the grid disturbances (i.e., load change and intermittent DER), the required controller drives a voltage source converter (VSC) to inject and/or absorb active and reactive power, respectively. The detailed description of testbed models of the microgrid samples and their control systems developed in MATLAB®/Simulink® environment along with the obtained system responses have been provided in this chapter.

10.2   STRUCTURE OF THE HYBRID MICROGRID

The conventional form of AC and DC microgrids have been separately developed over the past years. HMGs are only the new generations of combined AC and DC subgrids which are integrated through interlinking converters (ILCs) [6]. In this way, DC loads (e.g. electronic devices, the computer servers etc.) and DERs with DC current (e.g. Photovoltaic Panels (PV), fuel cells etc.) can be directly connected to DC subgrid while the AC subgrid serves the AC loads and DERs with AC currents (e.g. wind turbines, microturbines, diesel generators etc.). Nevertheless, AC and DC systems can be connected to DC and AC subgrids, respectively through power converters as required. Figure 10.1 shows a typical structure of an HMG. DC-DC

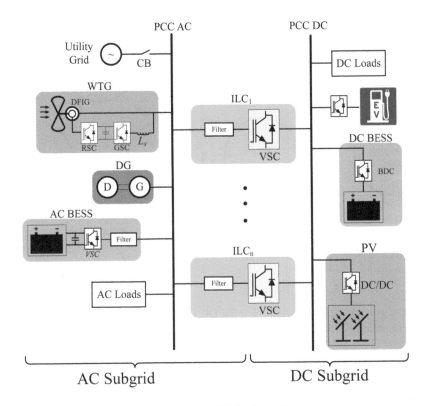

FIGURE 10.1 Typical structure of hybrid AC/DC microgrid.

converters and VSCs are two integral parts of every HMG to control the flow of power between various voltage levels and forms (AC/DC) as well as tracking the maximum power point in renewable energy sources (RES). Later, we will discuss how these power electronic-based converters are exploited to control the dynamic behavior of the HMG during the grid disturbances. HMGs are reasonable solutions to connect microgrids with different ratings to integrate verity of loads and DERs with dissimilar ratings. Figure 10.2 shows HMG topologies in which a subgrid is interfaced between two subgrids with unequal ratings. In addition, depending on the application, location and extent of the microgrid coverage, it might have several HMG subgrids (called clusters) which are interconnected throughout the AC or DC power distribution system. Figure 10.3(a) displays a microgrid cluster formed by AC distribution system [7] while Figure 10.3(b) shows a cluster based on the DC distribution system [8]. Each subgrid inside the cluster is capable of autonomous operation. Besides, a cluster can host any type of AC, DC or hybrid subgrid.

Although HMGs operate autonomously in a remote location, they can operate in a grid-connected mode where there is access to the main grid to either support the grid or get supported by the grid. They should be able to island and continue the service as quickly as possible when there is a fault in the main grid. In this context, HMGs are promising solutions for high quality and reliable electric power when there are uncertainties with the main grid.

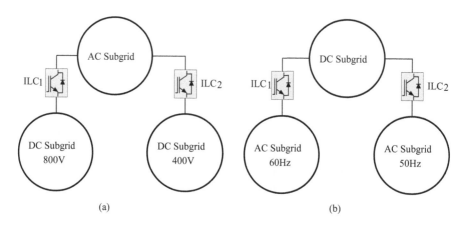

(a)   (b)

FIGURE 10.2 Hybrid microgrid systems including subgrids with different ratings: (a) inter-faced with AC subgrid (b) interfaced with DC subgrid.

10.2.1 POWER CONVERTERS

DC–DC converters usually interface the DC DERs and energy storage with the DC subgrid to control the flow of power and DC subgrid voltage. For instance, a boost converter (Figure 10.4a) is used to interface a PV with DC subgrid through a step-up conversion. In this case, the PV terminal voltage is lower than the DC subgrid voltage. If the PV voltage is higher than the DC subgrid voltage, a buck converter (Figure 10.4b) would be used to step-down the voltage. Energy storage like BESS needs a bidirectional DC–DC converter (BDC) (Figure 10.4c) which is controlled to absorb the power and charge the battery during DC subgrid overvoltage (or oversupply of renewable DERs) and to inject the power and discharge the battery during undervoltage (or undersupply of renewable DERs). In practice, the battery voltage is less than the DC subgrid voltage, which needs a boost (step-up) operation for the discharge process and a buck (step-down) operation for the charging process. Specific loads like electric vehicle (EV) charging station need DC–DC converter to control the charging process and match the DC subgrid voltage with the vehicle battery voltage. In Figure 10.4, S_{boost} and S_{buck} are power electronic switches (IGBT/MOSFET) driven by high-frequency pulse width modulation (PWM) switching technique [9]. The pulse width or the duty ratio (d) defines the converter gain in each operation mode. Depending on the application, the value of duty ratio is controlled based on the control scheme by voltage/current loops to control the grid voltage, flow of power or the charge/discharge process of the energy storage.

The VSC connects the DC DERs, energy storage and ILC to the AC subgrid [10–12]. When transferring power from DC link to AC side, it works as an inverter, and when exchanging power from the AC side to DC link, it works as a recti-fier. Figure 10.5 shows the power stage of a two-level, three-phase VSC along with the AC filter and DC link capacitive filter. It is called two-level because the output AC terminals can have either of the $-V_{dc}$ or V_{dc} values at a time. The power electronic switches are driven by a sinusoidal PWM (SPWM) to generate a sinusoidal waveform at the AC side. The pulse width is controlled by a modulation reference

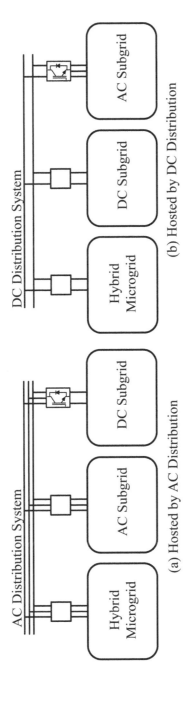

FIGURE 10.3 Microgrid clusters based on (a) AC or (b) DC distribution systems.

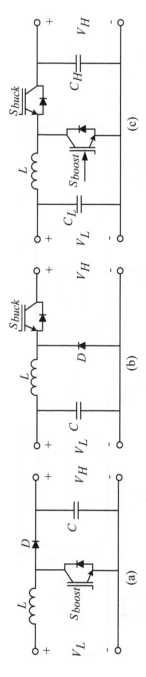

FIGURE 10.4 DC–DC converter topologies for microgrid applications: (a) boost, (b) buck, and (c) bidirectional.

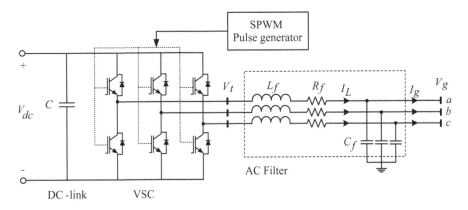

FIGURE 10.5 A two-level three-phase voltage source converter.

waveform (MRW), while the output pulse amplitude is equal to the DC link volt-age. During the time that power switches are OFF the current flows from AC side to DC side through anti-parallel diodes, if the AC side amplitude is greater than the DC link voltage at the same time. The MRW is generated by the modulation index, which varies in the range [0,1], and carrier frequency, which is the SPWM switching frequency. The DC link may be connected to DERs with DC voltage, energy storage or the DC bus of the DC subgrid in case if used in ILC. Due to the switching opera-tion, the input/output currents have switching harmonics. The AC filter restricts the harmonics in AC side current while the capacitor (C) filters the ripples in the DC link voltage. In order to increase the harmonic performance, the VSC is equipped with more than two levels, and the LC filter has another inductor in series with the output current I_g which makes it an LCL filter. In the next part we will discuss more details about the VSC.

10.3 CONTROL STRATEGIES IN HMG

Various control strategies have been proposed for microgrid control. Multi-level hier-archical control, which has been a standard approach, is suitable for the operability of the grids with distributed generations as it allows control of local and global variables such as frequency and voltage to guarantee power-sharing among DERs to generate a desired steady-state power [13]. The traditional hierarchical architecture has three con-trol levels: primary, secondary and tertiary. Each level has its own control objectives and specific bandwidth to ensure decoupled performance. The primary layer performs load sharing and provides ancillary protection and stability control to the local devices. An example of the primary level is the most common decentralized droop control tech-nique [14–16]. All the local measurements and controllers are in this control level. This level has the highest speed of response (bandwidth) since it is immediately attached to the devices with no communication interfaces.

The secondary level is responsible for compensating voltage/frequency deviations made by the primary control. This level has a relatively slower dynamic in com-parison with primary level as it requires an intermediate timescale to respond to frequency/voltage deviations and is usually implemented in control centers linked with communication channels.

Lastly, the tertiary control level, which is also called HMG power management system (PMS) performs supervisory management on solving the optimal power flow (OPF) problem and calculating power references for DERs considering economic objectives. This level has much slower dynamics as it needs to receive data from local measurement devices, perform the control objectives in longer timescale (compared with lower levels) and send back the reference setpoints through the low-bandwidth communication links.

In practice, the tertiary level is merged into the secondary level to form an equivalent two-level control architecture [17]. The upper level is usually implemented in a centralized and distributed configuration. Each topology has its own application and advantages.

The traditional centralized model of a converter-based HMG control is shown in Figure 10.6 that has two equivalent layers. All the DERs and devices are directly connected to the control center through communication links forming a star topology. Although relatively simple to implement, it has several drawbacks and limitations, as listed below:

- Requires substantial communication system between the control center and all DERs and devices which incurs a considerable cost in geographically scattered HMGs with several critical buses.
- Its security and accuracy are compromised as the entire HMG relies on one control center.
- This approach is suitable for HMGs with critical demand-supply balance and a fixed structure with no plug-and-play capability [18].
- It is not ideal for distributed energy storage application as it does not consider storage capacity and state of charge (SoC) [17].

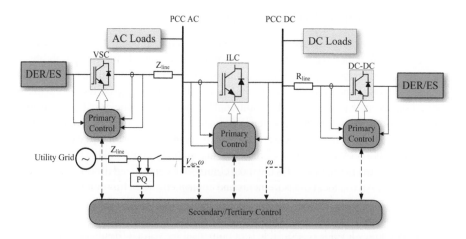

FIGURE 10.6 Centralized hierarchical control of HMG; dashed lines represent communication links.

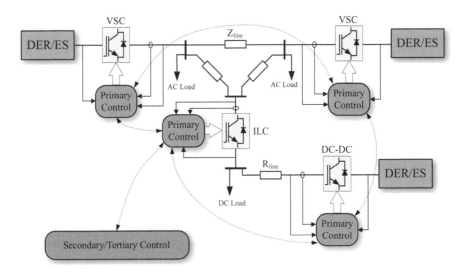

FIGURE 10.7   An example of a distributed control scheme.

The distributed control architecture has been recently proposed to provide a flexible, efficient and reliable operation based on a sparse communication network eliminating mere reliance on the control center. Figure 10.7 illustrates an example of a distributed control scheme for a converter-based HMG. This scheme is more suitable for spatially dispersed HMGs with several critical buses. In a distributed control architecture, the local controller can communicate with each other to define or decide the secondary control actions. The communication system is flexible from all-to-all link to neighbor-to-neighbor connection. The secondary control action is determined based on the averaging or consensus techniques in a multi-agent systems (MAS) framework. A MAS is a composition of multiple intelligent agents that have access to the local information and exchange information with each other to pursue several global and local objectives. In this scheme, the physical secondary/tertiary layer can be totally removed as the agents can establish the virtual secondary layer in the primary control level.

In the secondary averaging technique, each DER determines the secondary compensating terms based on the average value of some or all the other DERs. The real-time global average asymptotically tracks the control reference within a certain timescale. For instance, the average value of the frequency in an AC subgrid is the arithmetic mean of all other DERs. This process is a little different in case of the local variables like voltage. In this case, the output reactive power of other DERs also participates in the averaging [13].

In the cooperative consensus algorithm, the autonomous agents (i.e. DERs) share the information with other agents in a neighbor-to-neighbor configuration within a sparse communication structure. In the secondary level DERs achieve the accurate load sharing through global voltage/frequency restoration while in the tertiary level agents solve the problem of optimal power flow by iteratively solving limited size subproblems and share the results with neighbors [17]. Consensus technique reduces the communication links and can get adapted to different microgrid topologies. However, even with the sparse communication links, this topology mainly depends on the communication system implying that the system gets more complicated as the number of DERs grows.

10.3.1 PRIMARY CONTROL TECHNIQUES

The primary control is based on local measurement and information. DERs use the voltage, frequency, power etc. signals and primary set-points to establish a decentralized control with no information received from other HMG devices. The most common type of primary control for load sharing is droop control which has been used in conventional synchronous based power systems for a long time. It generates the reference signals applied to controllers, and the idea is that generating sources drop their frequency and voltage from the rated value to share a common load. The load sharing depends on the droop coefficients, which are determined based on the DRE rating. For DERs operating in AC subgrid, the reference values are determined by $P - \omega$ and $Q - V$ relationships:

$$\omega_i = \omega^* - m_i P_{oi}$$
$$V_i = V^* - n_i Q_{oi}$$

(10.1)

where ω_i and ω^* are the ith DER's reference angular frequency and its set-point value, V_i and V^* are the ith DER's reference voltage magnitude and its set-point value, m_i and n_i are active and reactive power droop coefficients respectively. P_{oi} and Q_{oi} are the output active and reactive power of each DER given in Equation (10.2) based on network parameters [19, 20]:

$$P_{oi} = \frac{V_i}{R_i^2 + X_i^2} \left[R_i \left(V_i - V_g \cos \varphi_i \right) + X_i V_g \sin \varphi_i \right]$$

$$Q_{oi} = \frac{V_i}{R_i^2 + X_i^2} \left[-R_i V_g \sin \varphi_i + X_i (V_i - V_g \cos \varphi_i) \right]$$

(10.2)

where R and X are components of the line impedance connecting the DER to the bus with the voltage V_g [see Figure 10.8(a) for a two DER system], V_i is the DER output voltage, ϕ_i is the phase difference between V_g and V_i. Figure 10.8(b) shows this simple concept for sharing a common load $P_L = P_1 + P_2$ between two parallel DERs displayed in Figure 10.8(a). In this case $m_1 P_1 = m_2 P_2$ and $\frac{m_1}{m_2} = \frac{P_1}{P_2}$ which implies that the load is proportionally shared to droop coefficients. The $P - \omega$ and $Q - V$ droop technique work based on the fact that a good decoupling exists between active and reactive powers such that active and reactive powers are proportional to DER's phase angle (or angular frequency) and voltage magnitude. In the practical medium and high voltage distribution systems, the lines connecting the DERs are mostly inductive implying that the ratio of $\frac{X}{R} > 1$ holds for lines reactance (X) and resistances (R). This ensures that a good decoupling exists between active and reactive powers. If the

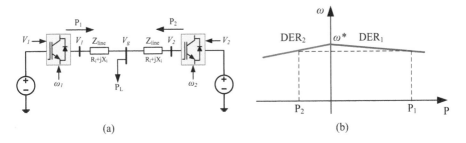

FIGURE 10.8 Droop control for active power-sharing in a two DER AC microgrid: (a) simplified circuit and (b) P-ω droop representation.

inductive part of the line is higher than resistive part such that the resistive part can be neglected, the active and reactive powers are simplified to:

$$P_{oi} \simeq \frac{1}{X_i} V_i V_g \sin\varphi_i$$

$$Q_{oi} \simeq \frac{V_i}{X_i}(V_i - V_g \cos\varphi_i)$$

(10.3)

However, if the inductive content of the distribution lines is less than the resistive part (i.e., $\frac{X}{R} < 1$), as in low voltage distribution systems, the $P - \omega$ and $Q - V$ droop would not make an accurate power-sharing. The active and reactive powers in a resistive distribution network will approximately be simplified to:

$$P_{oi} \simeq \frac{V_i}{R_i}(V_i - V_g \cos\varphi_i)$$

$$Q_{oi} \simeq \frac{-1}{R_i} V_i V_g \sin\varphi_i$$

(10.4)

In this case, a $P - V$ and $Q - \omega$ droop technique might be used. In general, to eliminate the dependency of the power sharing on line impedance, a virtual output impedance method is employed, which is briefly discussed later in this section.

A similar droop control technique is employed for the DC subgrid primary control, which is called $V - I$ droop by applying a virtual resistance:

$$V_{dcj} = V_{dc}^* - R_{vj} I_{oj}$$

(10.5)

where V_{dcj} and V_{dc}^* are the jth DER's output reference and DC subgrid's nominal voltage, R_{vj} is the virtual impedance and I_{oj} is the DER output current.

10.3.2 VSC Control System in DQ Reference Frame

In a VSC-based HMG, the DER power converters in the AC subgrid and ILC can be controlled to achieve certain control objectives. In fact, this capability is a remarkable competency of the modern HMGs (and AC microgrids) compared to the conventional power systems as they flexibly integrate the distributed RES and energy storage. The control scheme can be implemented in the natural reference frame (NRF) or *abc*, stationary reference frame (SRF) or *αβ* and rotational reference frame (RRF) or *dq*. The NRF is challenging to implement as it deals with complex sinusoidal signals and decoupling the active and reactive currents is impossible. SRF deals with decoupled sinusoidal signals (*αβ*) and hence can be used for harmonic control using proportional resonance (PR) controllers and virtual impedance implementation. However, the RRF converts three-phase sinusoidal signals to two DC direct and quadrature (dq) axes which can be easily used in linear control type controllers like Proportional-Integral (PI) compensators. If decoupled effectively, the *dq* control provides accurate active and reactive power control along with droop control. The VSC can be operated in grid-forming, grid-following (or grid-feeding) and DC voltage control modes depending on the application and the control objectives in HMG. In the grid-forming mode, VSC controls the AC subgrid voltage and frequency and operates in the islanded mode of the microgrid. The reference values of voltage magnitude and angular frequency (or phase angle) are generated by droop technique if more than one grid-forming VSC operate in the grid. The dynamic modeling of the power stage of a VSC starts by writing the KVL equations for the inductor current in *abc* frame and then converting them into RRF to acquire a *dq* decoupled control stage. It is assumed that VSC terminal voltage V_t is an averaged value of the DC side and PWM switching operation. From Figure 10.5 we have:

$$\mathbf{v}_t = L_f \frac{d\mathbf{i}_L}{dt} + R_f \mathbf{i}_L + \mathbf{v}_g \tag{10.6}$$

where

$$\mathbf{v}_t = \begin{bmatrix} V_{ta} \\ V_{tb} \\ V_{tc} \end{bmatrix} \quad \mathbf{i}_L = \begin{bmatrix} I_{La} \\ I_{Lb} \\ I_{Lc} \end{bmatrix} \quad \mathbf{v}_g = \begin{bmatrix} V_{ga} \\ V_{gb} \\ V_{gc} \end{bmatrix} \tag{10.7}$$

Transforming Equation (10.6) directly from *abc* frame to *dq* (RRF) it yields:

$$\mathbf{v}_{tdq} = L_f \frac{d\mathbf{i}_{Ldq}}{dt} + j\omega L_f \mathbf{i}_{Ldq} + R_f \mathbf{i}_{Ldq} + \mathbf{v}_{gdq}$$

$$\mathbf{v}_{tdq} = \begin{bmatrix} v_{td} \\ v_{tq} \end{bmatrix} \quad \mathbf{i}_{Ldq} = \begin{bmatrix} i_{Ld} \\ i_{Lq} \end{bmatrix} \quad \mathbf{v}_{gdq} = \begin{bmatrix} v_{gd} \\ v_{dq} \end{bmatrix} \tag{10.8}$$

assuming that *d*-axis in *dq* frame is aligned with *a*-phase in abc frame. The transformation matrix for any variable X from *abc* frame to *dq* frame is given by the Park transformation as follows [21]:

$$X_{dq} = TX_{abc}$$

$$T = \frac{3}{2} \begin{bmatrix} \cos\theta_s & \cos\left(\theta_s - \frac{2\pi}{3}\right) & \cos\left(\theta_s + \frac{2\pi}{3}\right) \\ -\sin\theta_s & -\sin\left(\theta_s - \frac{2\pi}{3}\right) & -\sin\left(\theta_s + \frac{2\pi}{3}\right) \end{bmatrix} \tag{10.9}$$

where θ_s is the angle difference between dq and abc frames. T^{-1} is used to transform dq frame variables to abc frame. It should be noted that by the operation of the phase locked loop (PLL) in the steady-state, $\theta_s = 0$ and d-axis is fully aligned with a-phase. Assuming that $X_{dq} = X_d + jX_q$, Equation (10.8) can be expressed in d and q axes for inductor current dynamics by decomposing into real and imaginary terms we obtain:

$$L_f \frac{di_{Ld}}{dt} = v_{td} + \omega L_f i_{Lq} - R_f i_{Ld} - v_{gd}$$
$$L_f \frac{di_{Lq}}{dt} = v_{tq} - \omega L_f i_{Ld} - R_f i_{Ld} - v_{gd} \tag{10.10}$$

The PLL functionality will be briefly discussed in the next parts. Equation (10.10) expresses the mathematical model of the inductor dynamics in terms of converter terminal and grid voltages as converter reference generating and disturbance signals. The dynamics of capacitor voltage can also be derived similarly by writing the KCL for capacitor node in Figure 10.5:

$$C_f \frac{d\mathbf{v}_g}{dt} = \mathbf{i}_L - \mathbf{i}_g \tag{10.11}$$

where $\mathbf{i}_g = \begin{bmatrix} I_{ga} & I_{gb} & I_{gc} \end{bmatrix}^T$. Transforming to dq space and decoupling the real and imaginary terms it holds:

$$C_f \frac{dv_{gd}}{dt} = \omega C_f v_{gq} + i_{Ld} - i_{gd}$$
$$C_f \frac{dv_{gq}}{dt} = -\omega C_f v_{gd} + i_{Lq} - i_{gq} \tag{10.12}$$

Combining (10.10) and (10.12) equations results in averaged dynamic model of the power stage of the VSC shown in Figure 10.9. v_{td} and v_{tq} are the averaged dq values of VSC terminal voltage at AC side that are generated by the modulation index u:

$$\mathbf{v}_t = \mathbf{u} \frac{V_{dc}}{2}, \quad \mathbf{u}^T = \begin{bmatrix} u_a & u_b & u_c \end{bmatrix} \tag{10.13}$$

$$u_a = u(t) \cos\left(\omega t + \theta_0\right)$$

$$u_b = u(t) \cos\left(\omega t + \theta_0 - \frac{2\pi}{3}\right)$$

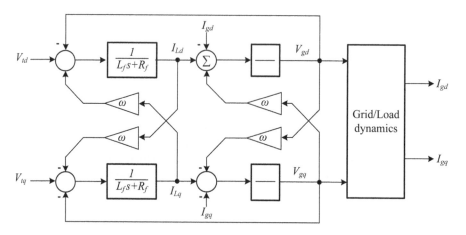

FIGURE 10.9 Dynamic model of the VSC power stage.

$$u_c = u(t) \cos\left(\omega t + \theta_0 + \frac{2\pi}{3}\right)$$

$$0 \le u(t) \le 1$$ (10.14)

where θ_0 is the initial and/or arbitrary phase angle of the VSC and **u** is the vector of the variable modulation indexes generated by the control system.

$$u_{dq} = u_d + ju_q$$

$$u(t) = \sqrt{u_d^2 + u_q^2}$$ (10.15)

where u_d and u_q are respectively d- and q-axis reference generating signals from control system. They are generally DC signals in the steady-state operation that are perturbed during transients. ωt is the phase angle generated by either droop control for a grid-forming or by the PLL for a grid-following VSC. In general form we have:

$$\omega t = \theta_s = \int \omega(t)dt$$ (10.16)

where $\omega(t)$ is treated as the time varying angular frequency generated by droop equations or detected by the PLL. Grid\load dynamics in Figure 10.9 is the equivalent model of the VSC external network connected to V_g terminal. In case of a stand-alone VSC and its load, the load current is simply equal to I_g. While in the networked microgrids with the operation of several DERs and other grid components, the load is shared between DERs. In general form, load dynamics are modeled by the state-space representation in which the dynamics of different DERs and grid components can be combined to form a multi input multi output (MIMO) system [22–24].

Figure 10.10 shows a VSC control diagram in grid-forming mode implemented in the RRF. The VSC is modeled as an independent averaged AC source discussed earlier. The control has two inner loops of current and voltage control [10]. The voltage control

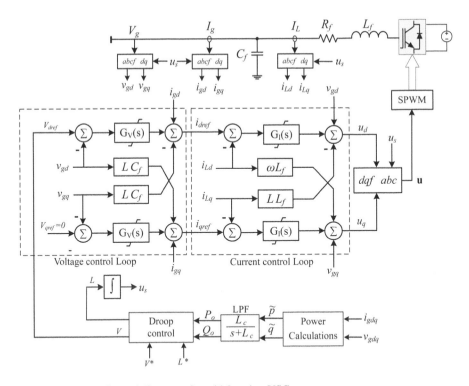

FIGURE 10.10   Control diagram of a grid forming VSC.

loop generates current references using the droop calculations. $G_I(s)$ and $G_V(s)$ are PI controllers tuned based on the converter time response and output filter dynamics. The current loop's bandwidth is assumed to be faster than the voltage control loop by at least one decade in frequency response measure. In this way, it can be ensured that during the current loop transients the PI outputs in voltage loop does not change significantly. The dq components of the measured voltage (V_g) and currents (I_L, I_g) are calculated using the Park transformation $(abc \rightarrow dq)$ in (10.9). Usually, low pass filters are used to suppress the switching harmonics from the measured signals which are neglected for simplicity. The current control loop has the following mathematical form:

$$u_d = k_{pI}\left(i_{dref} - i_{Ld}\right) + k_{iI}\int\left(i_{dref} - i_{Ld}\right)dt + v_{gd} - \omega L i_{Lq}$$

$$u_q = k_{pI}\left(i_{qref} - i_{Lq}\right) + k_{iI}\int\left(i_{qref} - i_{Lq}\right)dt + v_{gq} + \omega L i_{Ld}$$

(10.17)

where i_{dref} and i_{qref} are the reference values of inductor dq current generated by the voltage control loop [see (10.20)] and k_{pI} and k_{iI} are the proportional and integral coefficients of the controller, respectively. Feed-forward terms $\left(v_{gdq}, \omega L i_{Ld}, -\omega L i_{Lq}\right)$ are added to u_d and u_q to cancel the effect of V_g and I_L to decouple the d and q axes to be able to control two axes independently. However, perfect decoupling might

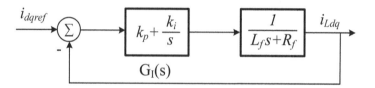

FIGURE 10.11   Equivalent model of the current closed-loop.

not be practically achieved with these feed-forward terms in (10.17) due to the harmonics and measurement errors. Reference [25] provides a method based on multivariable-PI current control with superior disturbance rejection approach to gain a fully decoupled control, specifically in asymmetrical loads situation. Assuming fully decoupled, the equivalent model of the current control closed-loop takes the form shown in Figure 10.11.

k_{pI} and k_{iI} can be selected based on the desired response time T of the converter in current loop. We have [10]:

$$k_{pI} = \frac{L_f}{T}, \; k_{iI} = \frac{R_f}{T} \tag{10.18}$$

Then current control closed-loop acts as a first order low pass filter with the time constant T i.e.:

$$i_{Ldq} = \frac{1}{Ts+1} i_{dqref} \tag{10.19}$$

The voltage control loop has the following mathematical form:

$$i_{dref} = K\left(v_{dref} - v_{gd}\right) + K\sigma \int \left(v_{dref} - v_{gd}\right)dt + i_{gd} - \omega C v_{gq}$$

$$i_{qref} = K\left(v_{qref} - v_{gq}\right) + K\sigma \int \left(v_{qref} - v_{gq}\right)dt + i_{gq} + \omega C v_{gd} \tag{10.20}$$

where v_{dref} is generated by droop control and $v_{qref} = 0$ corresponds to the fact that in steady-state the converter is driven to have $v_{gq} = 0$. Similar to current control loop, the feed-forward terms are added to cancel the effect of grid/load dynamics and to achieve a decoupled d and q axes. K and Kσ are the proportional and integral coefficients of the PI controller, respectively.

If T is small enough, the current control can be considered as an inner loop with a high bandwidth whose transients do not have significant impact on the outer loop (voltage control) over a wide range of frequencies. On the other hand, T should be large enough that the bandwidth of the closed-loop system $\left(\frac{1}{T}\right)$ is notably smaller (at least one decade) than the switching frequency $(2\pi f_s)$. Therefore the combination of the inductor current and capacitor voltage loop represent two linear and independent systems with i_{dqref} as input control and v_{gdq} as output variable shown in Figure 10.12.

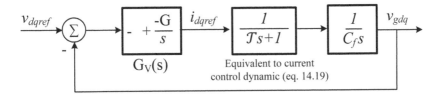

FIGURE 10.12 Simplified linearized system of voltage source converter control equal to voltage and current loops.

For such a system, that has two poles at origin (i.e. $s = 0$) and one real pole ($s = -1$) in its loop gain, control parameters of $G_V(s)$ are determined based on the phase margine criteria[10, 26]. The loop gain in Figure 10.12 has double pole at origin and one real pole at $s = -\frac{1}{T}$. The control objectives are based on the desired phase margin ρ_m that is kept in its maximum value at the frequency ω_m. If the gain crossover frequency of the loop is chosen as ω_m, then we have:

$$\rho_m = \arcsin\left(\frac{1 - T\sigma}{1 + T\sigma}\right)$$

$$\omega_m = \sqrt{\frac{\sigma}{T}}$$

(10.21)

Then the gain coefficient K must satisfy the unity gain of the loop at ω_m it holds:

$$K = \omega_m C_f$$

(10.22)

The phase margin is typically selected between 30° to 75°. In Figure 10.10, the power calculation block computes the unfiltered active and reactive power components directly from dq voltage and current measurements:

$$\tilde{p} = \frac{3}{2}\left[v_{gd}i_{gd} + v_{gq}i_{gq}\right]$$

$$\tilde{q} = \frac{3}{2}\left[-v_{gd}i_{gq} + v_{gq}i_{gd}\right]$$

(10.23)

In the steady state that the output tracks the references, $v_{gq} = 0$ and the second terms of (10.23) converge to zero. Then the active and reactive powers will be proportional to i_{gd} and i_{gq} respectively. The low pass filter (LPF) block suppresses switching harmonics. ω_c is the cut off frequency of the filter that is selected one decade below the power frequency.

$$P_o = \frac{\omega_c}{s + \omega_c}\tilde{p}$$

$$Q_o = \frac{\omega_c}{s + \omega_c}\tilde{q}$$

(10.24)

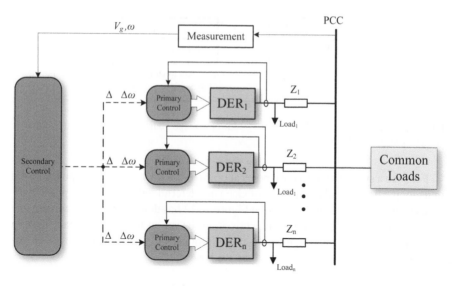

FIGURE 10.13   A droop-controlled AC microgrid equipped with centralized secondary control.

The grid forming VSC is suitable for DERs with controllable (or dispatchable) power sources like energy storage, fuel cells, conventional generators (AC or DC) or combination of RES and energy storage in islanded operation of AC subgrid. If several parallel DERs use grid forming VSCs with primary droop control, the values of grid voltage and frequency would have steady-state deviations from nominal ratings. Hence, the secondary control may restore the voltage and frequency in AC subgrid. There are several approachs to implement the secondary control based on the type of the architecture of the secondary layer hirerachy, i.e. the distributed or centralized considering various objectives [27–30]. As it was mentioned earlier, the centralized configuration collects the information from all DERs and communicates to the primary control of each DER to send corrective voltage and frequency signals. One approach is to measure the PCC voltage magnitude and frequency, and propagate a common corrective signal to all DERs[11, 31]. However, this configuration is only suitable for a grid of DERs with relatively equal ratings that share common loads located at a same bus as shown in Figure 10.13. In addition, local loads will experience voltage magnitudes a little higher than rated value. On the other hand, it is simple and does not rely on the configuration, and detailed model of the microgrid. The communication is relatively sparse and simple. Other approaches are based on the weighted sum of the voltage and frequency of all DERs[32] or to compensate the voltage magnitudes through accurate reactive power sharing. However, these topologies demand extensive communication links compared to the previous one.

The PCC voltage and frequency are compared with the reference values, and error signals are passed through PI compensators. The PI controller is the most popular among the other types[33]. The output signals are sent to primary control

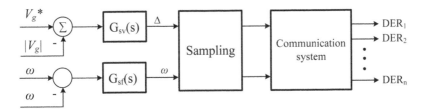

FIGURE 10.14 Centralized secondary voltage and frequency control.

of all DERs via low bandwidth communication medium in time intervals that are usually in the order of few milliseconds. Figure 10.14 shows a simple secondary control based on this topology. $G_{sv}(s)$ and $G_{sf}(s)$ are PI controllers and V_g^* and $\omega^* = 2\pi f^*$ are the reference voltage magnitude and frequency of the system. Sampling unit can be a sample and hold circuit implemented either in discretized or analog form depending on the communication system type. Modern communication systems use discretized values. Corrective signals have the mathematical form:

$$\Delta\omega = k_{sfp}\left(\omega^* - \omega\right) + k_{sfi}\int\left(\omega^* - \omega\right)$$

$$\Delta v = k_{svp}\left(V_g^* - \left|V_g\right|\right) + k_{svi}\int\left(V_g^* - \left|V_g\right|\right)$$

(10.25)

where k_{sfp}, k_{svp}, k_{sfi} and k_{svi} are the proportional and integral gains of the PI controller. Then the droop control equations in (10.1) takes the following form:

$$\omega_i = \omega^* - m_i P_{oi} + \Delta\omega$$

$$V_i = V^* - n_i Q_{oi} + \Delta v$$

(10.26)

Figure 10.15 shows the grid feeding VSC which has the similar current control loop to that of the grid forming VSC. The grid feeding mode is usually employed in grid-connected operation of microgrids. However, if it is operated in islanded mode, there must be another grid forming device in the AC subgrid (such as a grid forming VSC or conventional synchronous generator-based DER) to form the reference frame and control the voltage and frequency. The primary goal is to inject/absorb active and reactive setpoint powers, although it can be equipped with the reverse droop technique in islanded operation to support the AC subgrid during the grid voltage and frequency fluctuations as follows:

$$P_{ref} = P^* - m_r\left(\omega - \omega^*\right)$$

$$Q_{ref} = Q^* - n_r\left(V - V^*\right)$$

(10.27)

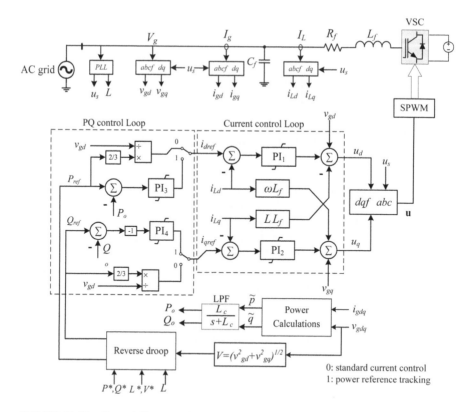

FIGURE 10.15   Control diagram of a grid feeding VSC.

where the superscript '*' represents the setpoint values and m_r and n_r are the reverse droop coefficients and are defined by:

$$m_r \geq \frac{P_{max} - P_{min}}{\omega^*}$$

$$n_r \geq \frac{V_{max} - V_{min}}{V^*} \tag{10.28}$$

and the subscript *max* and *min* refer to maximum and minimum values of power rating of VSC and voltage range at VSC output terminal. In grid connected mode of microgrid $m_r = n_r = 0$.

The grid feeding VSC is synchronized with the main grid (or AC subgrid) by the PLL [19]. The PLL has a park transformation block. It detects the phase angle θ_s between the *abc* and *dq* system and smoothly drives v_q to zero to make v_d aligned with the *abc* reference through a feedback control system. Figure 10.16 illustrates the standard structure of the PLL. The quality of PLL control is crucial for stable operation and an accurate power delivery. The grid feeding VSC is suitable for constant power exchange by the grid in applications such as energy storage and RES with

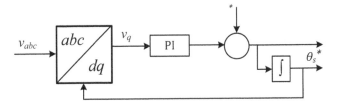

FIGURE 10.16 Standard structure of the PLL.

maximum power tracking (e.g. PV systems). If the PLL operates in steady-state, $v_q = 0$ then the dq current references for grid feeding VSC are defined as below:

$$i_{dref} = \frac{2}{3} \frac{P_{ref}}{v_{gd}}$$

$$i_{qref} = -\frac{2}{3} \frac{Q_{ref}}{v_{gd}}$$

(10.29)

If the current control loop is fast enough, it is guaranteed that output active and reactive powers track P_{ref} and Q_{ref} independently. Alternatively, the dq current references can be generated using PI controllers. In this case, the error between the output powers and reference values are passed through PI controllers. Current control loop must be fast enough to separate the transient operation of the current and power loops. Figure 10.15 depicts both schemes. The current control loop modeling and controller design are already presented earlier. The mathematical modeling of PQ control loop will lead to nonlinear dynamics as active/reactive powers are the product of voltages and currents. Therefore, the best method to determine the PQ loop PI gains is to tune the parameters by the standard method of tuning PI parameters.

The grid feeding mode can be modified to form a VSC for controlling the DC side terminal voltage. This control mode is suitable for ILC converter and also combined operation of DC RES and energy storage. Since the active power exchange between DC and AC subgrid is proportional to DC subgrid voltage, the d axis in the VSC decoupled control can perform the DC voltage reference tracking. Figure 10.17 shows the d axis reference current generating block diagram for a VSC with a DC voltage control scheme in which V_{dc} and V_{dcref} represent DC subgrid (or DC link) voltage and its reference value, respectively. The q axis can still track the reference

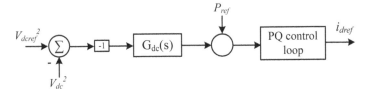

FIGURE 10.17 d axis reference current generating in DC link voltage control scheme.

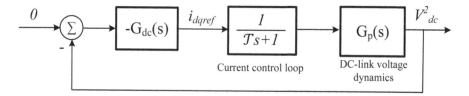

FIGURE 10.18   Equivalent linearized model of the DC link voltage control dynamics.

reactive power or be set to zero to maximize the active power capacity of the VSC and to operate at unity power factor, depending on the application.

Figure 10.18 shows the equivalent linearized model of the DC link voltage control dynamics. $G_{dc}(s)$ is the DC link voltage controller and $G_p(s)$ represents the DC link voltage dynamics as follows [10]:

$$G_p(s) = -\frac{2}{C_f} \frac{\dfrac{2L_f P_{ref0}}{3V_{gd}^2} s + 1}{s} \tag{10.30}$$

where P_{ref0} and V_{gd} are the steady-state values of active power flow reference and d-axis component of the grid voltage in AC side which is equal to the magnitude of the AC voltage since in steady-state $V_{gq} = 0$. Note that VSC does not control the AC side voltage. $G_{dc}(s)$ is, in general, a lead-integral type controller for the stability of the full-scale operation. However, a PI controller can also be implemented that works fine for a wide range of operating conditions. Reference [10] presents detailed design of the controller based on the phase margin criteria.

10.3.3   DC–DC CONVERTER CONTROL SCHEMES

As described earlier, the DC–DC converters are employed in DC subgrid to control the voltage and power in DC DERs and certain loads. Here, we briefly discus a conventional dual loop current/voltage control scheme based on PI compensators, that is widely used for different types. Also, the single loop current (power) control scheme is presented. The transfer functions of DC–DC converters are obtained using averaging approximation techniques over one switching period. References [34] and [35] provide detail of the converter transfer functions and different types of controllers for various applications. Buck and boost converters are two basic unidirectional topologies that are widely used in microgrid application. The half bridge bidirectional type is also composed of two buck and boost converters [see Figure 10.4]. The averaging approximation results in a two-input two-output small-signal model of the converter represented by the following system of transfer function equations in the s-domain applicable to both buck and boost topologies:

$$\begin{bmatrix} e_2(s) \\ i_L(s) \end{bmatrix} = \begin{bmatrix} G_{11} & G_{12} \\ G_{21} & G_{22} \end{bmatrix} \begin{bmatrix} d(s) \\ e_1(s) \end{bmatrix} \tag{10.31}$$

where $e_1(s)$ and $e_2(s)$ are input and output voltage variables respectively, $i_L(s)$ and $d(t)$ are the inductor current and duty ratio. The duty ratio $d(s)$ and converter input voltage $e_1(s)$ are the independent input controls in a general form. However, here the input voltage is assumed to be constant during converter operation, hence the duty ratio is the only control input. G_{11} - G_{22} are converter small-signal transfer functions that are briefly discussed here.

G_{11} and G_{21} are respectively duty-to-output voltage and duty-to-inductor current transfer functions. G_{12} and G_{22} are transfer functions of input voltage to, respectively, output voltage and inductor currents that feed-forward $e_1(t)$, as a disturbance, to the output variable. However, they are neglected here, since $e_1(t)$ is assumed to be constant. For the boost converter we have:

$$G_{11} = \frac{e_2(s)}{d(s)}\bigg|_{e_1=0} = \alpha_0 \frac{\alpha_1 s^2 + \alpha_2 s + \alpha_3}{\alpha_4 \alpha_5 s^2 + \alpha_4 \alpha_6 s + \alpha_4^2} \tag{10.32}$$

$$G_{21} = \frac{i_L(s)}{d(s)}\bigg|_{e_1=0} = \alpha_7 \frac{C(R_l + r_c)s + 1}{\alpha_4 \alpha_5 s^2 + \alpha_4 \alpha_6 s + \alpha_4^2} \tag{10.33}$$

where:

$$\alpha_0 = -R_l E_1 (R_l + r_c), \; \alpha_1 = LCr_c (R_l + r_c),$$

$$\alpha_2 = Cr_L r_c^2 (1 + D') + \left[CR_l (r_L D' - R_l D'^2 + r_L) - Cr_L D' + L \right] r_c + LR_l$$

$$\alpha_3 = (R_l - D' + r_c + R_l D' + r_c D') r_L - R_l^2 D'^2, \; \alpha_4 = R_l D' (R_l D' + r_c) + r_L (R_l + r_c)$$

$$\alpha_5 = LC(R_l + r_c)^2, \; \alpha_6 = \left[L + C(R_l r_L + r_c r_L + R_l D' r_c) \right] (R_l + r_c)$$

$$\alpha_7 = E_1 (R_l + r_c) \left[(1 - r_c - R_l) r_L + r_c R_l + D' R_l^2 \right] \tag{10.34}$$

where $D = 1 - D' \in [0.1]$ and E_1 are the DC value of duty ratio and the input voltage e_1 in the steady-state operation. C, L, r_c and r_L are respectively the output capacitor, inductor, and their equivalent series resistance (ESR). R_l is the equivalent converter load resistance connected to output terminal. The ESR of the power switch and diode are neglected for simplicity. In boost converter V_L is the input and V_H is the output voltage. The voltage gain ratio of the boost converter is given as:

$$M_{boost} = \frac{V_H}{V_L} = \frac{1}{1 - D} \tag{10.35}$$

which indicates that $1 < M_{boost} < \infty$. In practice $D_{min} < D < D_{max}$ to protect the converter against overcurrent and overvoltage. For a buck converter we have:

$$G_{11} = \frac{e_2(s)}{d(s)}\bigg|_{e_1=0} = \beta_0 \frac{s + \dfrac{1}{Cr_c}}{s^2 + \beta_1 s + \beta_2} \tag{10.36}$$

$$G_{21} = \frac{i_L(s)}{d(s)}\bigg|_{e_1=0} = \frac{E_1}{L}\frac{s+\beta_3}{s^2+\beta_1 s+\beta_2} \tag{10.37}$$

$$\beta_0 = \frac{E_1 R_l r_c}{L(R_l+rc)}, \beta_1 = \frac{L+C(R_l r_L+r_L r_c+R_l r_c)}{LC(R+r_c)}, \beta_2 = \frac{(R+r_L)}{LC(R+r_c)}, \beta_3 = \frac{1}{C(R_l+r_c)} \tag{10.38}$$

In buck converter V_H is the input and V_L is the output voltage. The voltage gain ratio of the boost converter is given as:

$$M_{buck} = \frac{V_L}{V_H} = D_{buck} \tag{10.39}$$

which indicates that $0 < M_{buck} < 1$. In practice $D_{min} < D_{buck} < D_{max}$ to protect the converter against overcurrent and overvoltage.

The transfer functions are used to design the controllers for the different applications. Figure 10.19 show the plant model of the single loop current and voltage control schemes applicable to the DC–DC converters depicted in Figure 10.4. I_L is the inductor current, d is the duty ratio, S and S' are the switching pulses applied to power electronic switching devices in case of the bidirectional converter. If unidirectional (buck or boost) converter is used S' is terminated. $G_{cv}(s)$ and $G_{ci}(s)$ are the compensators and H_v and H_c are feedback sensor transfer functions, respectively that are usually combined with an LPF to suppress the switching harmonics. V^* and I^* are voltage and current reference values. Power reference determines the current set-point I^*:

$$I^* = \frac{P_{ref}}{V_{in}} \tag{10.40}$$

where P_{ref} is the reference power and V_{in} and V_{dc} are, respectively, the input and output voltages of the converter in boost, buck or bidirectional operation. PI (or lag type) controllers are widely used for $G_{cv}(s)$ and $G_{ci}(s)$ implementation due to the large disturbance rejection. Although the PI controller relatively reduces converter response (hence its bandwidth), it reduces the switching noise in the feedback signal. Reference [35] provides theoretical and practical basis to design controllers for DC-DC converters. Current control scheme is suitable for constant power control in energy storage applications, whereas the voltage control is used to control the DC bus voltage where no other devices are used to do so (like ILC).

Unidirectional converters are usually used for the power conversion from primary energy sources like solar PV, fuel cell in DC or hybrid microgrids. However, as mentioned earlier, BDC connects the energy storage to the grid. Figure 10.20 shows a dual loop current/voltage control which is a commonly used scheme for BDC applications. G_v and G_c are designed such that the overall dynamic of the current loop is faster than the voltage loop. Hence, the bandwidth of the voltage control loop is placed at least one decade below the current control loop. Then it can be guaranteed that I_L tracks the reference I_{Lref} during the voltage transients. G_v and G_c are usually PI controllers. R_v is the virtual resistance described earlier and H_v is the voltage feedback sensor's transfer

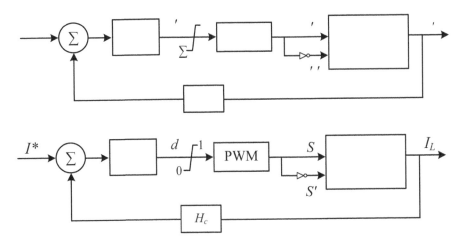

FIGURE 10.19 Plant model of the DC–DC converter control schemes; (a) voltage control, (b) current control.

function which is usually combined with an LPF to suppress the switching harmonics. V_{dc} and V_{dc-ref} are the converter output voltage and its reference value.

10.4 CASE STUDIES: HMG RESILIENCY EVALUATION AGAINST GRID UNCERTAINTIES

In this section the islanded HMG resiliency has been studied during renewable power fluctuations and other grid disturbances by showing some examples, emphasizing the battery energy storage system (BESS) role in maintaining the grid stability. The BESS can be utilized for several ancillary purposes in HMGs, i.e., peak shaving, dynamic local voltage support, short-term frequency smoothing, and grid contingency support. In this study, the BESS control system is adopted to contribute in oscillations damping during the grid disturbances.

Case I: HMG with AC BESS and LVRT scheme in Wind turbine

Figure 10.15 shows the structure of the islanded HMG under the study [5]. The complete system parameters are given in Table 10.1. The AC subgrid power system has a doubly fed induction generator (DFIG)-based variable speed wind generator.

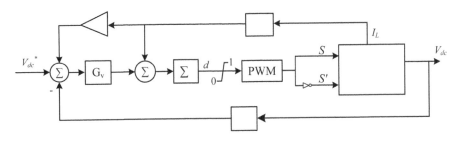

FIGURE 10.20 DC–DC converter dual loop control.

TABLE 10.1

HMG System Parameters for Case I and II

Wind Turbine

Rated power	1.67 MVA	Turbine inertia constant	4.32
Rated voltage	460 V	Shaft spring constant	1.11
Stator R_s, L_s	0.023, 0.18 pu	Shaft mutual damping	1.5
Rotor R'_r, L'_r	0.016, 0.16 pu	DC link capacitor	10 mF
Inertia constant	0.685 s	Rated DC link voltage	1150 V
Friction factor	0.01 pu	Pole pairs	3

PV

Module	SunPower	Model: SPR-315E-WHT-D	
Parallel and series strings	128, 8	Rating capacity	310 kW

Diesel Generator

Rated power:	800 KVA	X_d, X_q	2.59, 2.36 pu
Rated voltage	460 V	Inertia constant, Friction	0.1716s,0.0133pu
Stator resistance R_s	0.010 pu	Pole pairs	2

AC BESS

Battery voltage	950 V	AC filter: R_{f1}, L_{f1}, L'_{f1},C_{f1} 1.4mΩ, 0.4mH, 0.08mH, 1mF	
Rated capacity	500 Ah	DC link voltage, capacitor	750V, 5mF
VSC voltage (AC)	460 V	VSC carrier frequency	5 kHz

DC BESS (Case II)

Battery voltage	400 V	Rated capacity	1180 Ah
Switching frequency	10 kHz	Inductor: R, L	1m Ω, 50μH

ILC

DC voltage, C_{DC}	800V, 50 mF	VSC carrier frequency	5 kHz
AC filter: R_{f2}, L_{f2}, C_{f2} 1.9mΩ, 0.25mH, 830μF			

HMG Loads and Lines

Subgrid Voltage V_{ac},V_{dc}	460 V, 800V	Load 1 induction motor: 460V, 205KW, 1780 *RPM*, 1100*N.m*	
AC system frequency	60 Hz	Load 2,3 (Case I)	0.85pu, PF=0.83
Base power	1MVA	Load 2 (Case II)	0.57pu, PF=0.83
R1, R2, R3	0.2, 0.1384,0.2 Ω	Load 3,4 (Case II)	0.85pu, PF=0.83
L1, L2, L3	1.9, 1.3, 1.9 mH	Load DC	300 kW

The DFIG model used in this work is a widely reported model introduced in reference [36] with complete detailed switching modeling of rotor side converter (RSC) and grid side converter (GSC) control systems in *dq*-frame [10] that are VSC based converters. The RSC/GSC control, drive train and blade pitch angle regulator details can be found in [37]. The wound rotor induction machine, is modeled using fourth and second order for electrical and mechanical parts, respectively. A diesel generator (DG) is in parallel with wind turbine to provide dispatchable power requirements and is modeled considering the conventional salient pole synchronous generator [38], IEEE type 1 exciter [39] and speed governor control system. The DG establishes RRF for the DFIG and grid-following BESS.

The BESS includes a battery bank and a grid-feeding VSC discussed in earlier section and Figure 10.10 equipped with reverse droop. The battery model is a widely used dynamic charge/discharge model of Li-ion experimentally verified in reference [40]. The power management unit calculates BESS power set-point P_B^* based on load powers and power generation from DERs, as follows:

$$P_B^* = P_L - P_G \tag{10.41}$$

$$P_L = P_{Lac} + P_{loss} + P_{ILC}, P_G = P_{DFIG} + P_{DG}$$

where P_L and P_G are total load and generation active powers, P_{Lac} P_{loss}, P_{ILC}, P_{DFIG} and P_{DG} are respectively total AC load power, distribution losses, ILC power, DFIG and DG powers. Power signals are calculated and updated in the power management unit within time intervals that are longer than the simulation sampling frequency. Losses can be calculated by $\sum_{i=1}^{3} R_i |I_i|^2$, where I_i are distribution line currents in AC subgrid; losses in DC subgrid are neglected. Load$_2$ and load$_3$ are modeled as constant power loads. A dynamic (frequency-dependent) AC load (load$_1$) that is modeled as an induction motor [18] is also considered to emulate a more realistic HMG.

The DC subgrid comprises of PV solar panels and constant DC loads. A DC/DC boost converter along with a maximum power point tracking (MPPT) algorithm is integrated to extract the highest possible solar power in different ambient conditions and PV output voltages. The ILC performs as DC subgrid voltage controller (Figure 10.10 and 10.11) and tracks the zero reactive power generation $\left(i_{qref} = 0 \right)$. Also, its reference power is set to zero and standard current control is used.

During grid faults and severe voltage sags, the electromagnetic torque inside the DFIG declines to low levels. Inspired by [41], a coordinated low-voltage ride through (LVRT) control strategy of DFIG is adopted here to compensate for the reactive current requirement during the severe voltage dips. It is assumed that 1/3rd of output current is supplied by the GSC and 2/3rd of it is supplied by the stator. The reactive power set points Q_r^* in RSC are changed according to DFIG terminal voltage magnitude:

$$Q_r^* (pu) = \begin{cases} 0, \ V_t \geq 0.9 \\ -\dfrac{2}{3} \times (0.2)(0.8 - V_t), \ 0.5 < V_t < 0.9 \\ -\dfrac{2}{3} \times 0.8, \ V_t \leq 0.5 \end{cases} \tag{10.42}$$

The time domain simulations are implemented in MATLAB\Simulink environment considering detailed switching models for power electronic devices and discrete-time simulations with sample time $T_s = 5\mu s$. In the following, the dynamic performance of HMG shown in Figure 10.21 is analyzed against intermittent wind and irradiation levels with BESS connected to AC subgrid.

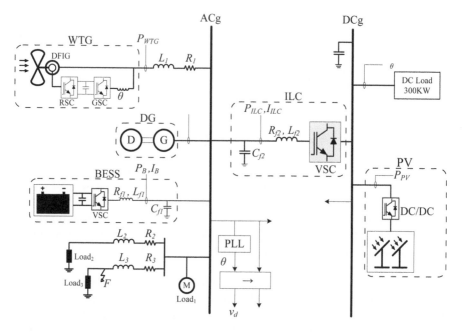

FIGURE 10.21   HMG for disturbance resiliency evaluation.

10.4.1 Disturbance In Wind Velocity And Solar Irradiance

In this part, the performance of the HMG is analyzed against intermittent wind and irradiation levels. Simulations are performed for a system with (**W**) and without (**WO**) BESS and the LVRT scheme discussed in previous part. First (case I-A), the wind speed changes from the initial value of 10m/s (which gives the rated output power) to following stochastic values in 2s intervals consecutively: [8.5, 11.2, 17, 12.5] m/s. The DFIG output power changes based on wind speed variation and pitch angle control action. In the next simulation (case I-B), solar irradiance in PV changes at 10s from the initial value of 1pu based on 1100 W/m² to the following values in 2-s intervals consecutively: [0.6, 1.4, 0.9] pu applying an appropriate ramp-rate. The PV power output changes based on irradiance values and the MPPT perturbations. Figures 10.22 and 10.23 depict the results of essential HMG parameters for these cases compared to the base case without energy storage and LVRT scheme. As it is expected, the BESS compensates for uncertain RES generation through a quick response of charge/discharge operation and thus keeps the load curve flat.

Case II: AC or DC BESS; influence of BESS location on HMG performance

This case makes a comparative analysis to investigate the influence of BESS location on the dynamic performance of the entire HMG against RES generation disturbance. A little modification is made to the case I to represent this case as shown in Figure 10.24 [42]. This study explores the dynamic performance of the HMG when operating with AC BESS (ACB) and DC BESS (DCB) separately under the same operating conditions. In order to emulate two comparison cases, during the ACB

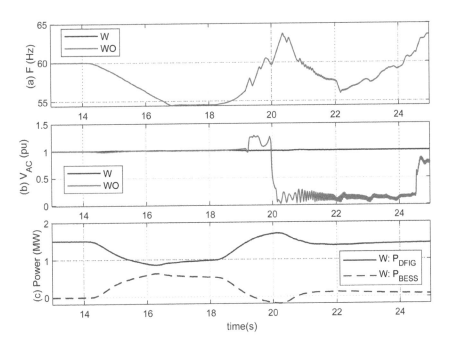

FIGURE 10.22 Wind speed fluctuation (case I-A): a) AC grid frequency, b) AC grid voltage magnitude and c) DFIG and BESS power in case of **W**.

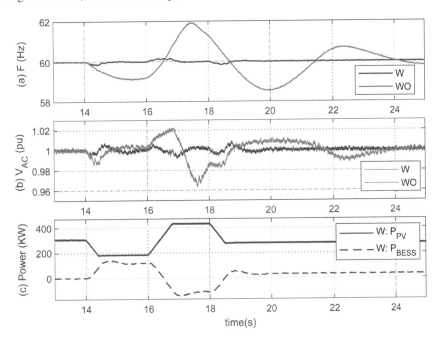

FIGURE 10.23 Solar irradiance variation (case I-B): (a) AC subgrid frequency, (b) AC subgrid voltage magnitude, (c) PV and BESS output power in case of **W**.

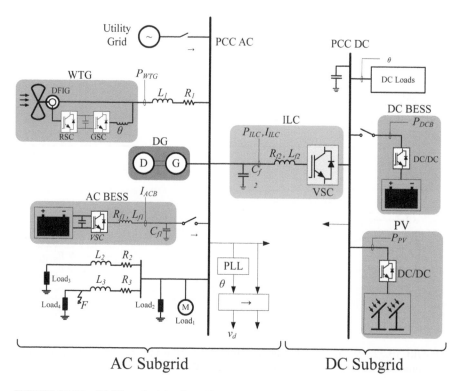

FIGURE 10.24   HMG testbed for Case II.

operation the switch B_1 is closed and the switch B_2 is open, while during the DCB operation B_2 is closed and B_1 is open (shown in Figure 10.24).

The ACB and DCB have the same capacity. The battery bank in DCB is connected to DC subgrid via a bidirectional DC–DC converter and its control system shown in Figure 10.20. The ILC power reference during islanding operation with DCB is given by:

$$P_{ILC}^* = \sum_{i \in S_L} P_{Lac,i} + \sum_{j \in S_I} P_{loss,j} - \sum_{k \in S_G} P_{G,k} \qquad (10.43)$$

where $P_{Lac,i}$, $P_{G,k}$ and $P_{loss,j}$ are active power of loads, active power generations and transmission losses in AC subgrid, respectively and $S_L = \{1,4\}$, $S_I = \{1,3\}$ and $S_G = \{WTG, DG\}$. When operating with ACB, the battery reference power is calculated by:

$$P_{ACB}^* = \sum_{i \in S_L} P_{Lac,i} + \sum_{j \in S_I} P_{loss,j} - \sum_{k \in S_G} P_{G,k} + P_{ILC} \qquad (10.44)$$

In this situation, the ILC controls V_{dc} and hence its power flow depends on the V_{dc} balance.

10.4.2 Disturbance In Wind Velocity (Case II-A)

In this part, a disruption in wind velocity is simulated which affects the power generation in wind turbine. The DC load is equal to 128kW, and the HMG works in oversupply mode, which means BESS operates in charging mode. The wind speed changes stepwise from the initial value of 14 m/s (rated value) to following values in 1s intervals consecutively: [9.8, 15, 18.2, and 14] m/s. Figure 10.25 depicts the dynamic response of the system to this disturbance. The wind turbine output changes

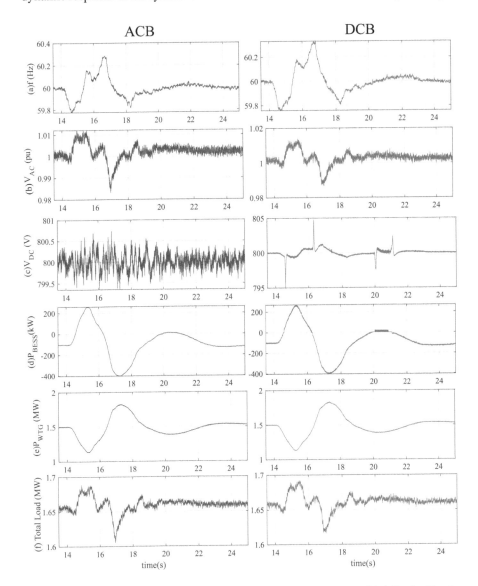

FIGURE 10.25 Wind velocity disturbance; (a) AC subgrid frequency, (b) AC subgrid voltage, (c) DC subgrid Voltage, (d) BESS power, (e) WTG power, (f) total AC load.

based on wind speed variation and pitch angle control action. As can be observed, both BESSs maintain the AC load profile similarly. Although V_{dc} fluctuations are within the acceptable range, unlike ACB, the DCB is unable to keep a flat V_{dc}.

10.4.3 DISTURBANCE IN SOLAR IRRADIANCE (CASE II-B)

In this simulation case, some solar variations affecting the power generation in PV are applied to evaluate the ACB and DCB dynamic performance. Initial system conditions are like cases II-A. The PV is initially generating 305kW with 1000 W/m² irradiance level, and the MPPT is 'ON'. A solar irradiance disturbance is applied to start from 10s with the following values in 2s intervals consecutively considering an appropriate slew rate: [1000 600 1000] W/m². Figure 10.26 compares the dynamic responses of essential system parameters of the HMG for both ACB and DCB operation. Although both BESS perfectly maintain system stability, DCB has better performance for DC subgrid dynamic.

10.4.4 UNPLANNED ISLANDING (CASE II-C)

Although this study mainly focuses on the islanded mode of HMG operation, this part shows the BESS capability in handing unplanned islanding incidence in the presence of DERs with various dynamic properties. Initially, in grid-connected mode, the BESS is set to absorb a constant power to charge the battery bank i.e., $P_{ACB}^* = P_{ILC}^* = -200kW$. This power is imported from the utility grid and the DG is working on droop-control power-sharing mode to generate 0.15 pu based on DG rated values. A summary of the initial values of parameters is given in Table 10.2. Once islanding is detected, power set-points are switched to (10.42) and (10.43). Figure 10.27 shows the essential parameters of the system in response to unplanned islanding incidence at 10s for ACB and DCB separately. The declined frequency at the islanding moment shows that the HMG runs in undersupply mode. The new operating points of both BESSs confirm this situation. Results show that both cases have almost a similar dynamic performance. The initial transient decline in V_{dc} (Figure 10.27 c) in DCB case originates from the load/generation power unbalance in AC subgrid. This power must be exchanged from DC subgrid to AC subgrid and causes a temporary reduction in V_{dc} that is adequately restored by DCB. Since the DC load is almost equal to PV output power, the ILC power exchange in the ACB case is around zero.

TABLE 10.2

Summary of Initial Parameters for Simulation in Case II

DG	P_{DG} = 0.15 pu (120kW)	
PV	P_{PV} = 305 kW	Irradiance: 1000W/m2
WTG	P_{WTG} = 1500kW	Wind vel.:10m/s
Load 1	P_M = 0.91 pu (at 205kW base)	
BESS SoC	%50	
P_{ACB}^*, P_{ILC}^*	-200kW prior to islanding	

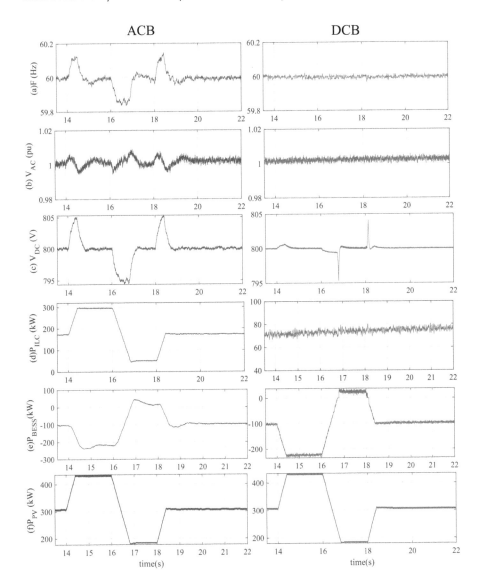

FIGURE 10.26 Dynamic response to solar irradiance disturbance; (a) AC subgrid frequency, (b) AC subgrid voltage, (c) DC subgrid voltage, (d) ILC power, (e) BESS power, (f) PV power.

10.4.5 STOCHASTIC LOAD CHANGE (CASE II-D)

This case analysis two random change in the AC grid loads and compares the HMG performance for both BESS.

Pulsed frequency-dependent load: This evaluation compares the response of the low inertia BESS to a pulsed, high inertia load. Initially, the system is working in oversupply mode and the BESS is getting charged. The $Load_1$ is a frequency

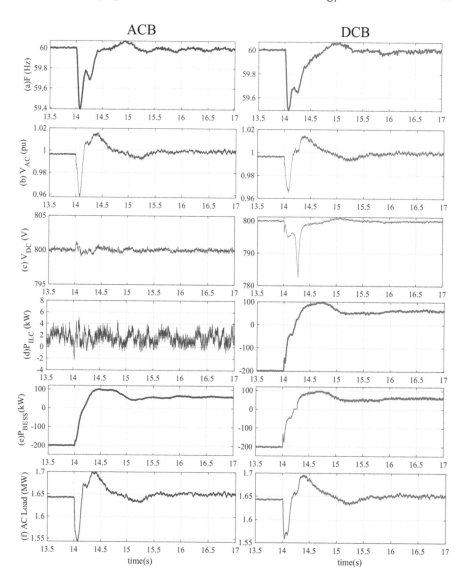

FIGURE 10.27 Islanding incidence simulation; (a) AC subgrid frequency, (b) AC subgrid voltage, (c) DC subgrid Voltage, (d) ILC power, (e) BESS power, (f) total AC load.

dependent load, that is a 205kW induction motor, is running with 0.91 pu (or 1000 Nm) of rated torque. At 10s, the following disturbances take place in motor torque in 1s intervals consecutively: [0.2, 1.2, and 0.91] pu. Figure 10.28 shows the HMG and BESS dynamic performance under this circumstance. Both BESSs almost perform similarly to keep the HMG stable. It should be noted that, since in this case, the DC load was changed from 305 to 128kW, large active power is exported from DC sub-grid to AC subgrid by ILC when operating with ACB.

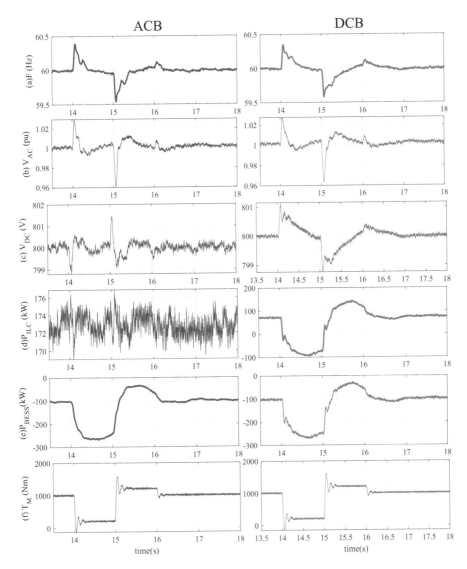

FIGURE 10.28 Dynamic AC load disturbance; (a) AC subgrid frequency, (b) AC subgrid voltage, (c) DC subgrid Voltage, (d) ILC power, (e) BESS power, (f) motor electromagnetic torque.

In DCB operation, a part of power is directly absorbed by DCB, and the rest is exported to AC subgrid to compensate for the power unbalance in AC subgrid.

Loss of large load: In this evaluation, a large-signal disturbance is analyzed. The Load$_3$ is disconnected in AC subgrid at $t_f = 10s$. This will result in losing $560kW + j112kVar$ of total rated load that will significantly cause voltage and frequency deviations in AC subgrid. Figure 10.29 shows the dynamic response of the essential parameters of the system to this disturbance. As it is clearly observable,

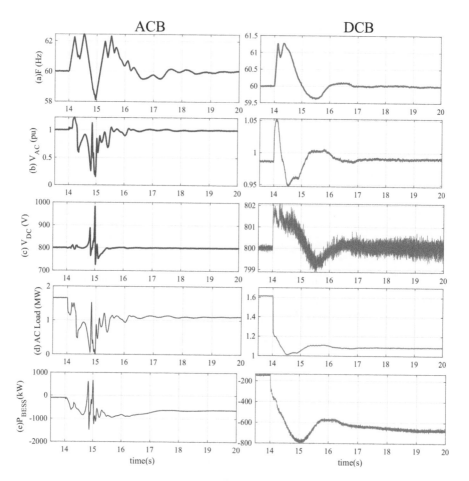

FIGURE 10.29 Dynamic response to loss of large load disturbance; (a) AC subgrid frequency, (b) AC subgrid voltage, (c) DC subgrid voltage, (d) Total AC load, and (e) BESS power.

the DCB has superior performance compared to the ACB operation. The system is restored to a new equilibrium point as reference set-points are adjusted to new values, and the BESS compensates for surplus power generation in steady-state. Although the system is recovered to a new operating point in case of ACB, in a practical HMG, the protection devices will disconnect the generating sources.

10.4.6   FAULT ANALYSIS (CASE II-E)

Although the BESS is not mainly employed to tackle the grid faults, here a performance evaluation is carried out to analyze the LVRT capability of BESS in HMG during and after fault events in the presence of RESs.

In this case, both AC subgrid and DC subgrid faults are considered. First, a 3-line-to-ground (3LG) short circuit is applied at $t_f = 10$ s at the end of feeder line 3

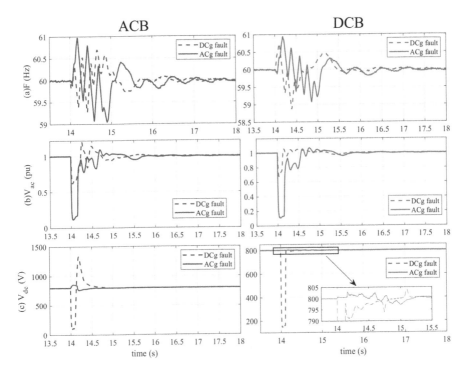

FIGURE 10.30 Dynamic response to fault events; (a) AC subgrid frequency, (b) AC subgrid voltage, (c) DC subgrid voltage.

in AC subgrid (location **F** in Figure 10.24). The fault is cleared after $150ms$ without opening any circuit breaker. In order to fulfill LVRT requirements, none of the DERs are isolated during or after the fault incidence. Next, a pole-to-pole short circuit with a fault resistance of $50m\Omega$ is applied at the DC PCC for the duration of $100ms$ at $t_f = 10s$. Figure 10.30 shows the dynamic response of the essential parameters of the system to these large-signal disturbances. Although in both cases the system can recover, the ACB cannot limit V_{dc} excursions in an acceptable range. In the case of DC subgrid fault, the DCB has superior performance compared to the ACB in controlling grid overvoltage. In ACB operation, the magnitude of V_{ac} and V_{dc} get the values higher than 1.2 and 1.66 pu that is far beyond the tolerable ranges. This means in case of ACB operation, the protection devices will isolate the faulted section.

10.5 CHAPTER SUMMARY

This chapter discusses new methods that provide overall stability and control improvements in hybrid AC/DC microgrids. Initially the chapter explains about structure of the HMG including their topologies, grid components, power converters such as DC–DC converters and VSCs, various control strategies such as centralized/distributed control, primary control, secondary/tertiary control, DC–DC converter control schemes, and interlinking converter control. Then the impact of

BESS dynamic control on hybrid AC/DC microgrid's transient stability during disturbed conditions has been analyzed. Detailed description on testbed models of the microgrid samples and their control systems developed in MATLAB\Simulink software along with obtained system responses have been provided in this chapter.

REFERENCES

1. A. Gupta, S. Doolla, and K. Chatterjee, "Hybrid AC–DC microgrid: systematic evaluation of control strategies," *IEEE Trans. Smart Grid*, vol. 9, no. 4, pp. 3830–3843, Jul. 2017.
2. A. Eisapour-Moarref, M. Kalantar, and M. Esmaili, "Power sharing in hybrid microgrids using a harmonic-based multi-dimensional droop," *IEEE Trans. Ind. Informatics*, vol. 16, no. 1, pp. 109–119, 2020.
3. R. H. Lasseter, "MicroGrids," *IEEE Power Eng. Soc. Winter Meet. Conf. Proc.*, pp. 305–308, 2002.
4. M. D. Keshavarzi and M. H. Ali, "FRT Capability Enhancement of Autonomous AC/DC Hybrid Microgrid by Coordinated MSDBR and Interlinking Converter Control Strategy," In *2019 IEEE Power & Energy Society Innovative Smart Grid Technologies Conference (ISGT)*, pp. 1–5, 2019.
5. M. Daviran Keshavarzi and M. H. Ali, "Disturbance Resilience Enhancement of Islanded Hybrid Microgrid Under High Penetration of Renewable Energy Resources by BESS," In *IEEE Transmission & Distribution*, pp. 1–5, 2020.
6. X. Lu, J. M. Guerrero, K. Sun, J. C. Vasquez, R. Teodorescu, and L. Huang, "Hierarchical control of parallel AC-DC converter interfaces for hybrid microgrids," *IEEE Trans. Smart Grid*, vol. 5, no. 2, pp. 683–692, 2010.
7. Z. Zhao, P. Yang, Y. Wang, Z. Xu, and J. M. Guerrero, "Dynamic characteristics analysis and stabilization of PV-based multiple microgrid clusters," *IEEE Trans. Smart Grid*, vol. 10, no. 1, pp. 805–818, 2019.
8. B. John, A. Ghosh, M. Goyal, and F. Zare, "A DC power exchange highway based power flow management for interconnected microgrid clusters," *IEEE Syst. J.*, vol. 13, no. 3, pp. 3347–3357, 2019.
9. M. K. Kazimierczuk, *Pulse-Width Modulated DC-DC Power Converters*, 2nd ed. Chichester, West Sussex: John Wiley & Sons, Ltd., 2015.
10. R. Yazdani, A. Iravani, *Voltage-Sourced Converters in Power Systems Part1*, Hoboken, NJ: John Wiley & Sons, Inc., 2010.
11. J. C. Vasquez, J. M. Guerrero, M. Savaghebi, J. Eloy-Garcia, and R. Teodorescu, "Modeling, analysis, and design of stationary-reference-frame droop-controlled parallel three-phase voltage source inverters," *IEEE Trans. Ind. Electron.*, vol. 60, no. 4, pp. 1271–1280, Apr. 2013.
12. M. Liserre, F. Blaabjerg, and S. Hansen, "Design and control of an LCL-filter-based three-phase active rectifier," *IEEE Trans. Ind. Appl.*, vol. 41, no. 5, pp. 1281–1291, 2005.
13. A. C. Zambroni de Souza and M. Castilla, *Microgrids Design and Implementation*, Cham: Springer, 2019.
14. X. Sun, Y. Hao, Q. Wu, X. Guo, and B. Wang, "A multifunctional and wireless droop control for distributed energy storage units in Islanded AC microgrid applications," *IEEE Trans. Power Electron.*, vol. 32, no.1, pp. 736–751, 2017.
15. P. H. Divshali, A. Alimardani, S. H. Hosseinian, and M. Abedi, "Decentralized cooperative control strategy of microsources for stabilizing autonomous VSC-based microgrids," *IEEE Trans. Power Syst.*, vol. 27, no. 4, pp. 1949–1959, 2012.
16. A. Bidram and A. Davoudi, "Hierarchical structure of microgrids control system," *IEEE Trans. Smart Grid*, vol. 3, no. 4, pp. 1963–1976, 2012.

17. T. Morstyn, B. Hredzak, and V. G. Agelidis, "Control strategies for microgrids with distributed energy storage systems: An overview," *IEEE Trans. Smart Grid*, vol. 9, no. 4, 3652–3666, 2016.

18. D. E. Olivares *et al.*, "Trends in microgrid control," *IEEE Trans. Smart Grid*, vol. 5, no. 4, pp. 1905–1919, 2010.

19. J. Rocabert, A. Luna, F. Blaabjerg, and P. Rodríguez, "Control of power converters in AC microgrids," *IEEE Trans. Power Electron.*, vol. 27, no. 11, pp. 4734–4749, Nov. 2012.

20. S. M. Ashabani and Y. A. I. Mohamed, "New family of microgrid control and management strategies in smart distribution grids analysis, comparison and testing," *IEEE Trans. Power Syst.*, vol. 29, no. 5, pp. 2257–2269, 2010.

21. C. J. O'Rourke, M. M. Qasim, M. R. Overlin, and J. L. Kirtley, "A geometric interpretation of reference frames and transformations: Dq0, Clarke, and Park," *IEEE Trans. Energy Convers.*, vol. 34, no. 4, pp. 2070–2083, 2019.

22. M. Naderi, Y. Khayat, Q. Shafiee, T. Dragicevic, H. Bevrani, and F. Blaabjerg, "Interconnected autonomous AC Microgrids via back-to-back converters—Part I: Small-signal modeling," *IEEE Trans. Power Electron.*, vol. 35, no. 5, pp. 4728–4740, May 2020.

23. F. D. Mohammadi and H. K. Vanashi, "State-space modeling, analysis, and distributed secondary frequency control of isolated microgrids," *IEEE Trans. Energy Convers.*, vol. 33, no. 1, pp. 155–165, 2018.

24. Z. Li and M. Shahidehpour, "Small-signal modeling and stability analysis of hybrid AC/DC microgrids," *IEEE Trans. Smart Grid*, vol. 10, no. 2, pp. 2080–2095, 2019.

25. B. Bahrani, S. Kenzelmann, and A. Rufer, "Multivariable-PI-Based dq current control of voltage source converters with superior axis," *IEEE Trans. Ind. Electron.*, vol. 58, no. 7, pp. 3016–3026, 2011.

26. W. Leonhard, *Control of Electrical Drives*, 3rd ed., Berlin, Heidelberg: Springer-Verlag, 2001.

27. A. Bidram, A. Davoudi, F. L. Lewis, and J. M. Guerrero, "Distributed cooperative secondary control of microgrids using feedback linearization," *IEEE Trans. Power Syst.*, vol. 28, no. 3, pp. 3462–3470, 2013.

28. M. A. Shahab, B. Mozafari, S. Soleymani, N. M. Dehkordi, H. M. Shourkaei, and J. M. Guerrero, "Stochastic consensus-based control of microgrids with communication delays and noises," *IEEE Trans. Power Syst.*, vol. 34, no. 5, pp. 3573–3581, 2019.

29. R. Zhang and B. Hredzak, "Distributed finite-time multiagent control for DC microgrids with time delays," *IEEE Trans. Smart Grid*, vol. 10, no. 3, pp. 2692–2701, 2019.

30. H. J. Yoo, T. T. Nguyen, and H. M. Kim, "Consensus-based distributed coordination control of hybrid AC/DC microgrids," *IEEE Trans. Sustain. Energy*, vol. 11, no. 2, pp. 629–639, 2020.

31. X. Zhao, J. M. Guerrero, M. Savaghebi, J. C. Vasquez, X. Wu, and K. Sun, "Low-voltage ride-through operation of power converters in grid-interactive microgrids by using negative-sequence droop control," *IEEE Trans. Power Electron.*, vol. 32, no. 4, pp. 3128–3102, 2017.

32. R. Heydari *et al.*, "Robust high-rate secondary control of microgrids with mitigation of communication impairments," *IEEE Trans. Power Electron.*, vol. 35, no. 11, pp. 12486–12496, Nov. 2020.

33. C. Ahumada, R. Cárdenas, D. Sáez, and J. M. Guerrero, "Secondary control strategies for frequency restoration in islanded microgrids with consideration of communication delays," *IEEE Trans. Smart Grid*, vol. 7, no. 3, pp. 1030–1041, 2016.

34. A. Ioinovici, *Power Electronics and Energy Conversion Systems, Fundamentals and Hard-switching Converters*, vol. 1. West Sussex: Wiley, 2013.

35. R. W. Erickson and D. Maksimović, *Fundamentals of Power Electronics*, 2nd ed. New York, NY: Kluwer, 2004.

36. G. Rashid and M. H. Ali, "Nonlinear control-based modified BFCL for LVRT capacity enhancement of DFIG-based wind farm," *IEEE Trans. Energy Convers.*, vol. 32, no. 1, pp. 284–295, 2017.

37. K. Clark, N. W. Miller, and J. J. Sanchez-Gasca, "Modeling of GE Wind Turbine-Generators for Grid Studies," 2010.

38. P. C. Krause, O. Wasynczuk, and S. D. Sudhoff, *Analysis of Electric Machinery and Drive Systems*, 3rd ed. Hoboken, New Jersey: John Wiley & Sons, Inc., 2013.

39. "IEEE Recommended Practice for Excitation System Models for Power System Stability Studies," In *IEEE Std 421.5-2005*. 2005.

40. O. Tremblay and L. A. Dessaint, "Experimental validation of a battery dynamic model for EV applications," *World Electr. Veh. J.*, vol. 3, no. 2, pp. 289–298, 2009.

41. B. B. Ambati, P. Kanjiya, and V. Khadkikar, "A low component count series voltage compensation scheme for DFIG WTs to enhance fault ride-through capability," *IEEE Trans. Energy Convers.*, vol. 30, no. 1, pp. 208–217, 2015.

42. M. Daviran Keshavarzi and M. Hasan Ali, "Influence of battery energy storage location on the dynamic performance of hybrid AC/DC microgrid," in *Smart Power and Internet Energy Systems*, pp. 1–6, 2020.

Index